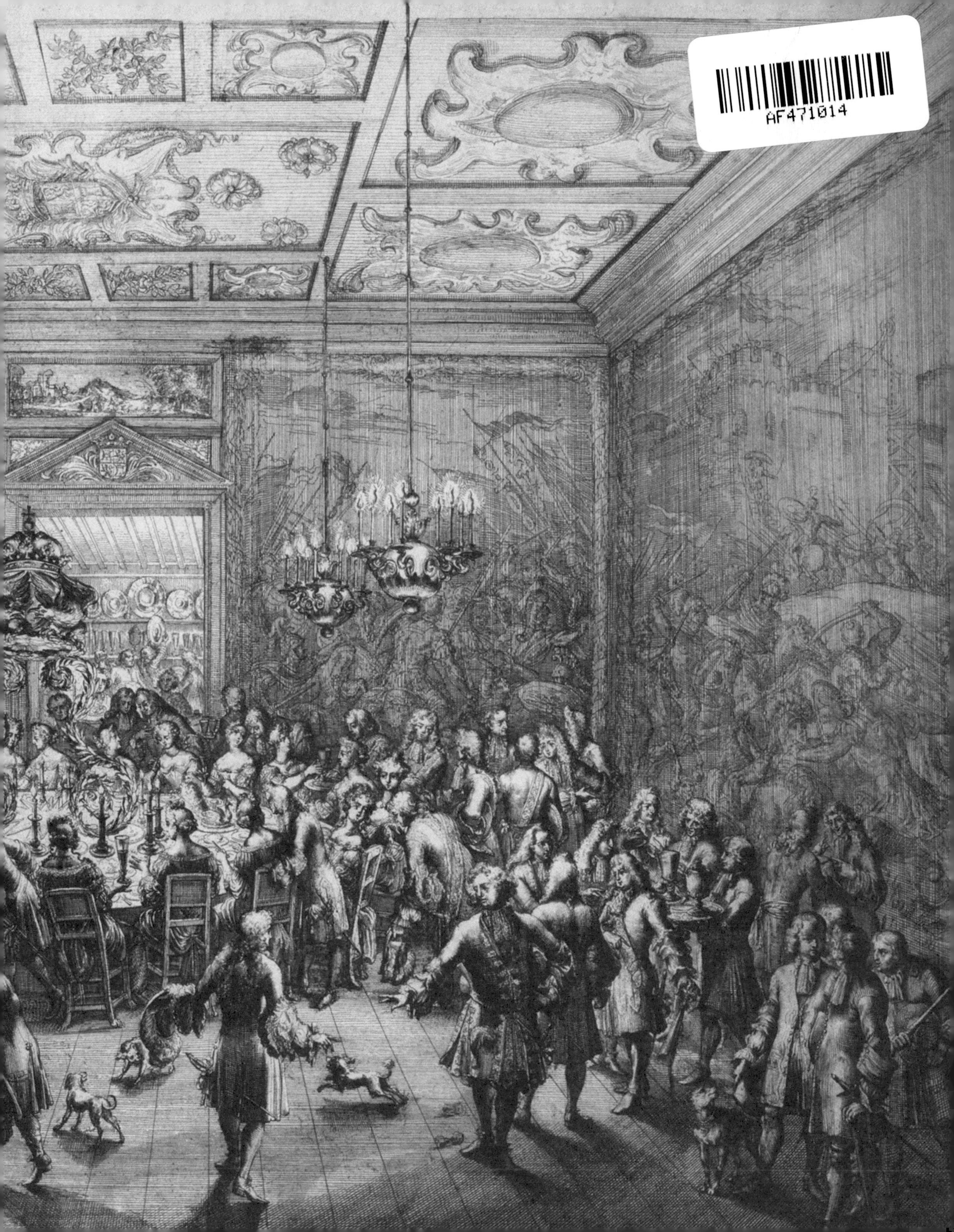

THE ART *of* DRINKING

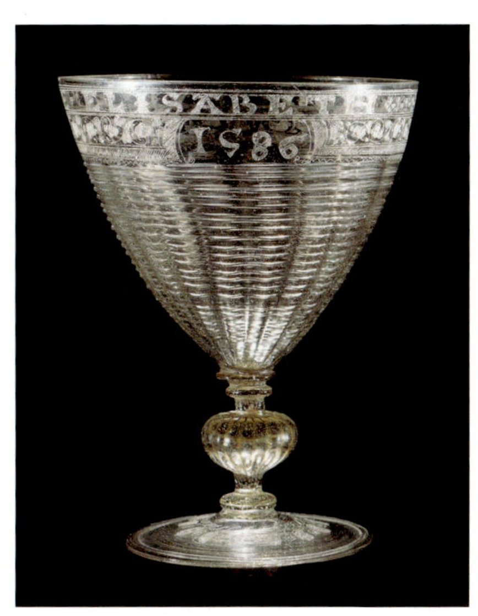

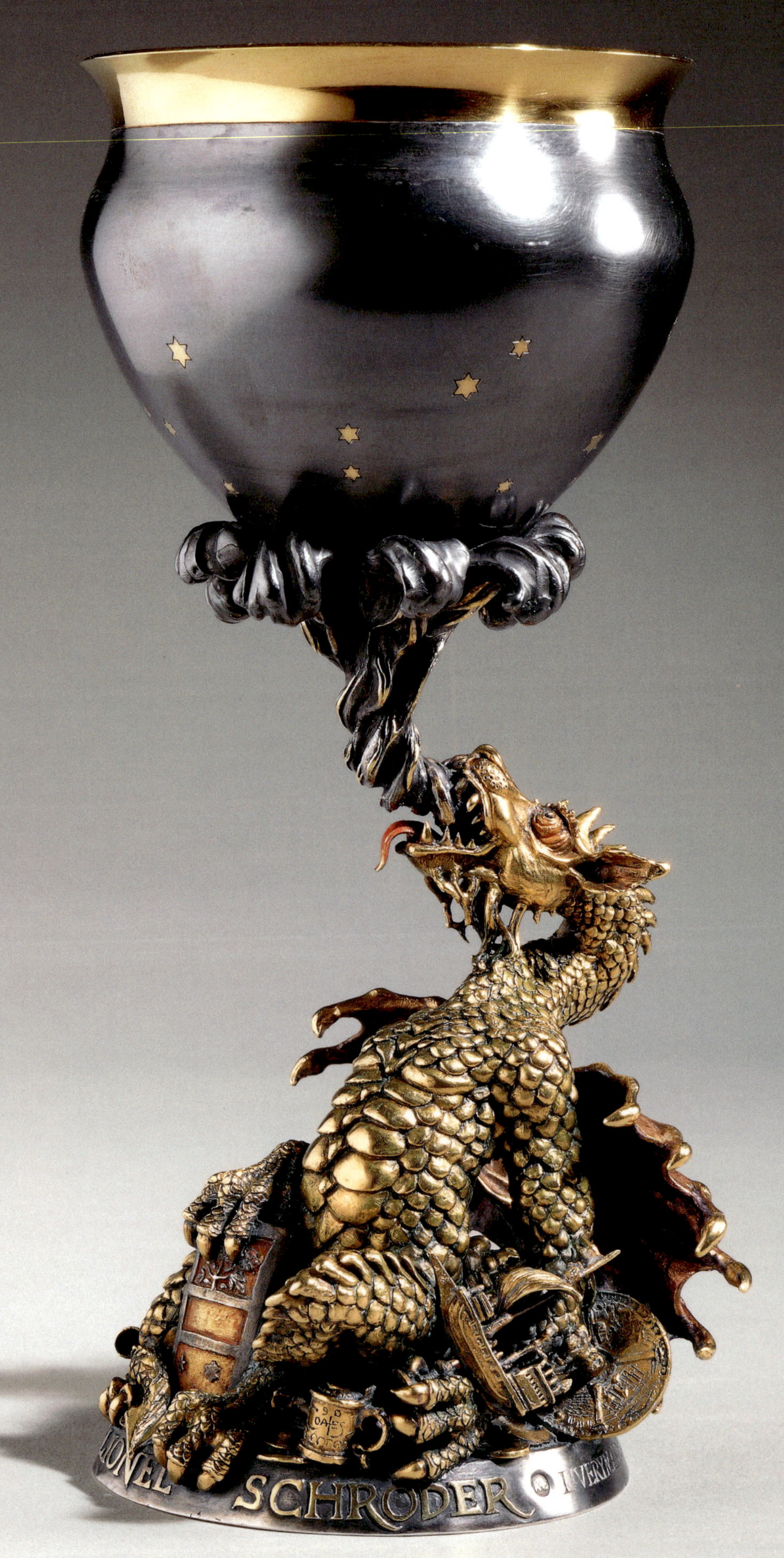
LIONEL SCHRODER LIVERPOOL

THE ART *of* DRINKING

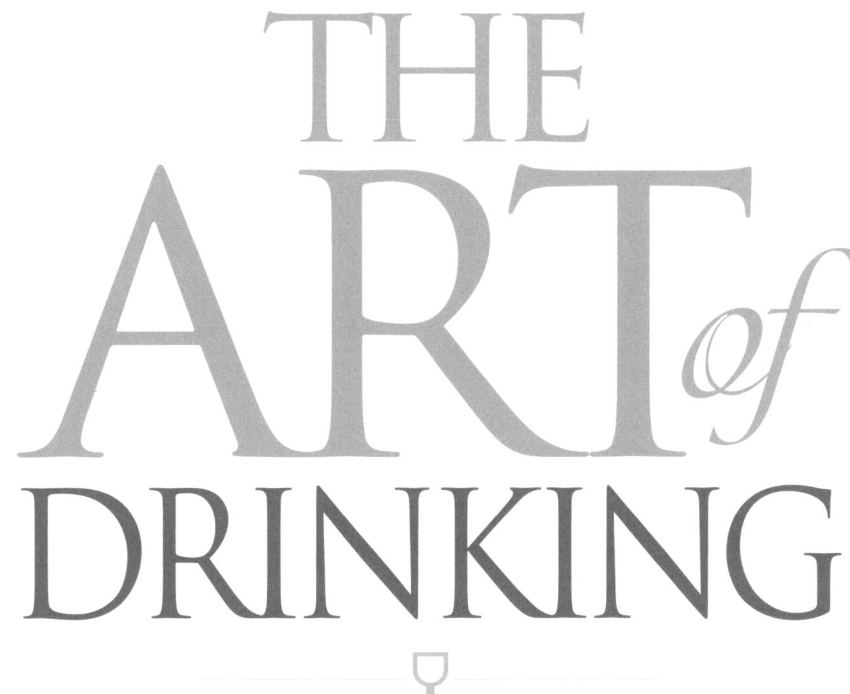

EDITED BY PHILIPPA GLANVILLE
AND SOPHIE LEE

V&A Publications

First published by V&A Publications, 2007
V&A Publications
Victoria and Albert Museum
South Kensington
London SW7 2RL

Distributed in North America by Harry N. Abrams, Inc., New York

Hardback edition
ISBN 978 1 85177 510 1
Library of Congress Control Number 2007924505

10 9 8 7 6 5 4 3 2 1
2011 2010 2009 2008 2007

Designed by Bernard Higton
New V&A photography by Mike Kitcatt, V&A Photographic Studio

Printed in Hong Kong

Acknowledgements
We are grateful for assistance with images from Rita Gans, Bruno
Schroder, Lord Rothschild, Anthony Marks, Mark Bills, James Lomax,
Timothy Lukes, Jonathan Horne, Melanie Aspey, Nicole Cartier,
Barbara Lasic and the staff at Waddesdon Manor. Peter Brears,
James Rothwell, Maja Houtman, Sue Laurence and Julie Banham
made valuable comments. Thanks are due also to Frances Ambler,
Asha Savjani, Ken Jackson, Eva White, Louise Hofman, Ethan
Kalemjian, Terry Bloxham, Christopher Maxwell, Hilary Young
and other V&A colleagues.

V&A
V&A Publications
Victoria and Albert Museum
South Kensington
London SW7 2RL
www.vam.ac.uk

Half title: Verzelini goblet, London, 1586 (p.35)
Title: Court Wine Cup by Kevin Coates for Bruno Schroder, 1990
Above: Milkmaid cup, silvergilt, late 17th century
Opposite: Memphis cocktail glass, 1983
Front jacket illustration: A *Vanitas* showing grapes, a German roemer glass
for white wine and a silver jug, oil on canvas by Willem van Aelst, 1659
Back jacket illustration: Cocktail equipment,1930s (p.137)
Endpaper: Royal dinner at the Hague, late 17th century (p.30)

Contents

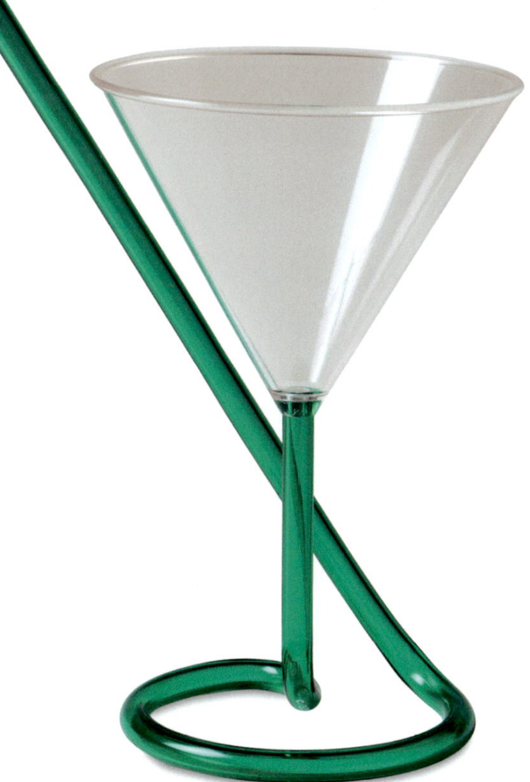

Introduction

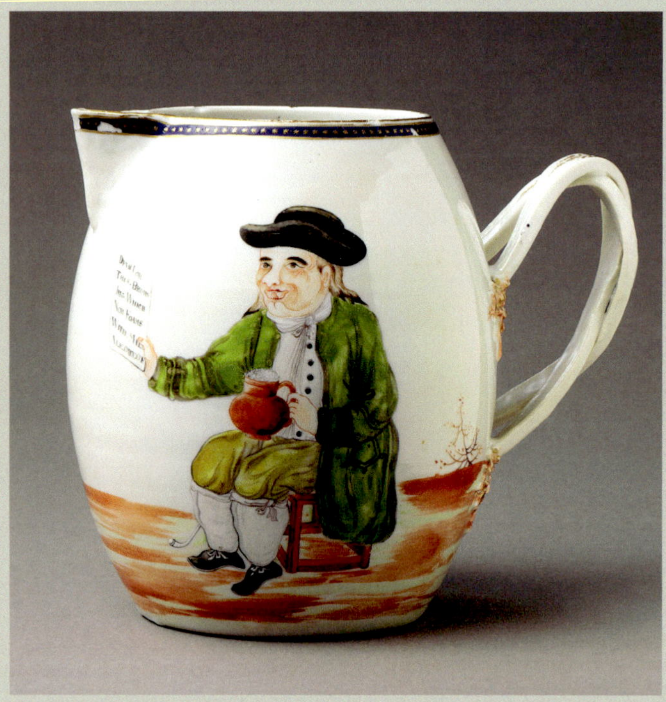

No heel-taps, taking down a peg or two, bottoms up, down the hatch, one for the road, a chaser?

Alcohol has many attributes: it brings pleasure, it alleviates pain, it builds good fellowship, it nourishes life and stimulates the creative juices; as Lord Byron said, 'Gin and water is the source of all my inspiration'. Its origins in the natural world make it an ancient and universally recognized metaphor for spiritual renewal, both in life and at death. When King Midas was buried in the early 8th century near Gordion (modern day central Turkey), residues recently analyzed from drinking vessels and libation bowls in his tomb show that the Phrygians drank a punch of mead, beer and wine. Molecular biologist Patrick McGovern at the University of Philadelphia identified fermented beeswax, calcium oxalate or beerstone, and tartaric acid – proof of grapes.

European drinking traditions have deep roots. Plato's *Symposium* set the pattern of men discussing ideas, playing games and drinking together – the origin of clubs in early modern Europe. Princes drank from a state glass or cup with a cover, which was ceremoniously raised to the sound of trumpets, just as the monarch kept his head covered when others bared theirs. Even today toasting is an intense ritual, a gender-determined reinforcement of hierarchy and personal bonds through an intimate pledge that still requires an exchange of glances. Women are often passive partners; in Sweden at formal dinners, a woman cannot drink at all until a man toasts her.

Drink has stimulated a rich material and visual culture. From caricatures by Cruikshank to Mouton-Rothschild wine labels, from gilded Bacchic wine wagons to sleek modern decanters, from porcelain punch pots and stoneware quart tankards to silver tumbler cups for spirits, an amazing diversity of objects and works of art have been devised to gratify the needs of drinkers. To attract attention in a competitive market, brewers, distillers and wine merchants have been inventive in exploiting design and colour for witty posters. Bass, the first company to register its trade labels of the red triangle and diamond, was particularly enterprising; note the exotic bottle of Bass prominently displayed with wine and spirits in Édouard Manet's *A Bar at the Folies-Bergère* and in the wall advertisement in John Henry Henshall's depiction of a Caledonian Road pub of the same year, 1882.

Across Europe showcases are crammed with a medley of glasses, cups and tankards in materials from the cheapest wood or pottery to gilded silver or gem-set rock crystal. Once, long ago, these were all called for because specific situations and drinks required particular vessels, graded by wealth and social class, as well as by local custom and personal idiosyncrasies. But their precise functions are forgotten; how are we to identify the playwright Thomas Heywood's 'mazers, noggins, whiskins, piggins, wassail bowls, ale bowls, Jacks tipped with silver', listed in 1635, not to mention tun cups, hottwater cups, College cups, Mounseer bowls and the many varieties of caudle, posset and punch vessels named in contemporary English inventories?

Some tankards and glasses seem too large for one and indeed sharing vessels was quite acceptable, although a Massachusetts magistrate in late Stuart Boston, Samuel Sewall, felt humiliated when he had to take communion from a tankard rather than the usual prestigious cup, because on that occasion the congregation outranked him. Costly porcelain or silver tankards for home-brewed beer, personalized with heraldry or scenes, were a daily table item, far removed from today's sturdy pub glasses or tavern mugs of pewter or stoneware. A reminder of hundreds of years of local regulations and national excise duties are the capacity marks, lines or stamps on these tavern pots.

Looking at the history of drinking, rather than at the objects alone, explains many anomalies. Glass wine bottles, for example, smaller than the Imperial measure introduced in 1824, were drained by the hundred but rarely turn up in rubbish pits, because a tax on glass made it worthwhile to recycle the cullet, or scrap. Wine was drunk not as the sole accompaniment to food but as an alternative to beer, ale or cider, reserved particularly for toasting and for after-dinner drinking, in small glasses, almost like a liqueur. Today's large glasses lead directly to greater consumption and problems of over-indulgence.

Drinking according to rules was a means to alleviate the boredom inherent in small, tight-knit societies. It relieved tedium, stimulated convivial behaviour and consolidated fellowship; 'to drink at table without drinking to somebodies health, especially among middling people, would be like drinking in a corner and would be reckoned a very rude action'.

Challenges for men abounded north of the Alps, sconcing (see p.73) vanished from Oxford and Cambridge only a decade ago and military establishments still exploit drink and rough games after dinner as a test of endurance. Guests at the Saxon court faced many occasions for drink, whether visiting the Dresden *Rüstkammer* (armoury) or the *Jägerhof* (hunting lodge), when special large vessels, such as the 'Old Woman with her 14 Boys' (beakers) had to be drained. In the 1640s John Evelyn, approaching the gates of a Swiss town, had to drink from a tall welcome cup carried by two guards, a traditional greeting to distinguished travellers.

Because a minimum intake of liquid is essential for survival, and because pure water was neither available nor indeed considered healthy ('water weakens a person' was the medical opinion), for centuries beer, cider and wine produced from whatever grew locally were central to the European diet. Extreme thirst drove many decisions. Why in November 1620 did the *Mayflower* pilgrims land on chilly Cape Cod instead of hospitable Virginia? 'We could not now take time for further search or consideration, our victuals being much spent, especially Beere.'

Alcohol was antiseptic, anaesthetic, preservative, and a valuable source of nutrients. Beer made from local barley was patriotic and healthier than drinking water, and the tavern was the focus of social life, often equipped with tankards and

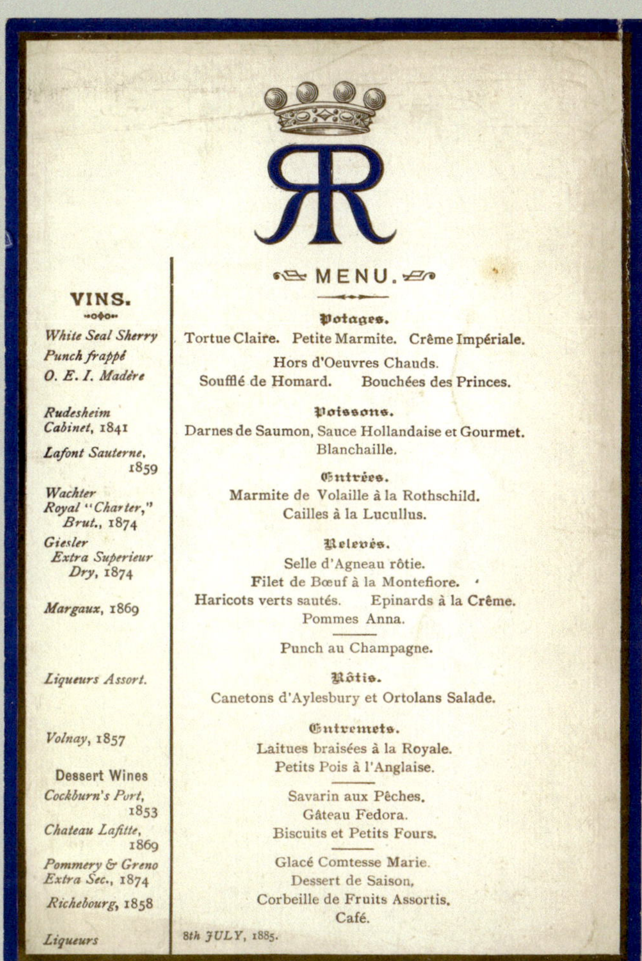

This lavish dinner, celebrating the first peerage for a British Rothschild, matches old, costly and rare wines to an eight-course menu.

punch bowls of precious metal, as witnessed by the many thefts of silver recorded in the Old Bailey Sessions Papers. Drinking at work was normal; 'strong beer made strong to labour' until the Industrial Revolution and the temperance movement allied to attack the practice, and drink was often supplied in lieu of cash wages, until the Truck Acts of 1831 and 1887 ruled it out. Beer was part of the diet for women, children and workhouse occupants. A tap for selling beer existed in every prison, although Daniel Defoe praised the rule for Bedlam in the second volume of his *Tour throu' the Whole Island of Great Britain* (1725): 'That no person be allowed to give the lunaticks strong drink, wine, tobacco or spirits, or to sell any such thing in the Hospital.'

A quick pick me up, 'Dutch courage', and an antidote against surfeit, spirits were a revolutionary innovation and created a whole new set of drinking practices and objects. On active service or out hunting, a swig of *genever*, *aqua vitae*, vodka or *brandij-wein* from a portable screwtop flask gave a quick lift, and required neither ceremony nor a servant with a corkscrew and a glass. Spirits did not spoil, and were lighter than beer or wine to transport. Tumbler cups with curved bases, lacking the foot or handle of wine and beer vessels, were devised for a quick private swallow, far from the long-established conviviality of the tavern or inn, where to drink alone was bad form. But spirits took on traditions too; rum and pink gin became the drinks of the British Navy, Scotsmen, as Dr Johnson noted, took a dram of whisky for breakfast and aquavit is still drunk with crayfish in August in Sweden.

Silver hottwater cups, shallow and small enough to slip in a pocket, were on sale in Cheapside, London by the 1620s, and Queen Elizabeth I owned a silver *aqua vitae* bottle in 1574. Spirits or hottwaters, at first a medicinal remedy and stimulant, were cheap, strong and addictive, triggering medical criticism as early as 1726.

Wine was always considered healthier than spirits, although the reinforced wines liked in Georgian England, especially port and Madeira, were recognized as risky. In 1767 Horace Walpole, suffering from the acute discomfort of gout, was unhelpfully advised by his friend Thomas Gray: 'Remember it is only the wine-drinking nations that know what the gout is; whereas those that even indulge themselves in distilled liquors, as well the laborious and hard-faring populace as the indolent and luxurious, are utter strangers to this.'

Wine might be considered safer than spirits, but from 1672 the English had to pay high import duties on it. In that year two hogsheads bought in Bordeaux cost four pounds, but by the time the Duke of Bedford took delivery, the cost had risen to fifteen pounds, with customs duties, freight, cooperage, cranage, porterage and cartage. This restrictive policy, devised to protect the distillers and brewers, had the inevitable consequences of adulteration, cheap spirits, smuggling and addiction. Thomas Jefferson understood its foolishness:

No nation is drunken where wine is cheap; and none sober, where the dearness of wine substitutes ardent spirits as the common beverage. Fix but the duty at the rate of other merchandise, and we can drink wine here as cheap as we do grog, and who will not prefer it? Its extended use will carry health and comfort to a much enlarged circle.

Drink history has its national myths. Champagne has had a special status since the late seventeenth century; for Madame de Pompadour it was 'the only drink which leaves you still beautiful after having drunk deeply'. Uniquely, Spanish bars retain, just, the tradition of *tapas*, simple slices of bread and Manchego cheese or smoked ham to cover (*tapar*) a drink, a custom abandoned by English tavern keepers as profits came under pressure in the Napoleonic Wars. In the 1820s an

500 years ago, a potter in the Rhineland made this jovial figure of a tavern keeper clutching a glass and a jug of beer.

English traveller in Italy was struck by the glass of *vermut* (wine flavoured with wormwood) he was offered as an aperitif, since this agreeable custom was unknown at home.

Crude stereotypes classify the French as wine drinkers, the Germans and Anglo-Saxons as beer drinkers and the higher social classes as preferring wine. In Germany even today, the average annual consumption of beer is 170 litres for every man, woman and child, more than milk, wine and soft drinks combined. And yet these stereotypes are not universally true, nor can they be projected backwards. Only 50 per cent of the French now drink wine. Verses beneath Hogarth's polemic engravings of 1751, *Beer Street* and *Gin Lane*, 'Beer, happy product of our Isle / Can sinewy strength impart / And wearied with fatigue and toil / Can cheer each manly heart', mock the water-drinking French. But the French did dilute their wine and despised Anglo-Saxon loss of control from intoxication, as did the Italians.

Matching wine to food is recent, a product of late

nineteenth-century bourgeois anxieties and the rise of the
competitive dinner party. As André Simon, the great wine
writer, commented, 'Wine is like sex, in that few men will
admit to not knowing all about it'.

In Paris in 1847 Eugène Delacroix deplored the new
attitude to dining, with menus that attempted to match food
and wines: 'How chilly and tiresome is this modern fashion
for dinner parties…. The fragile glasses – an idiotic
refinement! I cannot touch my glass without making it shake
and spilling half the contents over the cloth'. As philosopher
and historian Theodore Zeldin pointed out, analyzing the
roots of the burgeoning wine culture of nineteenth-century
France, 'Wine was seen as responsible for that part of the
French national character which was most admirable –
cordiality, frankness in human relations, good humour, the
gift for conversation and delicacy of taste.'

Visual evidence for the entrepreneurial energy of vintners,
brewers and distillers is rich and diverse. Peddlers of
pleasure, they have worked around the legislators, finding
ways to satisfy their customers' thirsts and sustain interest,
even as tourist attractions. Running Britain's largest brewery
in Chiswell Street, London, Samuel Whitbread I invested in a
Watts steam engine and built a capacious Porter Tun Room.
He received the honour of a royal visit from George III and
Queen Charlotte and the engine was published in the
Gentleman's Magazine. Vats for beer and wine could be
massive; at Heidelburg the Great Tun in the cellars of the
ruined castle was already an international tourist attraction
by 1600. This huge barrel was re-made in 1751 from 130 oak
tree trunks. Some 8.5 metres (28 feet) across and 7 metres
(23 feet) tall with a dance floor on top, it holds 220,017 litres
(58,124 gallons). A Prussian prince marvelled at the scale of
the tun in 1793: 'we were at least 20 people having breakfast
in a comfortable manner on top of it, I came down pretty
drunk in the end'.

Efforts to curb drinking have met with limited success.
Temperance was one thing, prohibition another; in fact,
denial simply bred illegality. Edward Marks, an English wine
merchant active in the USA in the 1920s, recalled devices
such as the American 'Cordial Shops'. They displayed soft
drinks, as at Rainbow Cordials on West 74th Street, which in
fact was selling Johnny Walker at $3 a quart and $30 a case.
As soon as prohibition was removed, the drinks trade rapidly
resumed, so that, for example, 79 imported port wines were
competing for business in the US market in the early 1930s.

For over 200 years public policy see-sawed between
encouraging and repressing drink. Attitudes remain
ambiguous; unease about addiction and excessive drinking
sits uncomfortably beside the economic clout and
inventiveness of the drinks industry and the revenue
generated from drinkers over the past three centuries. Excise
duties, customs and licences made up as much as 40 per
cent of government revenue in Britain until income tax
became a more significant income stream after
1900. Contemporary concern also
reflects the long-running struggle
between raising revenue and
protecting the health of the nation,
as against traditions of liberty.
Deaths from alcohol-related causes,
such as chronic liver disease, cirrhosis,
accidental poisoning and heart attacks,
rose to 8,380 in 2004, double the figure in
1991, and yet every Sunday supplement
offers discounted wines, in boxes, plus a
regular listing and comment, based on
trade PR. Attempts to achieve alcohol's
beneficial effects without the downside
of addiction or headache flourish; an
Observer article in August 2006
reported on research by David Nutt,
professor of psychopharmacology at
Bristol University, to devise a drink,
without alcohol, which would
induce euphoria without side
effects, exploiting drugs called
benzodiazepines.

So the social history of drinking
practices over the past four
centuries encompasses pleasure
and pain, tradition and innovation,
and extraordinary creativity in the
decorative arts, in architecture for
drinking and in advertising.

A post-War decanter from a
Scandinavian glasshouse
epitomizes the modern virtues
of simplicity and clarity, in an
easily grasped bottle shape.

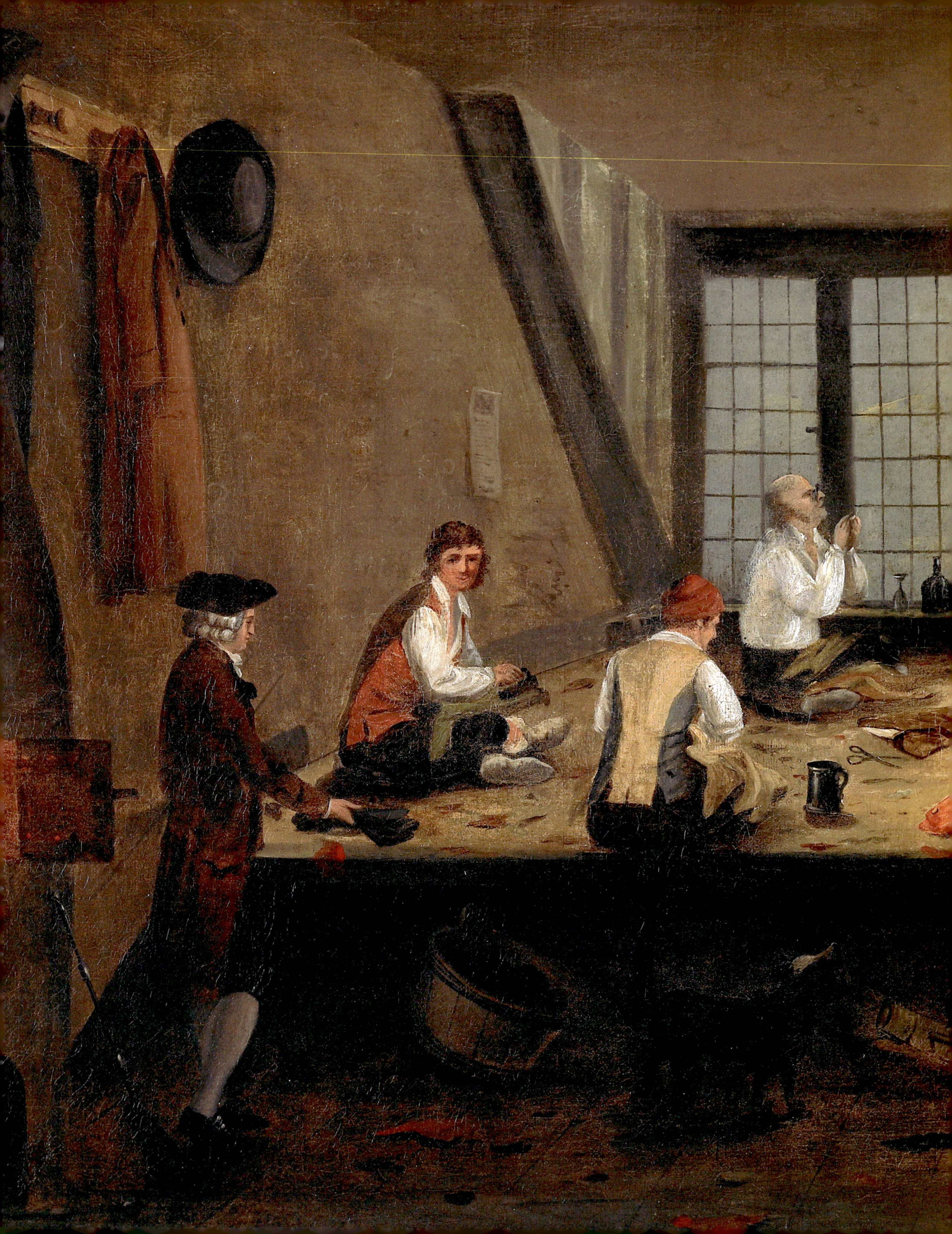

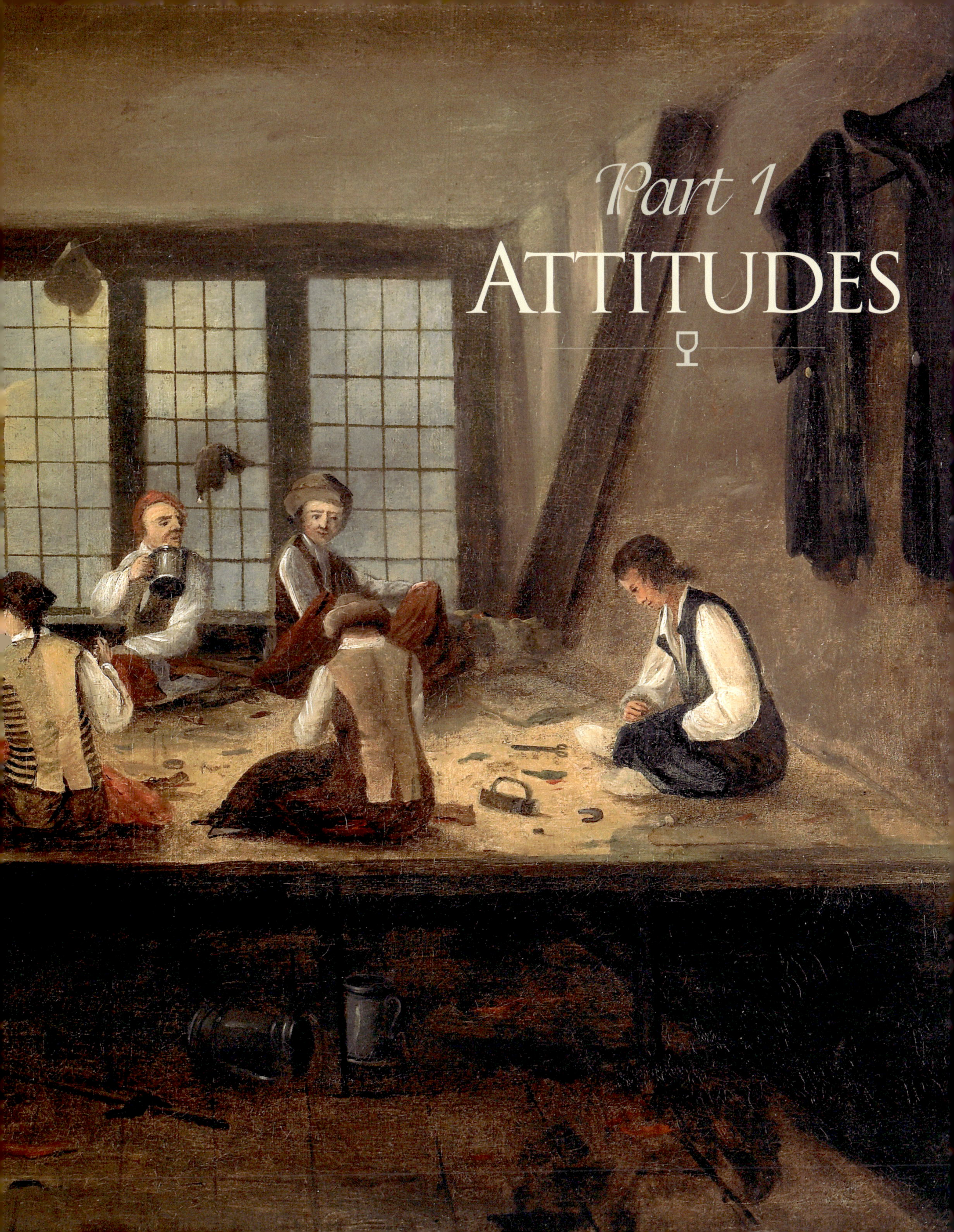

Part 1
ATTITUDES

Alcohol as Medicine

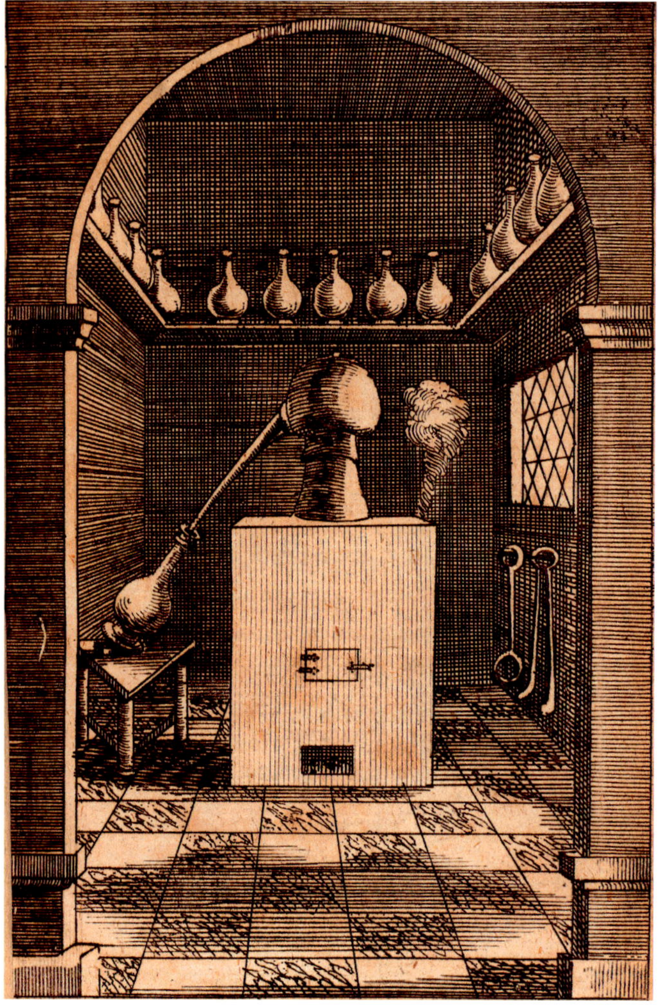

*A*lcohol has been praised and vilified in equal measure over the centuries. Its negative effects sadly are only too clearly recognizable, but its proven benefits as an antiseptic, anaesthetic, preservative and valuable source of nutrients have long been understood and appreciated by a surprisingly wide range of cultures for as long as records exist. Even so, it is alcohol's immediate future that heralds the brightest phase in its history so far; some alcohols, red wine in particular, seem to offer the potential to prevent and even cure many of the Western world's most insidious health afflictions.

Without doubt alcohol is a highly effective antimicrobial agent, particularly useful in conditions where access to

Previous pages: Drinking in a London workshop: tailors and printers refreshed themselves through the day with beer from quart pewter tankards, refilled at a local tavern.

water, clean or otherwise, is limited. So found the Good Samaritan who treated with oil and wine the wounds of the man injured beside the road. Florence Nightingale also came to respect alcohol's antiseptic value during the Crimean War. And Persian armies of the sixth century BC are reputed to have added red wine to water half an hour before drinking in order to render it safe. Wine, whatever its quality, harbours no harmful bacteria.

The Babylonian Talmud describes wine as 'the foremost of all medicines; wherever wine is lacking, medicine becomes necessary'. This quotation clearly underlines the ancient understanding of the very positive link between wine and health, despite the strict prohibition of alcohol for religious reasons. In the early eleventh century the Bukharan physician Ibn Sina (Avicenna) recommended the anaesthetic potential of alcohol: 'If it is desirable to get a person unconscious quickly, without his being harmed, add sweet smelling moss to the wine, or lignum aloes.'

Until recently it was thought that the Arabs were the originators of the distillation process, since they were distilling rosewater into perfumes and essences well before AD 1000. The first coded references in Europe to wine being distilled to spirit or *aqua ardens* occur in a Rhineland monastic manuscript, a copy of a Latin technical treatise, the *Mappae Clavicula*, of about 1190.

As a preservative of flesh, the role of brandy in conserving Nelson's corpse after the Battle of Trafalgar is particularly memorable. As a preservative of potions, spirits distilled from grain, grape or other fruits served as the base liquid or 'mistress of all medicines', as they were described by Hieronymus Brunschwig, a fifteenth-century German pharmacologist. Many religious orders developed their own therapeutic concoctions, particularly digestives to warm a cold stomach. Chartreuse, first commercially available in 1764, and Benedictine (1510) are two such liqueurs that are still celebrated; tin miners in China supposedly alleviate the worst symptoms of rheumatism with Benedictine. And as an aid to the human soul the spread of *aqua vitae* across Europe in the 1340s may have been hastened since it was believed that brandy, in particular, was a physical antidote to the Black Death, an affliction that was often believed to be a punishment for the guilty.

As a nutritional supplement, beer, which has been brewed domestically for thousands of years in almost all societies, has also played a valuable role, both as an additional source of calories and as a source of vitamins. Because the water for brewing was boiled, even polluted water could be transformed into something healthy and thirst quenching. A supply of clean drinking water was a luxury for most people in Britain until outbreaks of cholera from the 1830s in London and the Midlands were finally linked to polluted water, and public health reformers promoted piped supplies in larger cities. Currently researchers are investigating the health benefits of aspects of beer, and one recent study (2006) has even spotted a correlation between young males drinking beer to excess and their later achievements as entrepreneurs.

Cordial waters, home-distilled spirits for which many recipes appear in seventeenth-century household books, such as Anne Blencowe's of 1694, were recommended as good against 'any infections as ye plague, pox, measles, burning fever'. A caution followed, since their delicious warmth was recognized as tempting: 'But neither this nor any other hott waters should be used ordinarily, but when ye party hath need of such help.'

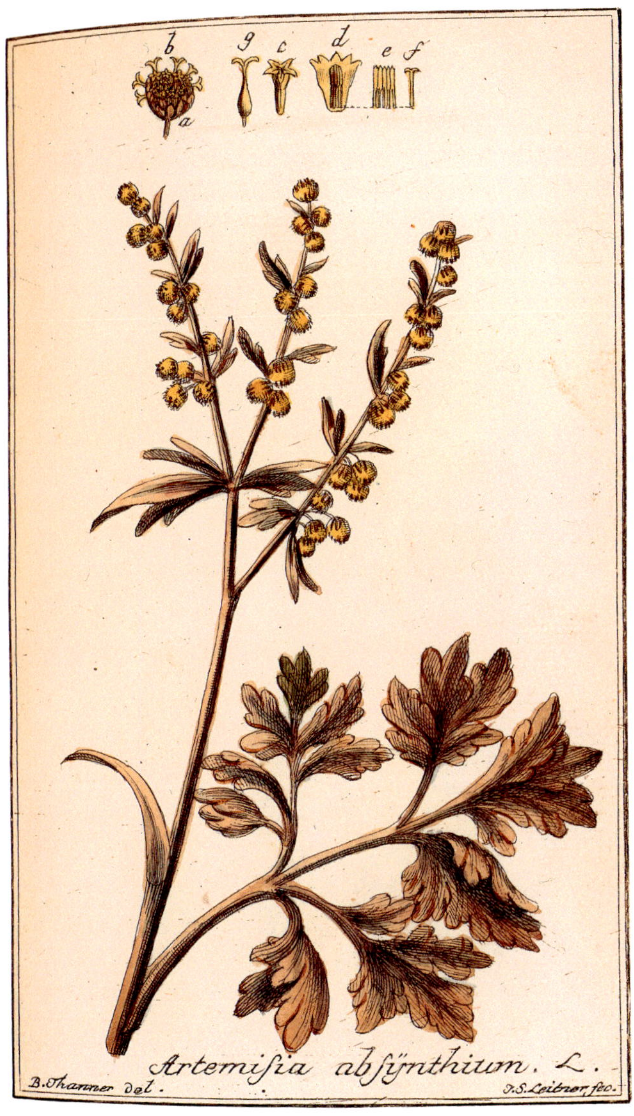

Distilling secrets

The distinguished food historian C. Anne Wilson has speculated that Gnostic Christians already knew the secret of distilling wine long before the 12th century; two Coptic texts refer to transforming wine into 'a baptism of fire' for initiates. If the Gnostic Christians did have the secret of distilling, they may have passed it to the Cathars, who also had a fire-baptism ritual. Attacks by the Inquisition eventually wiped out this heresy in Western Europe, and monks were initially prohibited from owning and practising distilling equipment, for example by a Dominican Provincial Chapter at Rimini in 1288. But this prohibition did not last. The distilling process was a powerful tool that enabled alchemists and monks to produce aqua vitae or 'water of life', a valuable preservative of flesh, potions and, potentially perhaps, souls. By the medieval doctrine of humours, spirits that were both moist and warm were an excellent antidote to balance the constitutions of those who had become cold and dry – that is the elderly. So their aches and pains were alleviated either through drinking a little diluted spirit or by rubbing spirits on the affected area as an embrocation.

In the seventeenth century Dutch Cape doctors believed that alcohol could help to cure scurvy; in the last century Guinness was often prescribed as a source of iron and Vitamin B for pregnant women, and in the Low Countries a nursing mother is still likely to be presented with six bottles of stout. During the term of Morarji Desai, the vegan, teetotal president of India (1977–79), alcohol, the local wheat-derived spirit, was permitted in the mountain states only as an essential source of vitamins A and B.

The medicinal benefits of alcohol lost ground in the later nineteenth-century, partly as the pharmaceutical industry devised pills and potions for many conditions and partly as the worst effects of addiction to alcohol became more

Water Weakens a Person

Strong beer made men strong to labour. Beer, rich in calories and vitamins, was a dietary staple until 150 years ago. The daily diet, a monotonous choice of salty preserved foods – bacon, fish, cheese and butter – stimulated thirst. At Kilmainham Hospital, Dublin in 1699 soldiers were allowed 3 pints of beer to moisten their daily bread, gruel, meat and cheese. London artisans clubbed together to contract for a regular supply of beer, often dark porter, from a local alehouse, with bread and cheese thrown in. Work started early and ended late, with no formal meal breaks. Children and women drank small (weak) beer, about 3 or 4 per cent proof.

By the mid-nineteenth century beer was losing its importance as clean piped water became widely available, taxes on tea and sugar fell and abstinence from all alcohol became admirable.

apparent. Campaigners for prohibition in Northern Europe and in the USA underlined all that was negative (p.133). Any positive attributes appeared to be largely forgotten until the 1990s when US doctors began to explore the 'French Paradox': the seeming contradiction of the French who typically consume a high fat diet yet report a low incidence of coronary disease. The magic ingredient appears to be red wine.

According to recent research, red wine appears to be able to offer at least some protection from a vast range of health problems, including heart disease and hypertension, cancers, Alzheimer's, asthma, herpes and HIV. In addition, it seems it can also protect and strengthen blood vessels damaged perhaps by diabetes, enhance hair growth, encourage more youthful-looking skin and even improve eyesight for limited periods of time.

The critical elements are polyphenols found in grape skins and other dark-coloured fruits, green tea and certain nuts. These are among the most potent plant antioxidants. Of the many polyphenols identified, resveratrol

(pronounced rez-VER-a-trol) is believed to be of especial importance. Resveratrol is a natural substance made by grapes and other plants in response to fungal infection. An antioxidant, it can reduce the damage by free radicals to arterial cells that might otherwise become clogged and restrict blood flow. In cancerous cells resveratrol helps to starve the cells by inhibiting the action of the key protein (known as nuclear factor-kappa B) that feeds them. A recent study demonstrated that 'resveratrol exerts antiproliferative effects on prostate cancer and … breast cancer cells'. Resveratrol may be the best cancer-preventative agent available today.

According to some researchers, the total amount of resveratrol in one glass of red wine three or four times a week is the right amount to block the protein from feeding cancer cells. But exactly how much resveratrol is in a glass of wine depends on how the grapes were grown. Grapes sprayed with pesticides that prevent fungal infection contain little, if any, resveratrol so, logically, organic wines have much to recommend them. Drinking more than the recommended amount of wine, however, would stop the beneficial effect and may actually lead to a greater risk of cancer. Moderation in all things, the ancient motto, seems still to be valid.

Clearly quantity is a critical issue but current research would genuinely seem to suggest that a little of what you fancy does indeed do you good. As the Apostle Paul famously advised Timothy, 'use a little wine for thy stomach's sake and thine often infirmities' (1 Timothy 5:23).

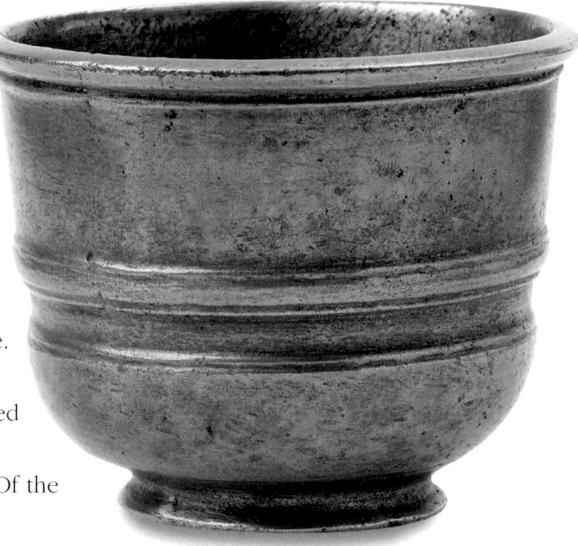

Left: White wine, poured into this antimony cup, took up some of the toxic metal and acted as a purgative or emetic, a popular treatment for many ailments before the 19th century.

Opposite: 'Nuisances' in water remained a Victorian problem; on an 1866 map, cholera cases clustered in the district supplied by the East London Water Board from the Old Ford reservoir.

OUR WATER SUPPLY.

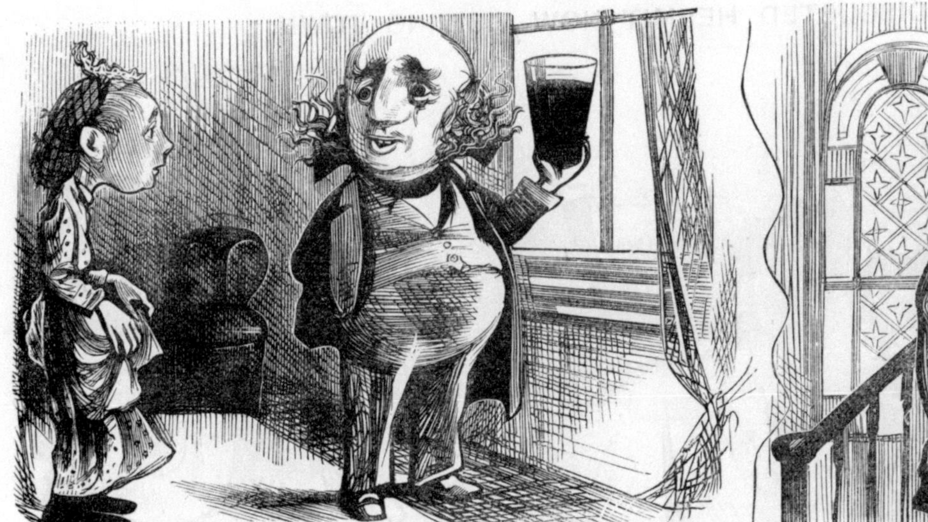

"Jane, this water is not at all clear.   I can see living organisms in it with the naked eye.   Go up and see if there's anything in the cistern."

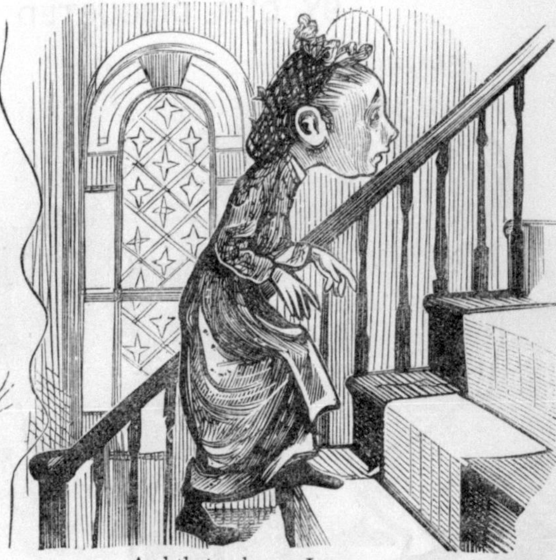

And that unhappy Jane went up.

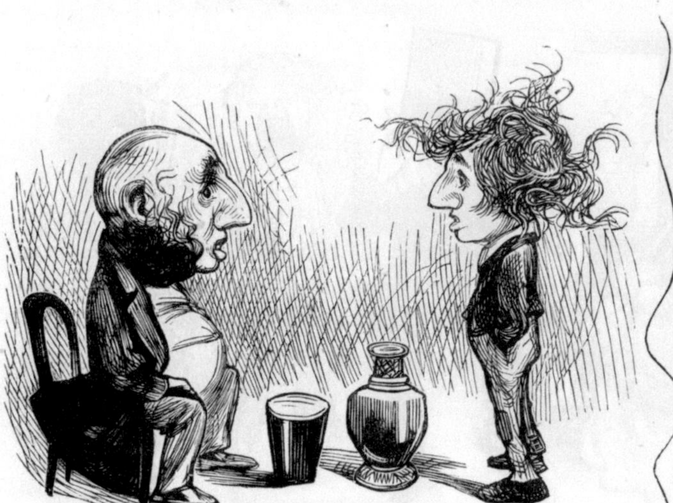

"John, my boy, it's very strange that Jane does not return from the cistern! Go up and see if there's anything the matter."

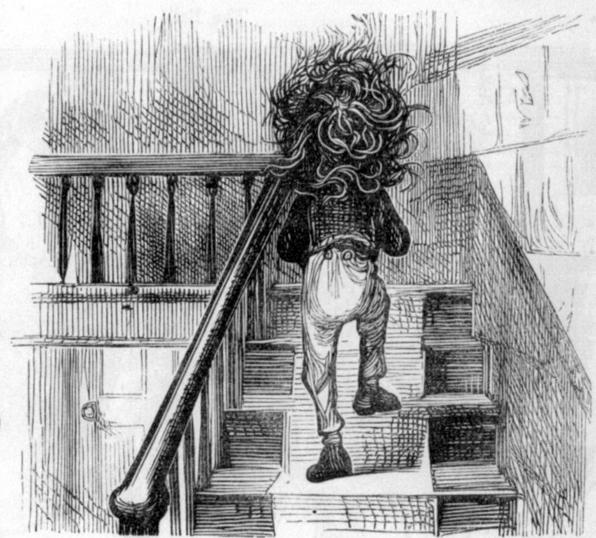

And the ill-starred John went up

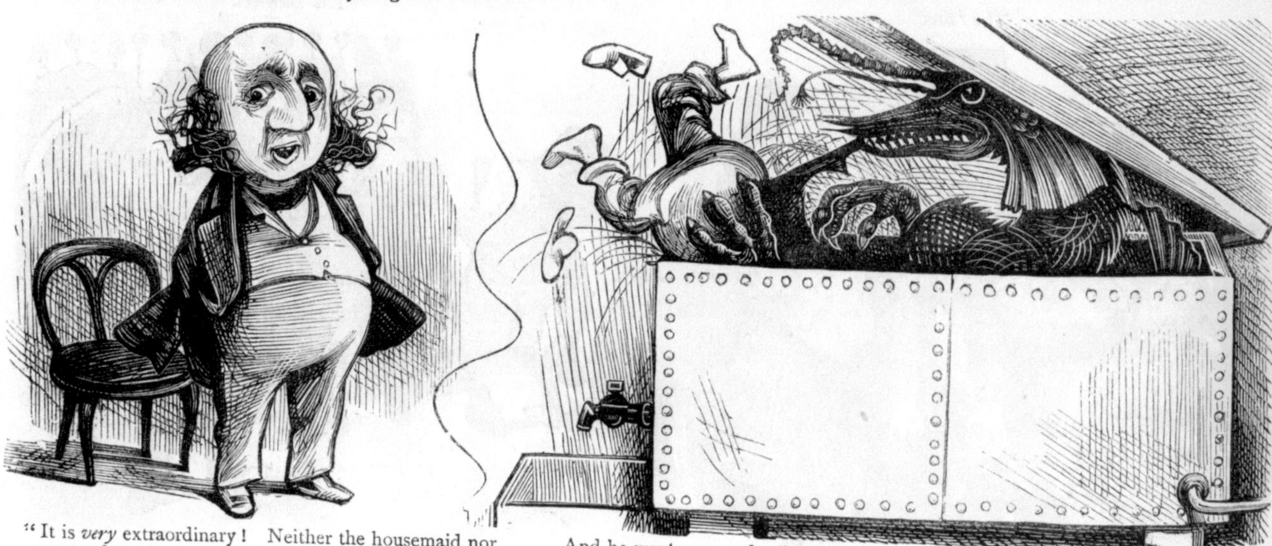

"It is *very* extraordinary!   Neither the housemaid nor my son has returned!   I will go myself and see!"

And he went up; and a Living Organism, which had eluded the careful filtration of the Water Company——

Evils of Alcohol

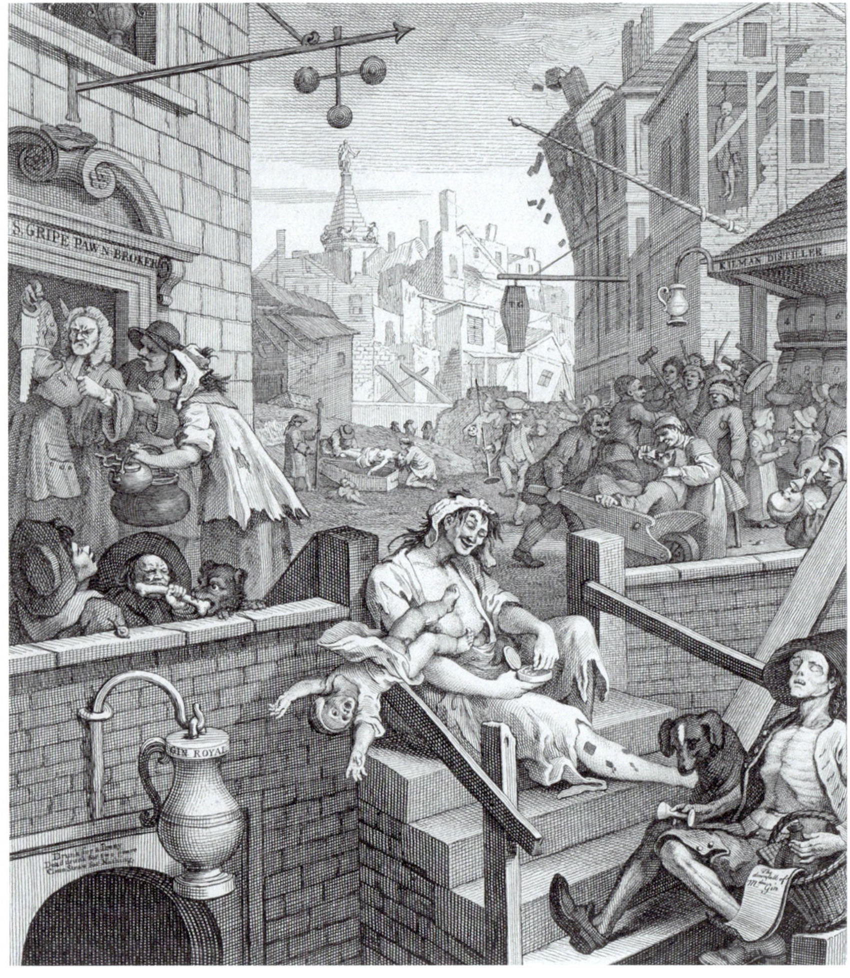

Depraved by gin, or Ladies' Delight, the mother lolls indecently as she drops her baby. In poverty-ridden St Giles, the only prosperous businesses are the distiller, the pawnbroker and the undertaker.

All, as well Brandy as Wine, and all our strong compounded Drinks, such as stout Ale, Punch, Double-Beer, Fine-Ale, are all drank to Excess, and that to such a Degree, as to become the Poison, as well of our Health as of our Morals; fatal to the Body; to Principles; and even to the Understanding; and we see daily Examples of Men of strong Bodies drinking themselves into the Grave ... Men of strong Heads and good Judgment, drinking themselves into Idiotism and Stupidity.

Daniel Defoe, *A Plan of the English Commerce* (1728)

In 2006 Hogarth's vivid *Gin Lane* image, reworked by Jas to depict binge-drinking by vicious youngsters, headed a *Daily Telegraph* article on drunkenness. Anglo-Saxons and Scots, with other Northern nations, seem to have a problem with alcohol, or with self-indulgence, which might be linked to climate as well as deeply-rooted social attitudes. In 2006, the anniversary of Sweden's and Norway's state monopolies for selling alcohol, exhibitions examined its history and sampled attitudes to their (by Anglo-Saxon standards) restrictive laws and the brown-bag culture for reducing the glamour of drink. This is a far cry from the clamorous, witty and appealing advertising campaigns financed by major drink companies in Britain.

Alcohol has always been recognized as a risk as well as a pleasure; self-denial is an element in most religious traditions. Rituals for limiting excessive drinking or censuring boorish behaviour have evolved in all societies. After about 1800, once the medical reality of addiction was acknowledged, caricaturists and temperance advocates emphasized the dangers of excess. Total abstinence, teetotalism, was urged by nineteenth-century reformers. But profits from selling drink cheaply – 'liquid madness at tenpence a quartern' as Thomas Carlyle described gin – and government dependence on revenue from excise duties meant that reformers were always struggling to get their message across. Moral force can operate effectively only if society is sensitive to its arguments, or alerted to a threat, as with a recent panic over under-age drinking, or female addiction rates.

Temperance was seen as a beneficial discipline for mind and body; when Margaret Cavendish described the spare diet of her 73-year-old husband William, Duke of Newcastle, she cited him as a model of a healthy life:

he is so sparing and temperate, that he never eats and drinks beyond his set proportion. He makes but one meal a day, at which he drinks two good glasses of small beer, one about the beginning, the other at the end, and a little glass of sack in the middle ... which glass of sack he also uses in the morning for his breakfast with a morsel of bread. His supper consists of an egg, and a draught of small beer. And by this temperance he finds himself very healthful....

A 1682 pamphlet, *A Warning Piece to All Drunkards and*

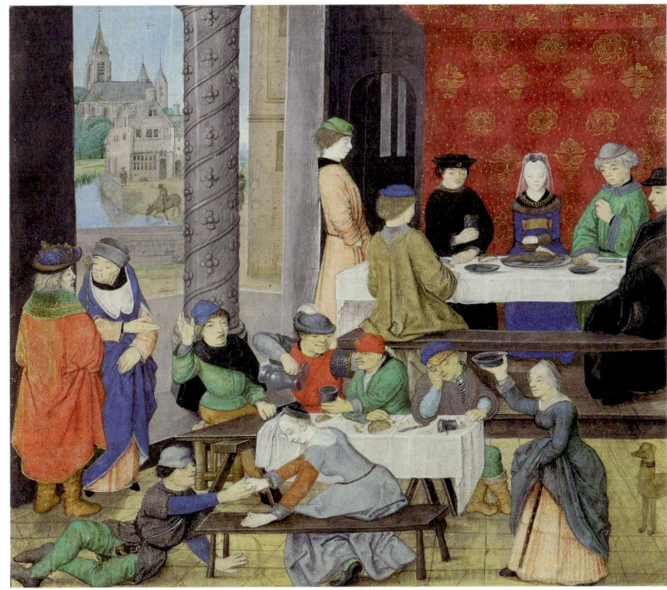

Health-Drinkers, expressed Puritan criticism of intemperance and sharing drinks. Concerns about the deleterious effects of gin-drinking on fertility and infant mortality were voiced in a College of Physicians report in 1726, a quarter of a century before Hogarth's lolling mother letting her infant fall from her arms captured public alarm. Lloyd George was able to undermine this tradition of conviviality only in the First World War, when treating, or buying rounds of drinks, was prohibited.

Doctors had few doubts about the damage done by hard drinking. Although the cause of gout, an excess of uric acid, was not understood, for centuries excessive alcohol consumption has been blamed. Horace Walpole sourly wrote to a friend George Montague after an attack in 1769, 'Consider that the physicians recommend wine, and then can you doubt of its being poison?'

Above: Temperate drinkers contrast with a riotous group. Platt's *Jewell House of Art & Nature* (1594) recommended 'Drink first a good large draught of sallet oil for that will float upon the wine … and suppress the spirits from ascending to the brain'.

Right: A broken tankard and overturned three-legged stool, symbol of the alehouse, contrast with domestic harmony and tea-drinking. Taxes on tea and sugar were reduced to encourage abstinence.

Erasmus Darwin, physician and grandfather of Charles Darwin, became a teetotaller, arguing 'a perpetual repetition of so powerful a poison must at length permanently affect' the drinker. The hard-drinking James Boswell made a New Year resolution in Utrecht in 1764, although his recipe for greater temperance still involved alcohol: 'Memo to self: You must take care of your stomach; eat always some toast and drink a little negus [punch] at night, so as not to clog the stomach at one meal'.

Although temperance benefits were understood, sociability and drinking together operated as social cement and was an obligation for those in authority, landlords and leaders of all kinds. In the 1730s, Anne Conolly deplored the fact that her husband William's abstemiousness was not understood by his Irish tenants. 'He has been so busy entertaining his tenants … that his head has not been enough settled to do anything for drinking you know does not agree with him 'tho he must practise it a little on his first coming into the country or he would not please'.

While neighbourly drinking in the tavern was recognized

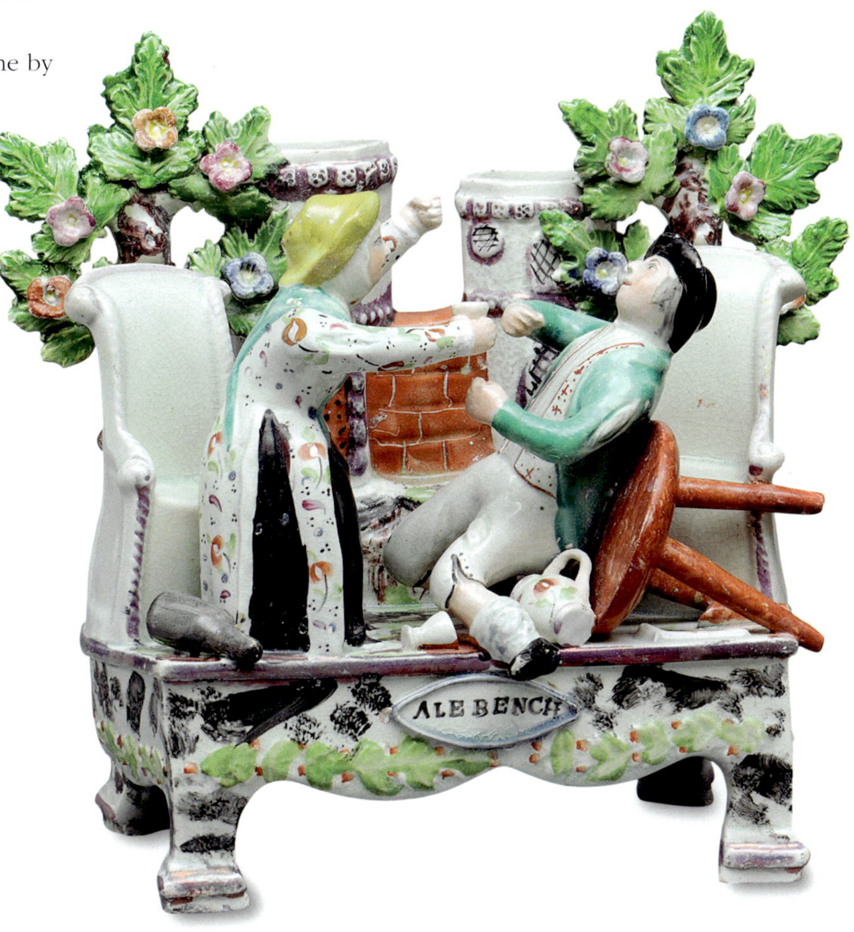

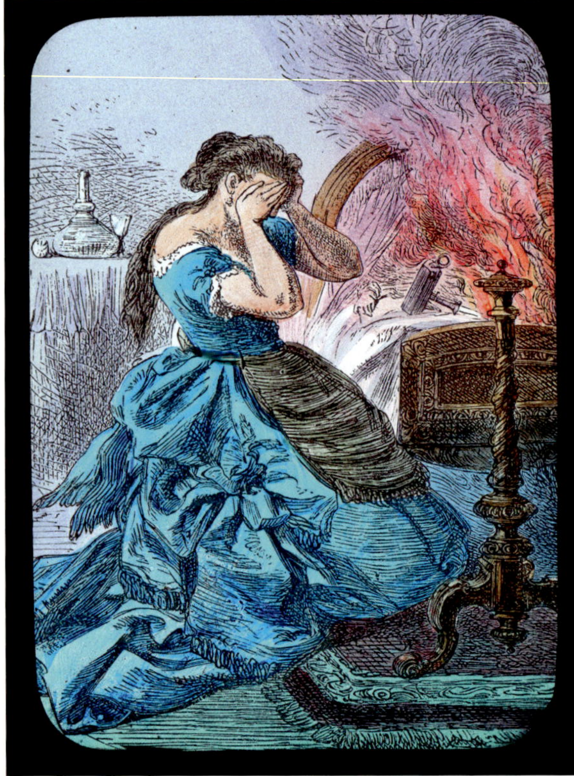

As the drunken mother starts awake, her baby burns to death. An emotional campaign of slide lectures and exhibitions urged abstinence although alcohol consumption reached an all time peak in the 1870s.

both as a need and as a benefit, drunkenness led to disorder and undermined local communities. The constables of Calne in Wiltshire complained that drinkers 'love [their] cup companions so well that no man will take upon him to be a sworn witness'. Feeding the families of incarcerated drunkards and impoverished gamblers created a financial burden. Once cheap spirits entered the picture, drunken behaviour, fuelled by mixing gin with the traditional longer quarts of beer, was seen as a threat to public order.

But the principle of drink as a social benefit continued, even while traditional working-class pleasures centring on the tavern were denigrated or suppressed. A visitor to Lancashire's manufacturing districts in 1842 commented, 'The very essence of our laws has been against the social meetings of the humble which have been called idleness and against the amusements of the poor which have been stigmatised as disorder'.

A strong clarion for the temperance cause was *The Bottle*, a series of eight inexpensive prints published in 1847. With Zola-like realism, George Cruikshank depicted how a mechanic and his family declined through drink via unemployment, the bailiffs and poverty, begging, child

neglect and death, domestic violence, to murder and finally insanity. Within weeks around 100,000 prints had been sold.

In the week that Cruikshank, a passionate Victorian campaigner for abstinence, died, *The Times* decried abstainers for their 'extreme denunciations of an article of food (intoxicating liquor) which has been hitherto valued the most highly by the most civilised and progressive races of mankind'. This argument for the nutritional as well as the social necessity of drink was constantly urged by a discomforting alliance of the drink business and successive Chancellors of the Exchequer. The influential British statesman Joseph Chamberlain put the other point of view in 1876 when he said: 'Temperance reform lies at the bottom of all further political, social and religious reform. The enemies of strong drink are the friends of mankind'.

Before the First World War temperance movements had a powerful impact. This 1920s Swiss poster extols the model family man, turning his back on spirits.

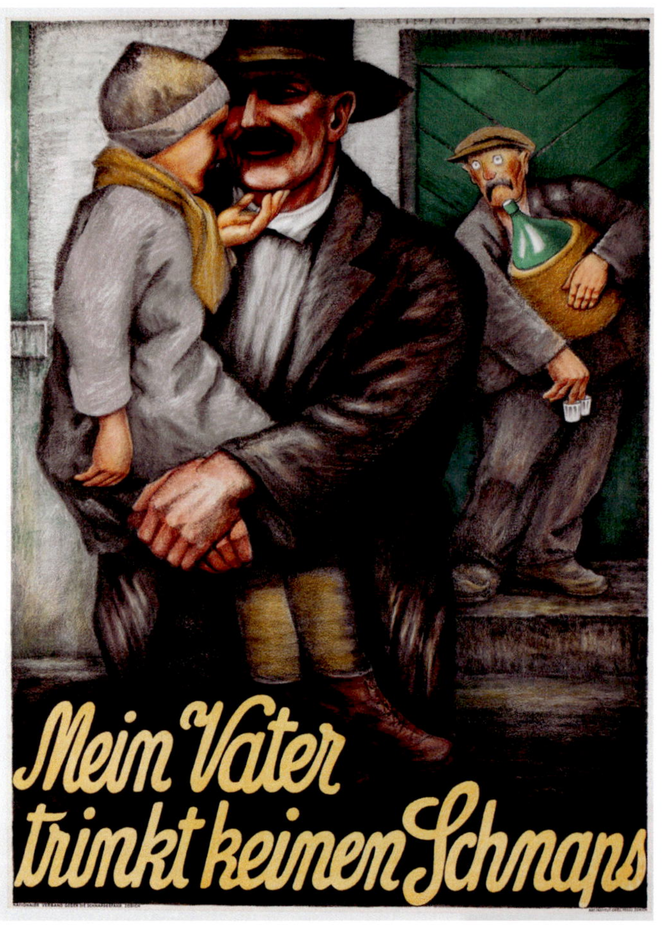

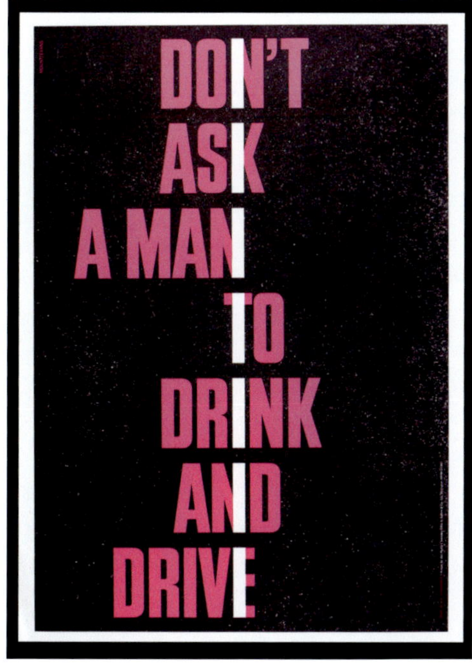

This memorable 1966 campaign changed popular perceptions about the joys of drink and altered behaviour in Britain for good.

Driven by its concern to limit drinking and fear that working women, especially vital munitions workers, would spend their wages on alcohol, the government ordered regular reports on female drinking practices throughout the First World War. In fact, reports from government observers show that these worries about extravagance and indulgence, drawn from the propaganda of pre-war prohibitionists, were misplaced. Overall, health problems from the bad effects of alcohol diminished. Convictions for drunkenness among women dropped by two-thirds from 1914 to 1917 (from 37,000 to 12,000). Cases of cirrhosis of the liver halved. Deaths from female alcoholism dropped from 680 to 220, a dramatic response to the national emergency.

An unholy alliance of the appetite for sugar, with its profitable by-product rum, and slavery created the greatest European shame associated with drink. Rum 'is a sovereign remedy against the grumbling of the guts, a kibe-heel [an ulcerated chilblain] or wounded conscience, which are three epidemical distempers that afflict the country' journalist Edward Ward commented, visiting New England in 1699. A clear-eyed look at the history of the Americas and alcohol is chilling. Rum proved as popular to quench the thirst as sugar to satisfy the palate. But as an immensely profitable by-product of sugar, it was always morally tainted. For the British Navy in the West Indies, rum offered a cheap and convenient alternative to French brandy or British corn-spirit for the daily ration, and as a basis for punch, it quickly appealed on both sides of the Atlantic.

In 1698 England imported only 207 gallons of rum; by 1770 some two million. George Washington demanded a barrel of Barbados rum at his 1789 inauguration. The first distillery in New England opened in 1700, and rum became the American colonies' major commercial industry and export. Sugar and rum helped make the British the richest people on earth, at a high moral price. During the eighteenth century over four million slaves were purchased by the sugar colonies. In a triangular trade, molasses was exported from the West Indies to Britain, France and America for distillation into rum. From America the rum was shipped to West Africa where it was traded for slaves. The slaves were shipped on the extremely brutal Middle Passage route to work the sugar plantations. It was a ring of iniquity.

Opening Hours

Buying drink to consume on the premises, or sending out for a tankard or jug of beer, could be controlled by limiting hours of opening, whereas the freedom for drinkers to take a bottle away for later, a mid nineteenth century innovation, made the authorities uneasy. This was a class and social divide, since those with large incomes and a private cellar, and employees in a large household with its own brew house could drink as they wanted, within reason.

Opening hours for taverns were controlled by local magistrates who issued licences, until the 1830 Beer House Act, which passed the licensing responsibility to the excise. A furious local magistrate in Norfolk complained of the 'Demoralisation of the Peasantry', who could now drink from 4am to 10pm, instead of 6am till 9pm. Legislation restricted Sunday opening until 6pm. Punch satirized the official attitude in 1855:

Mr. Hall, Chief Magistrate at Bow Street, cannot discover any 'inconvenience' in the present working of the Act, but recommends that the poor Sunday excursionist should 'Strap a Knapsack on his Back, with Two or Three Bottles of Beer, and the Child to boot, sooner than the Sunday should be desecrated by Opening the Public-House.'

However, bona fide travellers could buy a drink. In 1892 the thirsty Mr Pooter, epitome of the modestly aspiring City clerk, was too honest when, on a Sunday walk in Hampstead, he admitted that he lived nearby and so missed the refreshing brandy and soda enjoyed by his less scrupulous friends, who claimed to have walked all the way from Blackheath.

Political and Legal Controls of Drink

[I]f the State does not control the liquor traffic, the liquor traffic will control the State.

LORD ROSEBERY WHEN LIBERAL PRIME MINISTER IN 1894

From the late seventeenth century drinking was subject to government intervention. Prices were largely dictated by customs and excise duties; choice and availability by shifts in foreign policy; location by a system of licensing. The English government was not a free agent; its decisions were driven by groups of MPs who represented irreconcilable special interests such as the land, brewing, West India and temperance.

Following Charles II's restoration in 1660, the English court took to drinking French brandy and wines, particularly Champagne and Haut-Brion. When William III became king in 1688, Dutch influences predominated; England inherited a military and economic struggle with France that lasted on and off until 1815. William imposed very high customs duties on French wines and brandy, while promoting indigenous spirits by ending the Distillers' Company monopoly, and throwing distilling open to all who could pay the duty. The landed interest was squared by limiting, through excise duty, the use of any distillate save corn. With duties reduced on the finished product, a pint of gin was cheaper than a pint of beer. England, particularly London, embarked on an addiction to spirits that troubled Lloyd George as late as the First World War.

The Methuen Treaty of 1703 confirmed the exclusion of French products, stipulating that Portuguese wines would never pay more than two-thirds of the duty levied on French wines. Portuguese wines paid a duty of £7 per tun (252 gallons) and French £55 per tun. French wines became a rich man's pleasure and claret the Jacobite wine for toasting the 'king over the water'. Port, strengthened with the addition of Portuguese brandy, became the tipple of Whigs and Hanoverian Tories. The Methuen Treaty was rescinded only in 1831 when duty on all European wines was equalized at 5*s.* 6*d.* per gallon, initiating an age of wine for the mass of consumers accustomed to beer and spirits.

Gin-sodden London, depicted by Hogarth, reflected the inability of eighteenth-century governments, lacking an incorruptible excise force, to make their legislation effective. Consumption, rising from 500,000 gallons in 1688 to 7,160,000 gallons in 1742, was not easily cut back in face of the vested interest of distillers and landowners. In 1751 Parliament reduced dramatically the selling outlets and raised the duty. Consumption of spirits fell from just over 7,000,000 gallons in 1751 to 1,800,000 gallons in 1758.

A competitor to gin was rum, 'a hott hellish and terrible liquor' christened 'kill-devil' by English soldiers in Jamaica in 1651 because of its life-taking properties. From 1742 it was allowed to mature in bonded warehouses and Old Jamaica Rum formed the basis of rum punch. By the 1770s rum had a quarter of the spirit market. For those who could afford it, or have it smuggled, French brandy was the premier spirit. In the 1730s an anchor (4 gallons) could be bought in France for 16*s.* whereas in Britain with duty paid the price was £1. 12*s.* or £1. 5*s.* for smuggled brandy. By contrast smuggled rum sold at 5*s.* a gallon while a gallon with duty paid was 8*s.* 6*d.* Smuggled port and sherry were 2*s.* 6*d.* a gallon and with duty paid 4*s.* a gallon.

Smuggling and illicit distilling and brewing confuse the

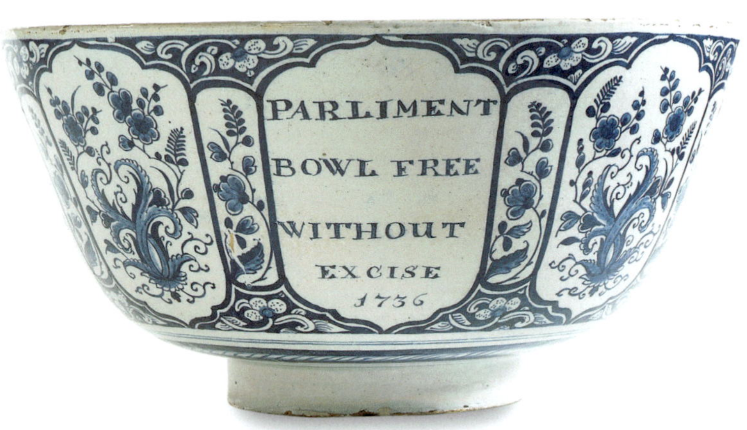

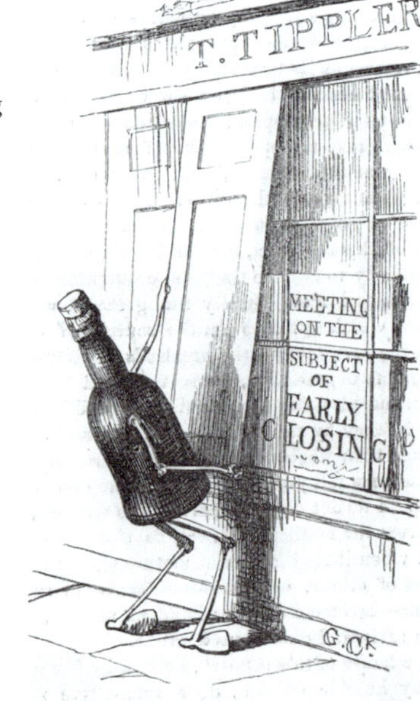

Left: By lobbying and bribes, the Distillers' Company had weakened the 1730 Gin Act. In 1736 the trade faced higher excise duties, imposed on spirits because of 'horrid excess'.

Right: Victorian reformers considered all Sunday pleasures as wicked and public house hours were cut back in 1853. But press criticism and the recent (1848) popular uprisings across the Continent, forced the government to relent.

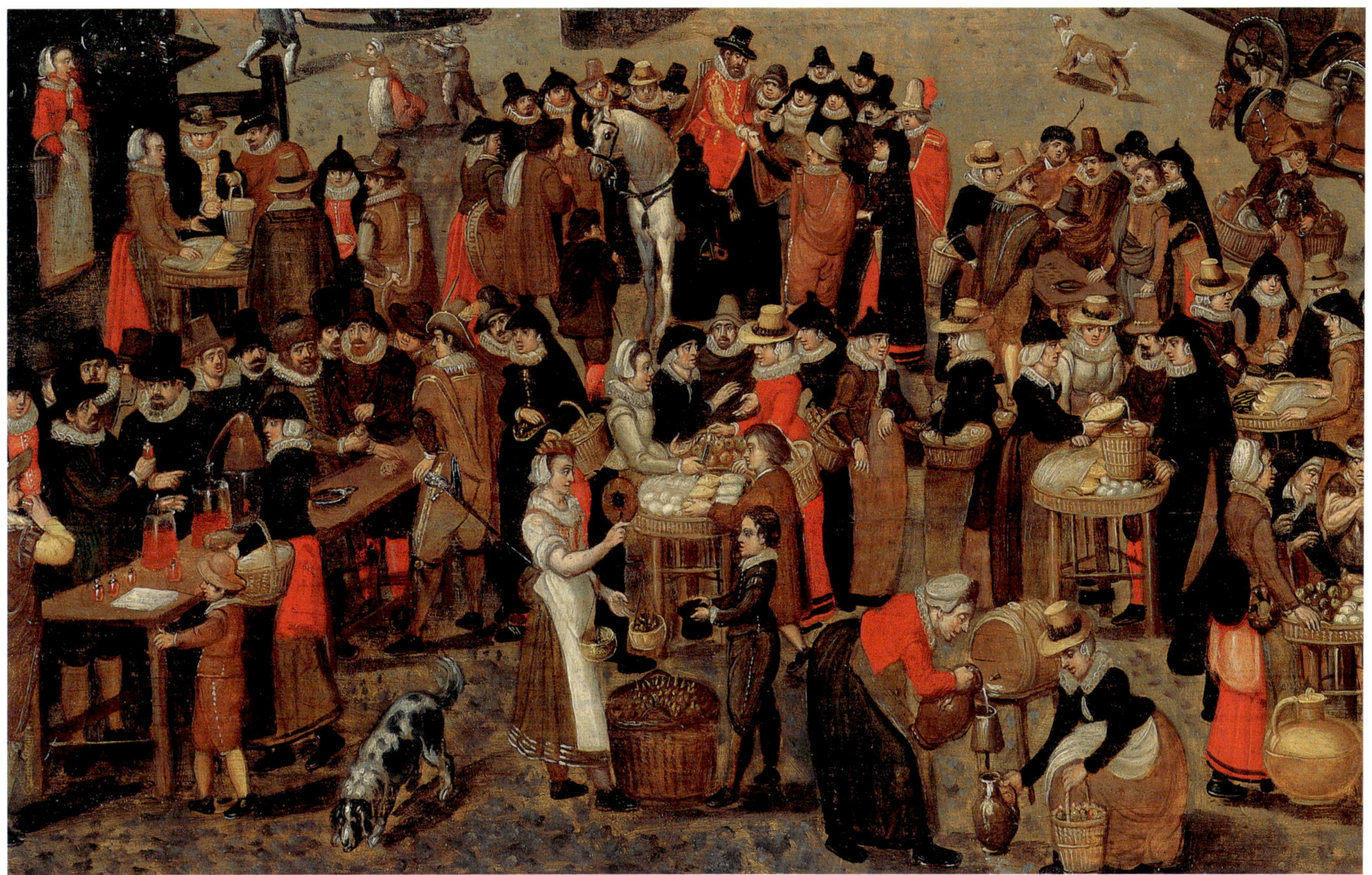

Selling home brewed beer was often an occupation for widows and other poor women. In this Antwerp market, a well-dressed man is selling a distillation of some kind, perhaps medicinal spirits.

statistics as to the overall consumption of alcohol and deny us an accurate figure of how much revenue was lost to the treasury. With a coastline of some 6,000 miles to be protected against smugglers, successive governments had to professionalize the preventative service until in 1856 it was finally handed over to the Navy.

Although spirits have attracted much attention, beer was the daily drink, widely available through inns, taverns, beer houses and home brewing. To the treasury, beer was the acme of revenue raising: duty was levied on the malt, the hops, the maltster's licence, the home-brewer, the brewer's licence and various licences for dealing in or retailing beer. Not surprisingly, in the era before income tax, the liquor trade provided up to 47 per cent of government revenue in any one year. Porter, the favourite tipple of Londoners, improved with keeping and was a capital-intensive operation. So by the late eighteenth century there emerged very large brewing companies from whom it was easier to collect revenue but which also had considerable political influence.

The huge costs of the Napoleonic Wars dictated that all liquors carried a heavier burden of taxation. Double duties were placed on malt and beer, raising the price of a pint (2*d.*) beyond the pocket of many labourers. Duty on a gallon of rum rose from 5*s.* 6*d.* in 1782 to 13*s.* 1¼*d.* in 1810 and was still 9*s.* in 1830. In 1825 the duty on home-distilled spirits was reduced from 11*s.* 9*d.* to 7*s.*; spirit production in England almost doubled. In Scotland and Ireland where whisky had become a regulated product, production almost trebled. At the same time the spirit licence for premises valued at under £20 was reduced from 5 guineas to 2 guineas. The collapse of the West Indian plantation system through soil exhaustion, natural disasters, competition from other centres of sugar production, and shortage of labour diminished the West Indian lobby in the House of Commons from 75 MPs in 1782 to 16 in 1832.

The future was to belong to native spirits – to gin and whisky – and the new challenge of French wine. Reducing the spirit duty saw the re-emergence of Hogarthian London and a steep rise in criminal activity. The solution legislated by Wellington's Tory administration was the Beer Act of 1830, coinciding with the first temperance campaigns. This Act radically reduced controls on beer-selling. As usual a medley

Selling Wine

Wine has always been regulated and sold differently from beer or spirits. Until the 1920s vintners in northern Europe imported most wine in barrels, and few wines were estate or château-bottled. Vintners judged quality, colour and sediment by pouring a small amount into shallow silver wine tasters; mudlarks have recovered at least two with names of London vintners from the river Thames, close to Vintners' Hall. Oliver Cromwell set legal maxima for prices in 1657, with Spanish wine the most expensive at 1*s*. 6*d*. a quart, Rhenish 1*s*., and French the cheapest at 7*d*.

Wealthy families and institutions bought wine by the barrel and had it bottled, whereas drinkers in taverns could buy only by the glass. Unlike alehouse laws, the licences for vintners (tavern keepers) who ran retail wine shops carried no obligation to maintain good behaviour, perhaps because wine cost much more than beer, a daily staple.

of influences were at work: a belief that greater access to beer would diminish the consumption of spirits and thus improve the health and morality of the nation; the removal of magistrates' power to grant licences would break up corrupt alliances between brewer and magistrate; overall it was held to be a freeing of trade. First the duties on beer and cider

were removed and then any person whose name was on the rate book might open his house as a beershop, free from justices' control, on payment of 2 guineas to the local excise office. Within six months 24,342 new licences had been taken out, and in two years there were 39,000 new alehouses.

The anticipated decline in spirit drinking did not take place and drunkenness increased dramatically. Gladstone's Liberal government of 1868–72 attempted to regain the control lost in 1830. Magistrates became the sole licensing authority, opening hours were reduced and adulteration of beer was checked. Disraeli's Conservative government removed in 1874 onerous regulations on licensed premises, gaining the political support of the licensed trade.

In 1860 Gladstone, as Chancellor of the Exchequer, determined to disprove the general conviction 'that an Englishman is not born to drink French wine', reduced the duty to 3*s*. a gallon, introducing a new system of duties differentiating between wines according to alcoholic strength. Duties on port and sherry were reduced to 2*s*. 6*d*. per gallon. This price reduction encouraged the working classes to drink these fortified wines, paralleled by a characteristic English middle-class taste for claret. From 1859 to 1876 French exports of wine to Britain rose from 695,911 gallons to an astonishing 6,745,710 gallons.

Statistics confirm that drinking more wine did not reduce

Lid of a 16th-century measure, probably for the 'Mitre' or 'Bishop's Head' tavern. The devices discouraged theft and pawning. A crowned 'HR' stamp recorded official local control of its capacity.

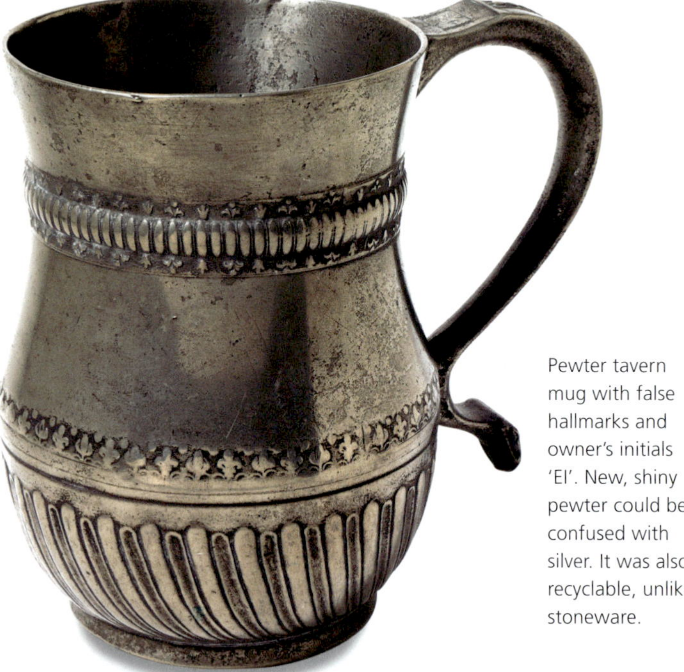

Pewter tavern mug with false hallmarks and owner's initials 'EI'. New, shiny pewter could be confused with silver. It was also recyclable, unlike stoneware.

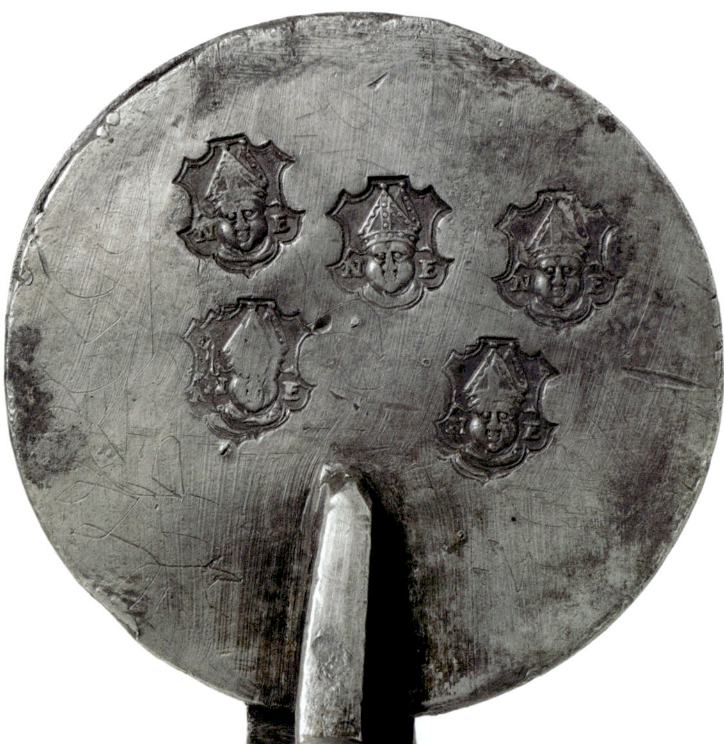

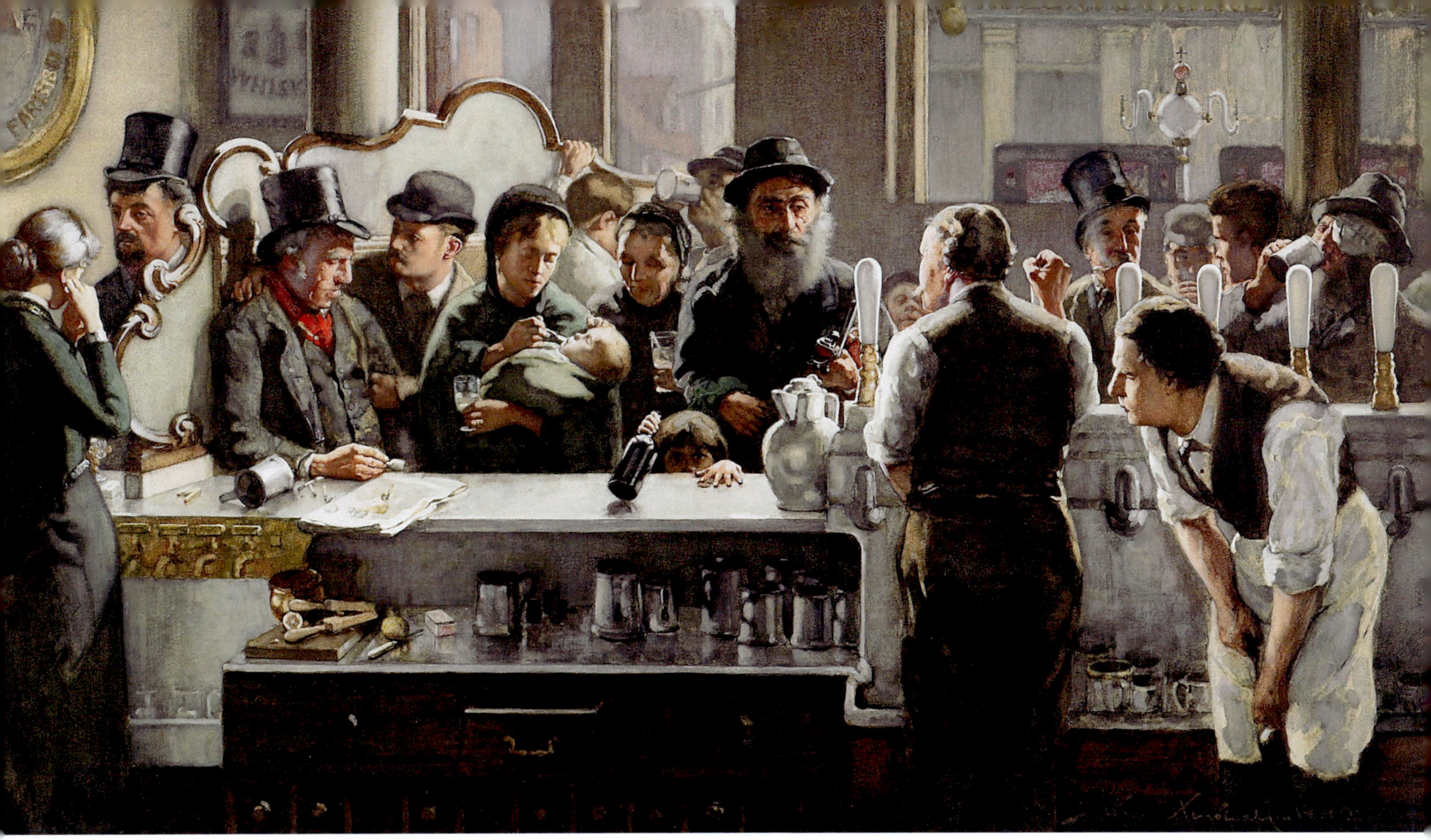

the demand for spirits. Spirit distillation reached a peak of 25,623,000 gallons in 1900. Battles continued between the vested interests of brewers, distillers and publicans and the Liberal Party, sensitive to their temperance supporters. With the First World War the state imposed upon the drinker a fiscal and organizational straitjacket that was loosened only in the 1980s.

Lloyd George, as Chancellor of the Exchequer, set the tone, declaring in March 1915 'drink is doing us more damage in the war than all the German submarines put together. We are fighting Germany, Austria and Drink, and the greatest of these deadly foes is drink'. The duty on a standard barrel of beer was raised from 7*s.* to £1. 3*s.* and to £2. 10*s.* in 1918; its strength was reduced. Spirit duty rose from 14*s.* to £1. 10*s.* per gallon and hours for drinking were dramatically reduced.

After the war was it the continuance of high prices, the missing generation of young men, the attraction of other leisure pursuits such as the cinema, which so dramatically reduced consumption? Men and women in the 1930s were consuming only 40 per cent of the beer consumed in the 1870s and only 15 per cent of the spirits. Britain had at last been weaned away from two addictions that had begun in the early eighteenth century – the people from liquor, the government from dependence on the revenue from making and selling liquor.

Part pub, part gin palace, this drinking place in the Caledonian Road attracts a social mix; the little boy with a bottle to be re-filled hints at the off-licence sales.

Below: A wreath of hops enclosing ears of barley, a maltster's shovel and a malt measure on a German brewer's sign advertise the tradition of wholesome beer, brewed from his own malt.

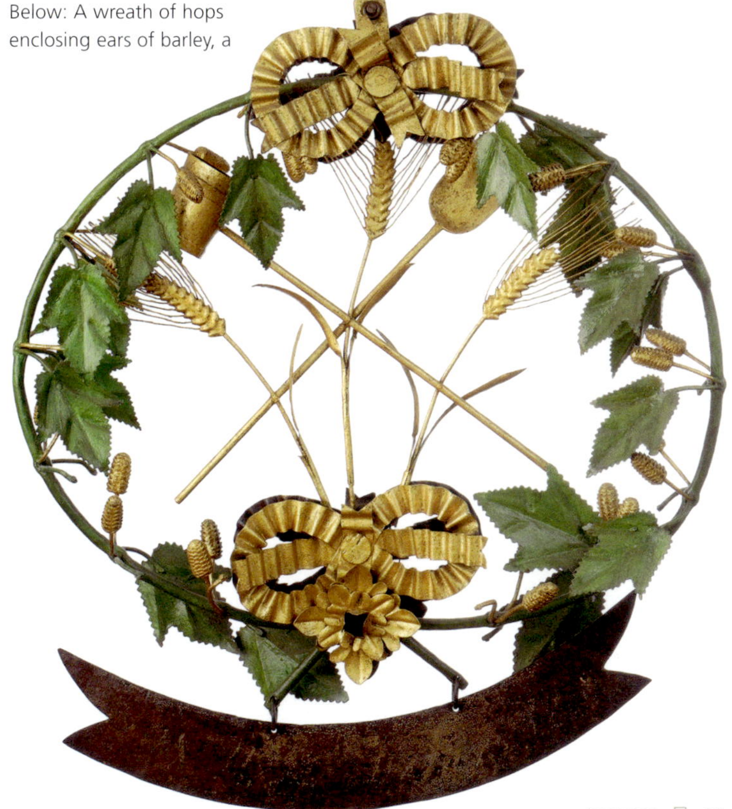

Wine and Religion

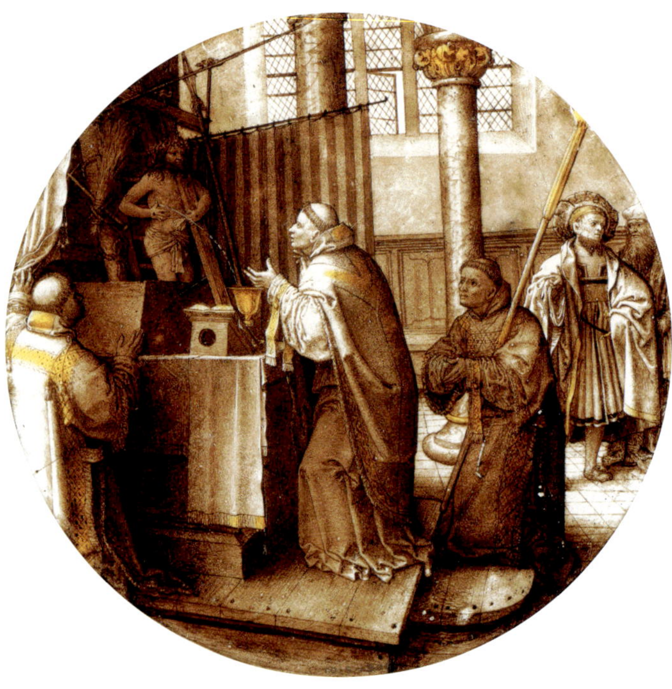

'I am the true vine'. (JOHN 15:1)

Mentioned far more often in the Old Testament than any other plant, the vine was considered one of the most significant of Divine gifts and its symbolism of renewal has conditioned beliefs and ritual practices from as long ago as ancient Mesopotamia. The ancient Israelites rejected the Rechabites, tent-dwellers and herders, as non-drinkers of wine, since their wandering way of life did not allow for seasonal cultivation of crops and vines. Although most practising Muslims do not drink wine because the Holy Koran considers it 'a satanic device', the devout are rewarded in the afterlife: in paradise there are 'rivers of wine, a delight to those who drink'.

Dionysus, god of wine for the Greeks, according to some accounts died each year along with the crushed grapes, journeyed to the underworld and was then resurrected in the spring, when the bare gnarled vines sprouted again. Regarded as a saviour figure with power to grant life after death, Dionysus, like Christ, was said to be the son of an immortal god (Zeus) and a human woman (Semele); he, Bacchus and Christ are all symbolized by the vine.

Eating Dionysus's flesh and drinking his blood brought the possibility of life after death, a belief in the supernatural power of wine echoed in the Christian Eucharist. From some perspectives, for the early Christians Christ became a new

wine god. This led to ambiguity and duality in pagan and Christian imagery; a fifth-century mosaic from Cyprus shows a child, his head lit by a halo, sitting on a lap and surrounded by worshippers. He represents Dionysus, god of wine, sitting on the lap of Eros, god of love, but this image could also be interpreted as a Christian nativity scene.

The Christian Eucharist originated in The Last Supper shared by Jesus and his disciples at the time of Passover, the Jewish spring festival that recalls and celebrates the release of the Israelites under Moses from Egyptian slavery. At the annual Passover meal, celebrated at home, unleavened bread and wine are shared. For early Christians, this Jewish Passover rite gradually moved from a domestic setting to special buildings, and the original meal became encased within layers of liturgical significance. Drinking wine and breaking bread together became symbols of the sacrifice of Christ, following His instructions at The Last Supper.

In the Jewish faith, wine is served in the home on every Sabbath and at all religious festivals. The ceremonial opens and closes with the Kiddush, a prayer of sanctification, which is said over the wine cup. In the thirteenth century the

Above left: The Mass of St Gregory is a graphic expression of the Roman Catholic belief that at the central act of the Mass, when the priest consecrates the wine, it becomes Christ's blood.

Below: A silver and enamel Kiddush cup made by Tamar de Vries Winter in 2005.

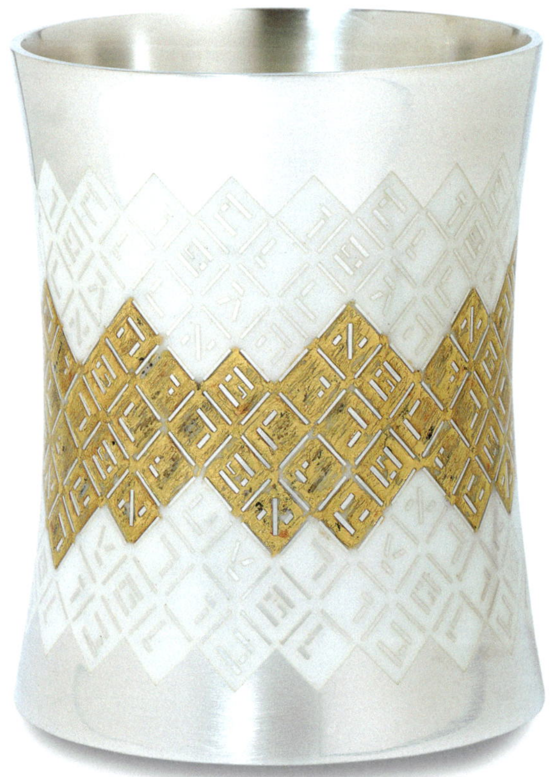

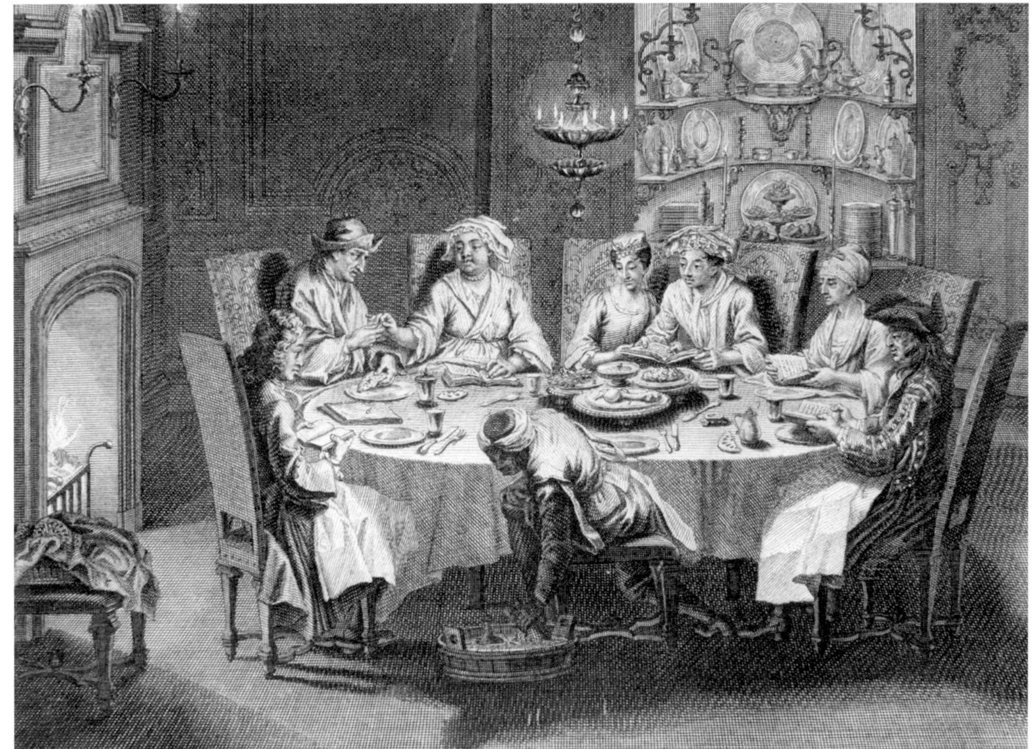

At this Passover meal, the young man, the lowest in the family hierarchy, reaches for a bottle to fill the ceremonial glass of wine before it is blessed and passed around.

Kiddush was also said in the synagogue in areas where Jews were migrating and had no fixed home. In 2005 Tamar de Vries Winter, a Jewish artist in enamels, created a Kiddush cup decorated with Hebrew letters that spell three words from the ancient benediction referring to God as 'creator of the fruit of the vine'. She explains that 'the uninterrupted repetition of these words 21 times encircling the vessel, with no space between them, reflects the continuity of the prayer's recitation at the beginning of Sabbath and holy days year after year for more than two millennia.'

In the Mass celebrated according to the Roman Catholic rite, the wine transforms into the blood of Christ. This belief is illustrated by the Miracle of Saint Gregory the Great (*c.*540–604), one of the four Latin fathers of the Church. He was celebrating Mass but finding a disbeliever in his congregation, prayed for a sign. His prayer was answered when the crucified Christ appeared above the altar with the instruments of His Passion. In the early Netherlandish stained glass representation of this scene Christ presses the wound in his side, from which blood spurts into a chalice.

Drinking wine together was itself a ritual, and belief conditioned practices such as the post-meal grace. It has a long ancestry. At the ancient Greek symposium, after the first course, a goblet of wine was passed around. Each poured a few drops on the ground as a libation, and then drank a little. At Jewish ritual meals, the grace after eating is also accompanied by a cup of wine. For Christians, this shared

cup of wine pointed to the afterlife, envisioned as a shared banquet. Jesus alluded to this: 'I will never again drink this wine until the day I drink the new wine in the Kingdom of God' (Mark 14:25). Sharing grace cups after eating together became part of Western monastic practice, a separate wine-based tradition from the transformation at the Mass.

The names of wines consumed today that include such words as Clos, Kloster, Hermitage, St, Abbaye, Preiure and Commanderie recall their origins in vineyards owned by prelates and monastic houses. In 1096 Pope Urban II planted the Marmoutiers vineyard which surrounded the holy relic of St Martin at Ligugé near Poitiers in France. The relics of St Vincent, the patron of wine merchants, were housed in the church St Vincent-hors-les-Murs, outside Paris, now Saint Germain-des-Prés. Champagne originated with the wine of *Ay* and *Vindey* (*vinum dei* – the wine of God) as a wine for the Mass. However, the quantity of wine produced by monastic vineyards greatly exceeded liturgical needs; as the celebrating priest alone drank from the chalice during Mass, the proportion of wine used was tiny. Most was consumed by bishops and higher-ranking clerics with their meals. Monks were given wine only on feast days.

As a result of fresh thinking about the roots of Christian belief in the early sixteenth century, Protestant Reformers reintroduced the early Christian practice of sharing the cup of wine with all adult worshippers. The reformer John Calvin accused the Roman Catholic hierarchy of snatching the wine 'from the greater part of God's people and giving a special property to a few shaven and anointed men'. Extreme Protestants denied any mystical significance in the bread and wine, emphasizing merely the act of commemoration through a shared meal. For this, as early as 1547, large broad bucket-shaped cups, adapted from a current form of wine goblet, replaced the small-bowled Mass chalices. At the first Presbyterian communion celebrated in Scotland, people did not take a small sip of wine, but took a large mouthful and 'refreshed themselves as it were'. Each strand of Protestantism promoted the consumption of wine by all adult members of the congregation, whether as a sign of their communion, or as a reminder of Christ's sacrifice by sharing his body and blood. Attendance even became a way of testing political allegiance.

By the eighteenth century Calvinists took communion seated at a long table. Many Protestant variants in sharing the wine flourish today, from small individual glasses or pottery cups to a cup passed along each pew to seated worshippers and refilled by the deacons as necessary. For some Nonconformists, the issue of the alcoholic content of communion wine became significant. Great sophistry was employed to suggest that biblical references were to unfermented grape juice and some teetotallers celebrated communion with grape juice or even tea. One outcome of temperance policies, allied to a new emphasis on ritual in the 1830s and 1840s, was the switch to taking a reverent sip of wine from the cup at the altar rail, so that the large flagons of the seventeenth and eighteenth centuries became redundant.

In the Roman Catholic rite water was mixed with wine in the Eucharist. Special cruets of glass or precious metal identified with initials A for *aqua* and V for *vinum* were made for this purpose. The ancient Greeks usually diluted their wine with water to reduce its density, strength and

A Profane Sacrament

After the Reformation of the sixteenth century, Lutheran and Calvinist communion plate followed domestic designs. Cups frequently carried signs of state rather than church and had to be capacious enough for all adults to share the Eucharistic wine. Flagons and tankards, along with drinking-cups, sweetmeat plates (for the Eucharistic bread) and rose water basins and charger plates (for collecting alms), were now part of both liturgical and secular life.

In the religious and political turmoil of the Stuart and Hanoverian eras, both taking communion and 'loyal health drinking' became vital shared public rituals demonstrating loyalty to church and state. Loyal health drinking, or disloyally drinking to 'The king over the water' offered unlimited opportunities for expressing religious and political divisions. Healths and drinking songs were couched in violent terms, offering painful or humiliating punishments, to 'cripple your shanks' or 'a full one in the face', to anyone who did not join in a 'loyal bumper'. While

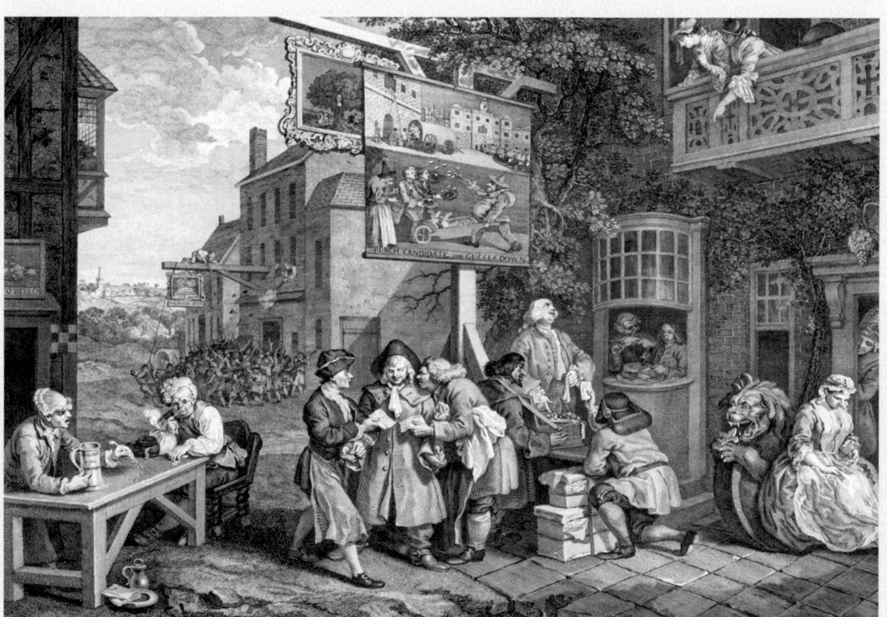

Two men, sharing a tankard, re-enact the naval victory at Portobello in 1739. Hogarth contrasts their nostalgia with drink-fuelled corrupt canvassing.

private societies could exclude unwanted participants, taverns and alehouses rang with the sounds of quarrelling men (and women) as they attacked those who had refused a loyal health.

acidity, but for Christians the drop of water represented the meeting of God and man in the person of Jesus, the blood representing the divine and the water the human. A set of silver-gilt cruets and dish were made by London-based German goldsmith Charles Kandler in about 1735 for the private Roman Catholic chapel of the 8th Lord Petre at Thorndon, Essex. The contemporary *Rules to be observ'd in the Chapple* at Thorndon list the special feast days on which these cruets and their matching chalice were used, including the Exposition days of Epiphany, Candlemas, Easter Sunday and Corpus Christi. Although purchased separately, the wine consumed at Mass in Thorndon chapel was the same as that served at table. In two Roman Catholic households, Arundel and Lulworth Castles, the private chapels later became the family dining room; at Lulworth the altar tabernacle frame remained as a reminder of the room's previous use.

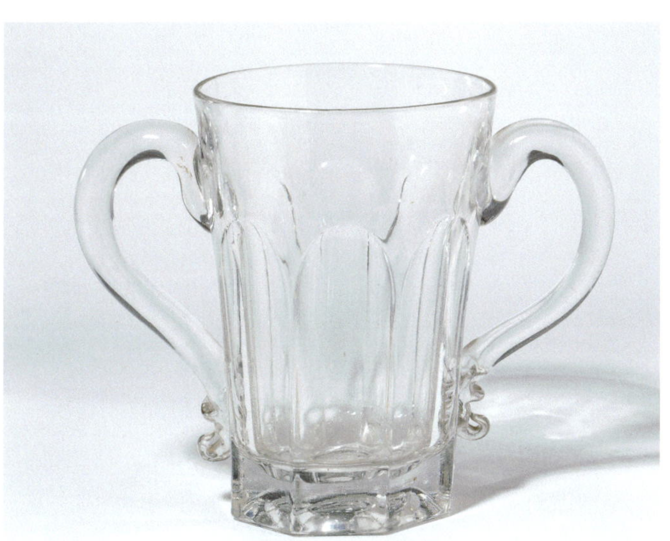

This 19th-century two-handled cup was passed along the row of communicants at a Nonconformist service, perhaps filled with unfermented grape juice.

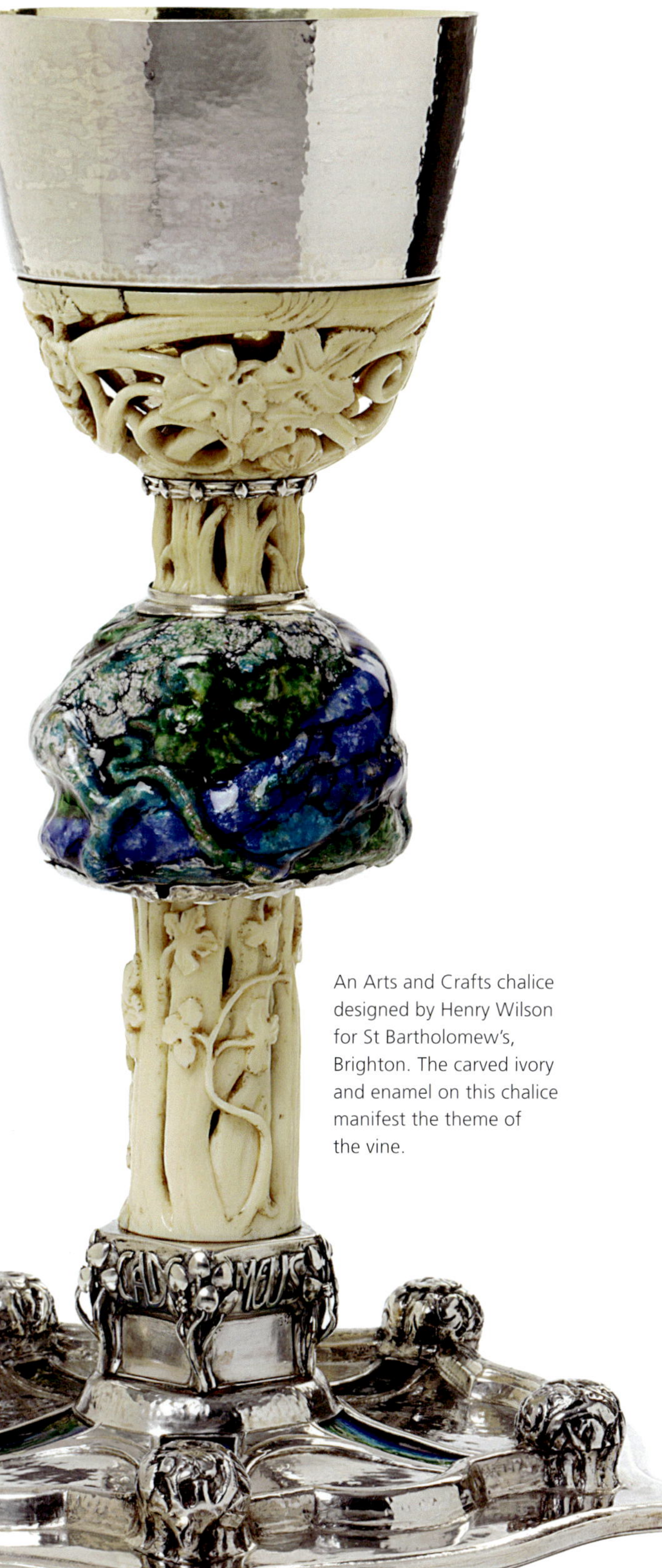

An Arts and Crafts chalice designed by Henry Wilson for St Bartholomew's, Brighton. The carved ivory and enamel on this chalice manifest the theme of the vine.

Holding Your Drink

The English are too apt to reproach Foreigners, particularly the Germans and Dutch with the vice of Drunkenness, yet none more guilty of it than themselves.'

ALI MOHAMED HADGI, *A BRIEF AND MERRY HISTORY OF GREAT BRITAIN* (C.1730)

Offering and receiving alcohol at dinner was an essential element of hospitality. A complex social etiquette dictated what to drink, how to drink and when. Manuals by French and Italian authors such as Giovanni della Casa (1503–56) and Antoine de Courtin (1622–85) brought the manners and customs of the most fashionable European courts to a wider audience.

Some ideas of proper behaviour remained constant in courtesy books from the sixteenth to the twentieth centuries. For example, it was important to keep your mouth, drink and glass clean. One medieval treatise advises, 'Before drinking, wipe your mouth so that you do not dirty the drink.' Some 400 years later, *Etiquette for Women* in 1919 advised, 'do not forget to wipe your lips before drinking and after.' However, instructions to remove one's helmet before giving wine to a lady had a much shorter vogue.

Keeping the vessel clean was good hygiene as well as aesthetically pleasing, important since vessels were often shared. This sharing of glasses dismayed some commentators. Germans and other Northern Europeans were thought to be most at fault, not surprisingly since these were the beer-drinking regions, with a tradition of large, shared beakers and tankards. In 1717, according to a French book of manners, 'this manner of drinking from the same glass, and still less of drinking what the ladies have left, it has an air of impropriety'. The emerging French etiquette of dining politely from the mid-seventeenth century demanded that each diner should be supplied with his or her own flatware, plate and glass rather than sharing communal dishes, tankards and cups.

Placing glasses at dinner acquired specific rules. In 1530, Erasmus of Rotterdam instructed that 'You should have your goblet and knife on the right of your plate', but another sixteenth-century writer commented that 'it might happen that the glass could not be left on the table or on his right without being in someone's way.' In fact, glasses stood normally on a side table, ready for service, just as their metal predecessors, cups, tankards and bowls, had been stacked

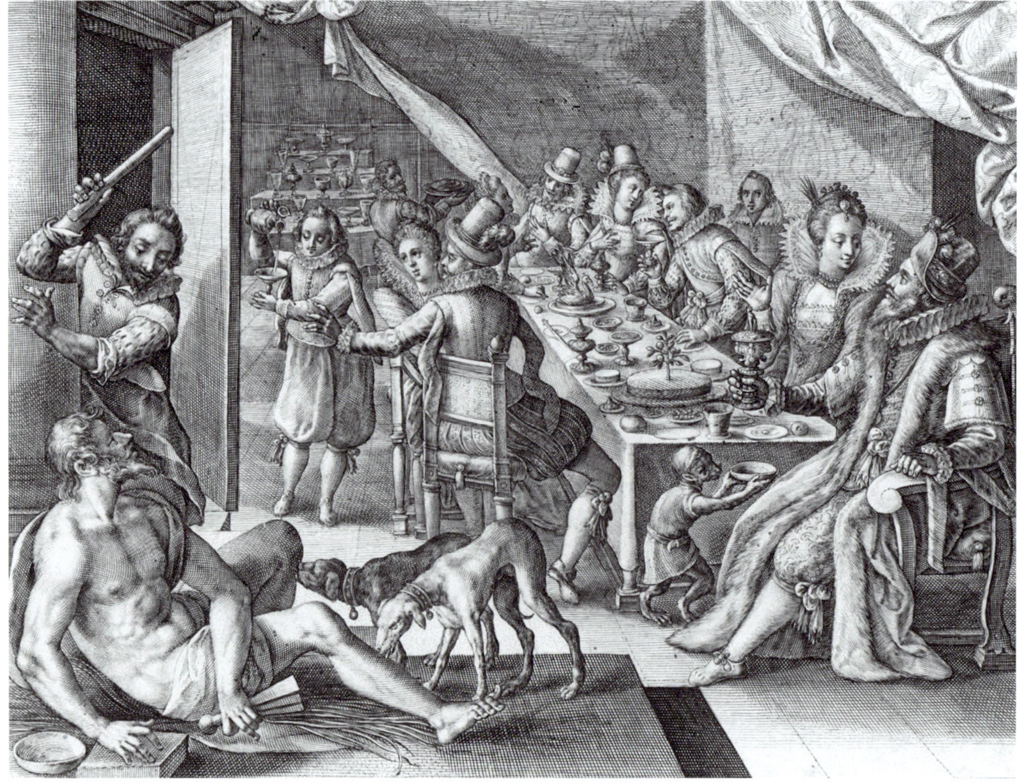

At this luxurious dinner, diverse drinking vessels are graded by rank and gender. In England, vessels were shared, 'not considering it any inconvenience for 3 or 4 persons to drink out of the same cup', as a Venetian visitor commented.

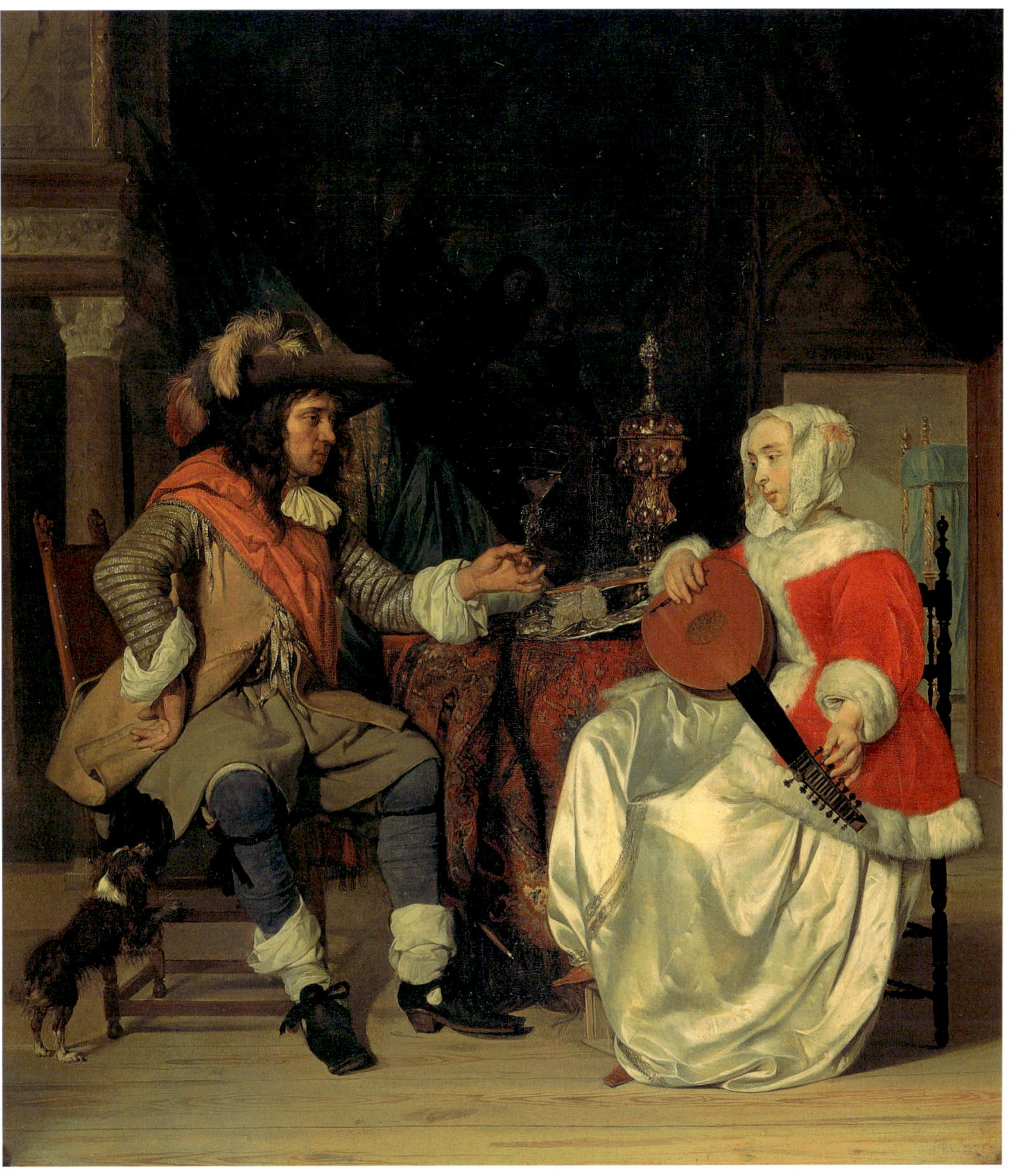

decoratively on the medieval cupboard. In 1786 the German
visitor Sophie von la Roche described seeing at Seddon's
London showroom a new kind of sideboard:

*an invention for use in a dining room – with both a
practical and a decorative purpose – that most appealed … a
table made of mahogany … against the wall is a set of shelves
for glasses and salvers … the touch of a spring opens a
pewterlined apartment with partitions for wine bottles to
stand in cold water, while on the other side a swivelling basin
is fitted'.*

Good manners required, especially at formal dinners with
many attendants, that wine be brought to the table in the
glass, drained by the diner and the empty glass returned
immediately. Madame de La Tour du Pin recalled dinners
with her uncle the Archbishop of Toulouse in the late
eighteenth century:

*In those days, everyone with a decently dressed servant was
waited on by him at table. No decanters or wine glasses on
the table. At big dinners, silver buckets were set on a
sideboard with wines for various courses. There was also a
stand of a dozen glasses and anyone wishing for a glass of
one of the wines sent his servant to fetch it.*

Glasses were always presented on a silver salver.
Instructions for servants advised: 'When handling wine, put
your thumb on the foot of the glass and on receiving it back
do the same, this will prevent its overthrow.' Diners called as
they wished for beer, cider, ale or wine, until it came to the
formal toasts, which were normally drunk in wine.

Service changed at the dessert course. Attendants
withdrew, leaving decanters of wine or spirits to be
circulated by the host to his male guests, while women
withdrew after taking one glass. From the early nineteenth

Holding Your Drink

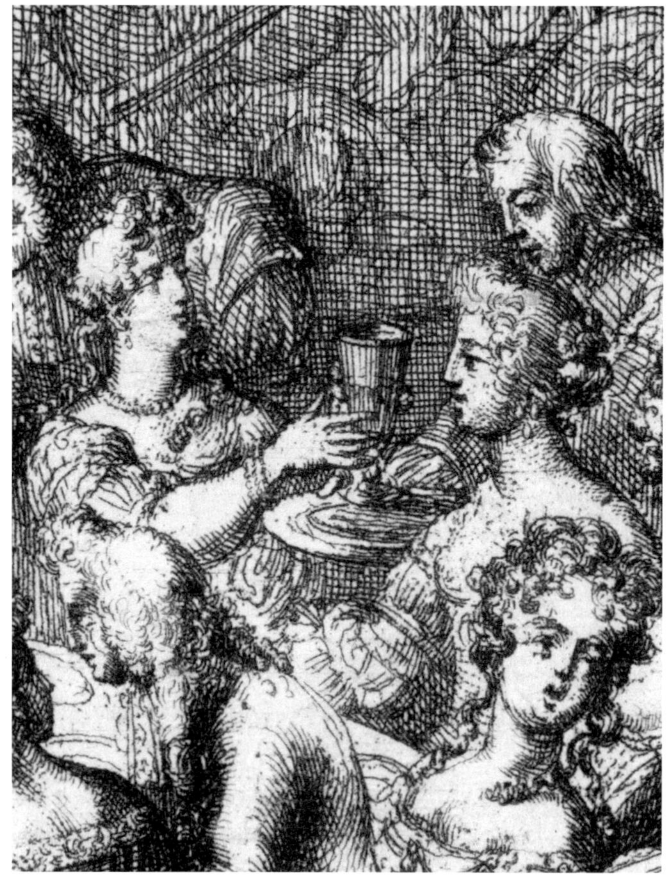

Drinking at a court dinner in The Hague; each woman is offered a large glass, presumably to toast the men before passing the glass on for sharing.

century it became more common to allow men to pour wine for themselves and for female guests, and decanters remained on the table throughout the meal.

Basic rules for drinking in company emerge from the etiquette books. Do not drink or speak with your mouth full. Do not spit, cough or sneeze. Do not smell the wine. Drink quietly, 'not pouring like a gutter'. Do not leave wine in the glass. Do not call too loudly for wine from the attendants. Hierarchy played an important role. You could not ask for a drink until the 'persons of quality have drank before you.' If you are speaking to a more senior companion and he takes a drink then you must stop talking until he has finished drinking.

For centuries, at dinners given for or by princes, a fanfare of drums and trumpets added to the ceremonial nature of the moment of formal drinking. Proposing healths or toasting was common throughout Europe, although the custom could be excessive in England. A French visitor recorded in 1700:

''Tis here so the Custom to Drink to everyone at the Table, that by the time a Gentleman has done his duty to the Whole Company, he is ready to fall asleep, whereas with us, we salute the Whole Table with a single glass only.' Glasses were far smaller, but had to be drained at a toast. A complicated etiquette surrounded the giving and receiving (or pledging) of the health. Diners were warned to remove their hats or bow if a gentleman of higher status drank their health. They might return the toast by saying 'Sir, my service to you. A good health to my lord.'

Ladies could propose a health and certainly received them, but in Britain by 1750 toasting became an essential part of the heavy drinking after dinner in exclusively male company. After the wine had circulated:

and mere thirst has become inadequate reason for drinking, a fresh stimulus is supplied by the drinking of 'toasts', that is to say, the host begins by giving the name of a lady, he drinks to her health and everyone is obliged to do likewise. After the host someone else gives a toast and everyone drinks to the health of everyone else's lady. Then each member of the party names some man and the whole ceremony begins again. If more drinks are required, fresh toasts are always ready to hand; politics can supply plenty – one drinks to the health of Mr Pitt or Mr Fox or Lord North....

From 1800, the drinking of healths during dinner was falling from fashion. The young Queen Victoria, travelling in Germany with Prince Albert, was shocked by the German custom of chinking glasses together when drinking healths. Count D'Orsay commented in 1843: 'The custom of drinking toasts and of forcing people to drink bumper after bumper of wine until drunkenness results, is quite banished from gentlemanly society to its proper place – the tavern'. Toasting continued to be practised in clubs, at military dinners and on other communal occasions and is part of formalities at state or civic functions and at weddings.

By the early nineteenth century, the French and Russians offered their guests aperitifs such as brandy with hors-d'oeuvres before dinner, but this would have been considered unhealthy and even offensive earlier. *The School of Good Manners,* which was translated from French in 1595, explained, 'To begin the meal with drinke, is a fashion of Drunkards, which drinke for custome and not for thirst; and that custome is both unwholesome and unseemly.'

As the type and quantity of alcohol drunk at dinner became more highly codified in Victorian Britain, so it became harder

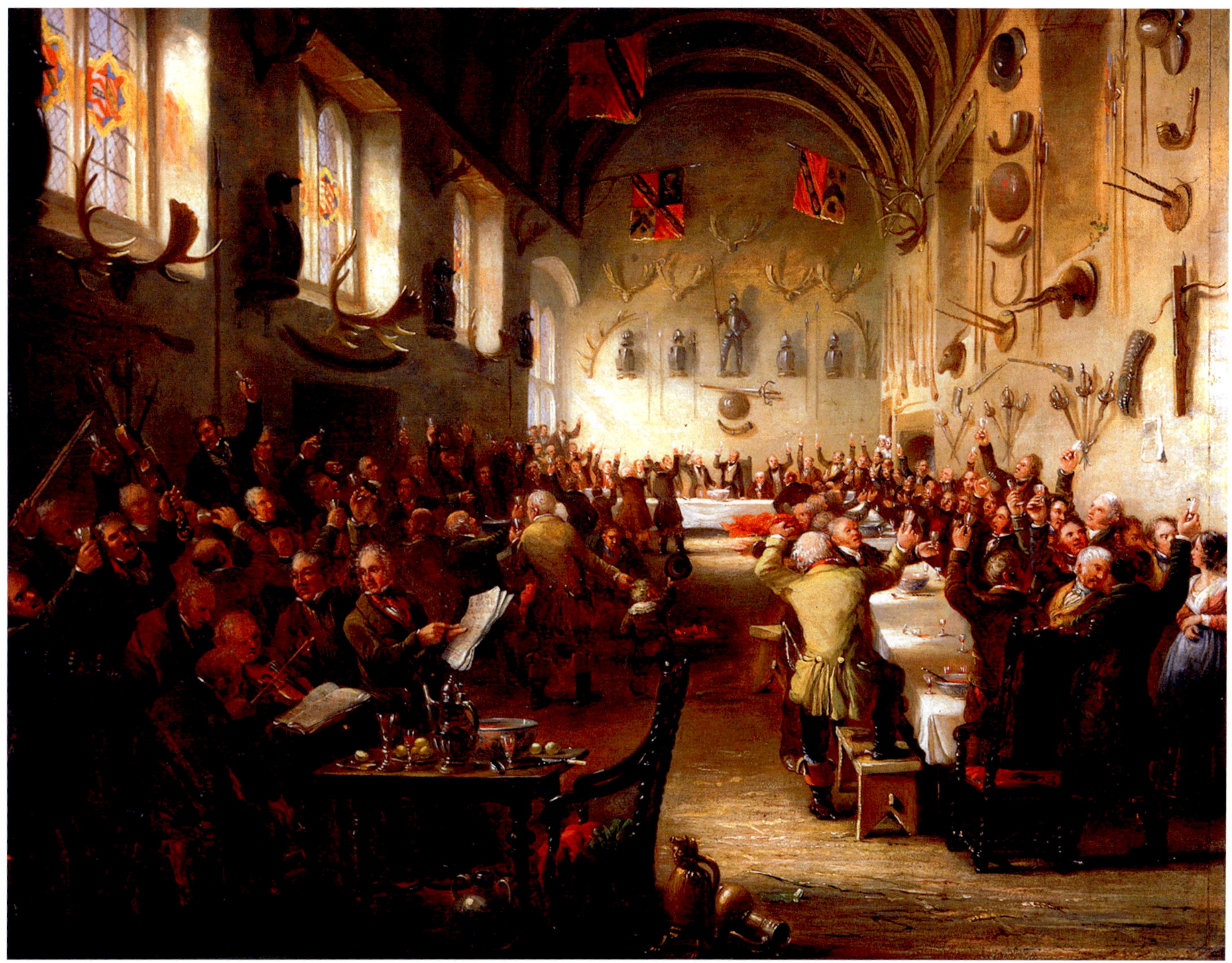

At this Cornish tenants' dinner, punch has been consumed throughout the meal, a practice that was becoming unacceptable in polite circles, under the influence of temperance reformers.

to disentangle drinking etiquette. Each dish was becoming associated with a distinct wine, a process assisted by growing wine expertise after the Napoleonic Wars. Wines became heavier and fuller bodied as the meal progressed while the dishes that were eaten became progressively lighter and sweeter, although the sherry drunk with soup was far darker and stronger than would be to today's taste.

Menus for meals catered by the Parisian chef Antonin Carême for the Prince Regent and Tsar Alexander I show this novel matching of wines to specific foods, as the traditional medley of *à la française* dining gave way to a linear meal structure. Alcohol was not given to guests in Britain until after the soup had been served. Then diners could expect sherry or dry Madeira, but with turtle soup the English drank punch. The fish course necessitated one glass of hock or

Moselle if the fish was white, but salmon required claret cup, claret or port. With the *entrées* could come Champagne although some male commentators were disparaging: 'In order to pander to the prevailing weakness of the day, and assuming that the champagne is choice in quality and perfectly iced, this much overrated but now favourite with the ladies may be introduced and continued throughout the dinner.'

The Champagne would be offered three or four times. Claret or port could be drunk with game and by the dessert claret, Madeira or sherry would be offered with cherry brandy and liqueurs after the ices. This variety of wines complicated the table layout of the glasses. Up to four different glasses could be set out (for sherry, claret, Champagne and hock), although three were typically recommended and dessert glasses arrived separately, at the end of the dinner, as part of the dessert. For those new to society, the number of glasses must have been intimidating, but as the etiquette books reassured, 'If you have any difficulty in remembering, just say which of the two wines offered you, you will take, and the servant will give it to you in the proper glass.'

While it was important for the host to offer a wide variety of wines, guests, especially women, were advised to limit their drinking to one or two types of wine during the meal. Nor were ladies expected to take a second glass of wine with the dessert. Tumblers for water or a small carafe with a glass were sometimes on the table although, until the twentieth century, it was more usual to serve water and beer from the sideboard as requested. In 1874 Walsh's *Manual of Domestic Economy* estimated the relative cost of dinner wines for middle-class families. An economical dinner party for eight could cost £5 for the food but twice as much, at between £5 and £10, for the wines.

After the dessert the ladies, in Britain, left the men to continue drinking in the dining room and withdrew for tea and coffee. The rest of Europe found this custom frankly shocking. It was well established by 1700 when a French traveller observed that he had left the dinner table to sit with the ladies while the host remained drinking below. Later, a drunken fight spilt into the withdrawing room from the men's party, breaking a looking glass as well as some fine porcelain vases and frightening the ladies. Foreign visitors

continued to highlight the difference. In 1725, César de Saussure wrote:

After these toasts the women rise and leave the room, the men paying them no attention or asking them to stay; the men remain together for a longer or lesser time. This custom surprises foreigners, especially Frenchmen, who are infinitely more polite with regard to women than are Englishmen; but it is the custom, and one must submit.

The gentlemen drank prodigious amounts and the conversation was frank. François de la Rochefoucauld reported in 1784 that

Very often I have heard things mentioned in good society which would be in the grossest taste in France. The sideboard too is furnished with a number of chamber pots and it is a common practice to relieve oneself whilst the rest are drinking; one has no kind of concealment and the practice strikes me as most indecent.

By the Victorian period, British commentators themselves could be quite critical. Separating the sexes after dinner was unfavourably compared to the arrangements in France: 'the dreary interregnum which still occurs in this country, whilst mine host is circulating the bottle below and ladies are discussing the servants … in the drawing room above is unknown in the salons of Paris.' The length of time men

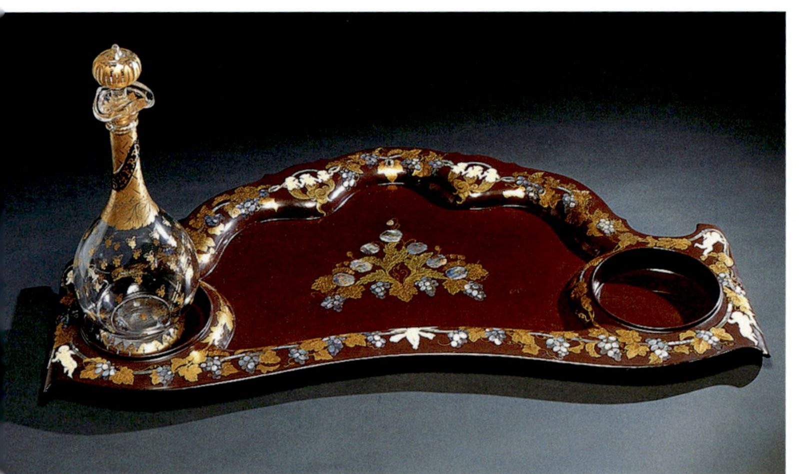

Above: Decanters on trays appeared after dinner. At a Scottish dinner in c.1830, an American was offered half a dozen wines, 'old Madeira from his friends in the United States, and a round of whiskey, the usual finale of a Scotch dinner'.

Right: At a Wynnstay family celebration, a young child, dressed in costume and disguised in a wig and moustache, is drinking a toast, perhaps to the New Year.

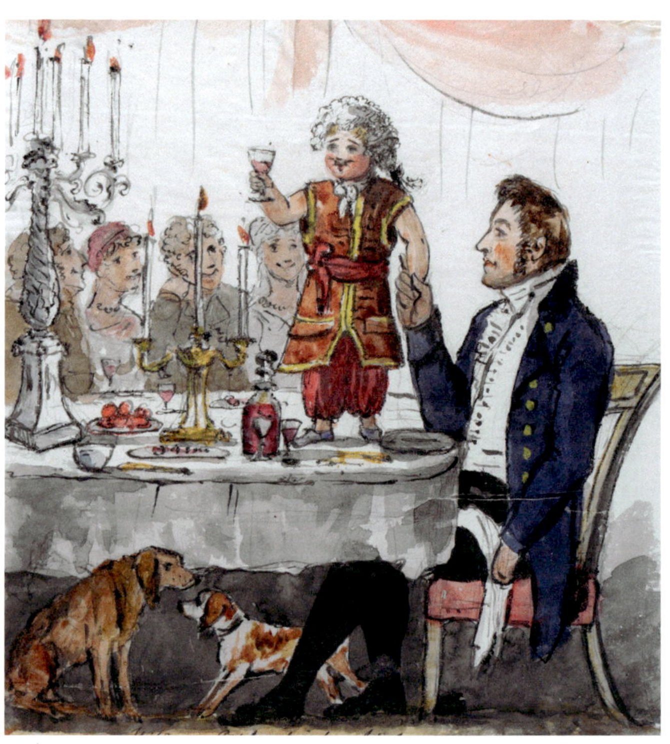

Hot or Cold?

Although most wines, spirits and ales were taken chilled, there was a recurrent vogue for hot alcoholic drinks made by mixing wine or spirits with spices and sugar. Sweet, warming brews, beneficial in the winter, could be drunk at any time. An early concoction, hypocras, was prepared hot to amalgamate the sugar and spice with the red wine, but was actually drunk cold. Wassail, or spiced ale, a medieval winter drink, was taken warm, as were posset and caudle. Flip was beer, sweetened and strengthened with rum; a red-hot poker thrust into the mixture made it foamy and gave it a bitter, burnt flavour. Punch, so popular for 250 years, could be drunk hot or cold.

Wine, on the other hand, both red and white, was drunk chilled, the bottles cooled in ice or iced water in large cisterns that stood close to the table, although the single bottle wine cooler emerged in the late seventeenth century. Glasses too were chilled, either in individual coolers or in scalloped-edged verrières of porcelain or silver in which the bowl could be up-ended in ice. Before refrigeration, ice itself was a costly commodity, cut from ponds and painstakingly preserved through the year in ice-houses cut deep into the ground. Ice was later imported from Scandinavia and the Baltic, and by the 1840s the

Glass made for the Prince of Wales, imitating the refreshing effect of ice. Polite hosts provided iced water for cooling wine glasses.

North Americans were breaking into the market. The Wenham Lake Ice Company, suppliers of choice to Queen Victoria, set up its main office in the Strand in London, complete with a window displaying a block of ice a foot thick. In hot weather many drinkers still prefer to drink red wine chilled.

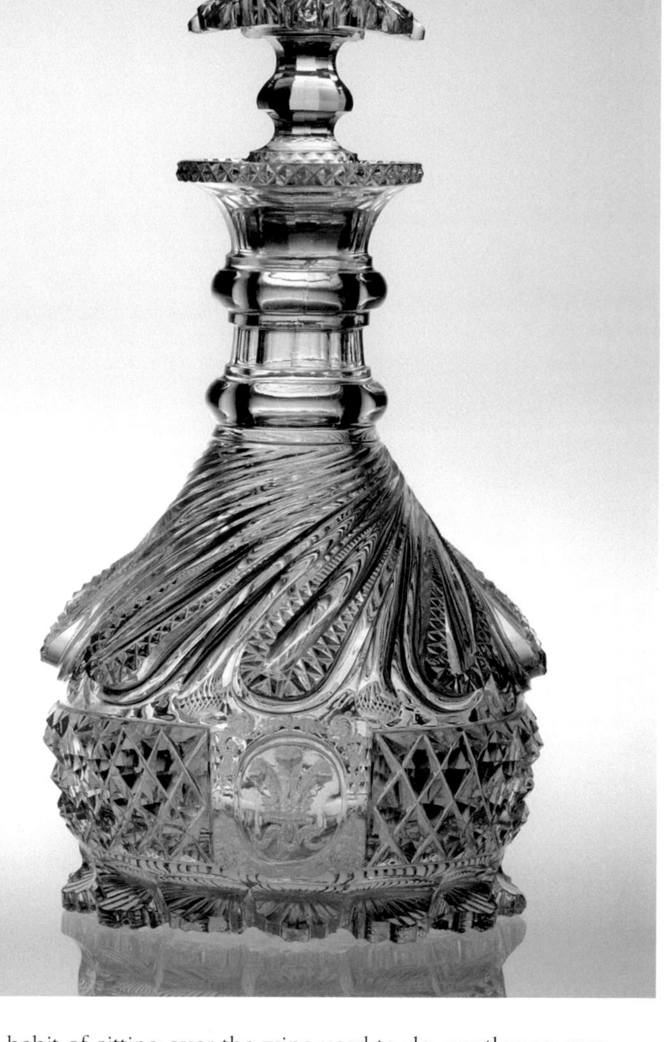

spent drinking after dinner was condemned, but by the 1880s the reduction from an hour or two to a more perfunctory 15 minutes could be commended. Smoking had by now begun to replace alcohol as a postprandial activity. In 1892, Lady Greville lamented 'owing to the prevailing habit of cigarette smoking in the dining room after dinner, which keeps men downstairs as long as the old fashioned habit of sitting over the wine used to do, gentlemen may come upstairs a few minutes before the carriages are announced and the party disperses.' Despite the frequent recommendation to follow continental etiquette, in many circles in Britain ladies continued to be segregated from gentlemen after dinner until the general loosening of social convention in the 1970s.

Vintages and Vintners

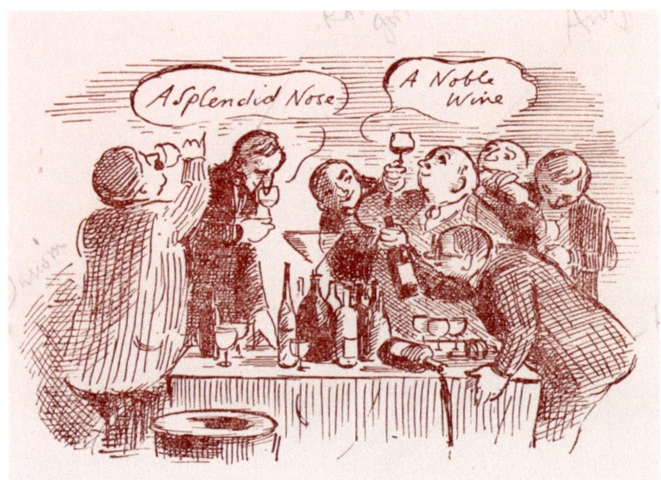

Wine, defined as 'the fermented juice of the grape', has been with us for thousands of years. Fermenting vessels complete with grape pips found in the Republic of Georgia date back to at least 4,000 years BC. Research has shown that the vine was cultivated in Macedonia one thousand years before. Egyptian princes drank wine imported from Syria in the second millennium BC. A shipwreck off the Tuscan coast foundered with 130 amphorae of Greek wine intended for Gallic aristocrats around 580 BC.

Vintners have been with us since those far off days, yet vintages, in the sense of a comparison of quality and ageing potential between the produce of different years, have been possible only since the advent of a glass bottle to store the wine and, more importantly, the cork to offer a hermetic seal. In this sense, true vintages date only from the early years of the eighteenth century. Wines may well have been aged in containers great and small in the cool cellars of the estate where they were grown, but once exported in barrels, a slow decline through oxidation was inevitable. Thus the young wine, *le vin de l'année,* was more prized than the old, and for the old to be good it had to be stored and served correctly.

Nobody understood this better than François-Auguste de Pontac of Château Haut-Brion in the Bordeaux region known as 'the Graves' due to its pebbly soil, whose vineyards are today totally engulfed by the city of Bordeaux itself. The origins of Haut-Brion as a vineyard date back to 1525, when Jean de Pontac, First President of the Bordeaux Parliament, married Jeanne de Bellon, daughter of the mayor of Libourne, whose dowry included lands already known under the Haut-Brion name. A vineyard was planted and later a fine house was built, some of which still stands. When Jean de Pontac died aged 101, the oldest and richest Bordelais, his grandson, Arnaud de Pontac, inherited. François-Auguste, his son, travelled to London to promote their wine. In 1666 he opened the Pontac's 'Head Tavern', the original wine bar and an immediately fashionable eating house which was to remain open for over 100 years. François-Auguste was an active salesman and Samuel Pepys drank 'a sort of French wine called Ho Bryan' at the 'Royal Oak Tavern' in Lombard Street. The de Pontacs were largely responsible for opening up the trade with England in fine clarets; well into the 1700s, the price of Haut-Brion far surpassed that of other Bordeaux wines, even Latour and Lafite.

Yet even the highest quality, recognized in the Official Classification of 1855, which placed Haut-Brion as the single First Growth in the Graves, alongside Margaux, Lafite-Rothschild and Latour in the Medoc, could not withstand French vineyards' difficulties in the late nineteenth century through the scourge of phylloxera and the 1930s depression.

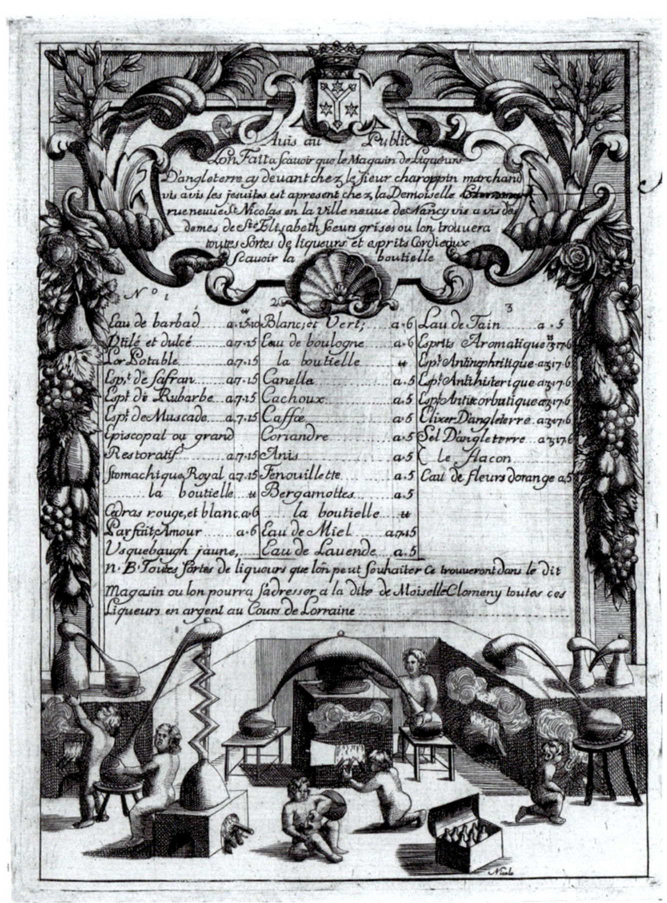

"

Opposite above: The Royal College of Art Senior Common Room at a tasting. Wine education flourishes in Britain. The Vintners' Company set up classes and an examination on wine in 1952; now there are more than 150 Masters of Wine and a rich vocabulary of descriptive terms.

Opposite below: Trade card of Demoiselle Clomeny of Nancy, seller of liqueurs and *esprits cordials*. Because of transport costs and regional tariffs, those who could afford it largely drank the products of their region.

Right: The Venetian glassmaker Jacomo Verzelini established a workshop in Elizabethan London producing fine wine goblets. At dinner with the Lord Mayor in the 1590s, a German visitor drank 'the best beer and all manner of heavy and light wines … Greek, Spanish, Malmsey, Languedoc, French and German, for all kinds of wine can be had for relatively little money because of the low freightage by sea'.

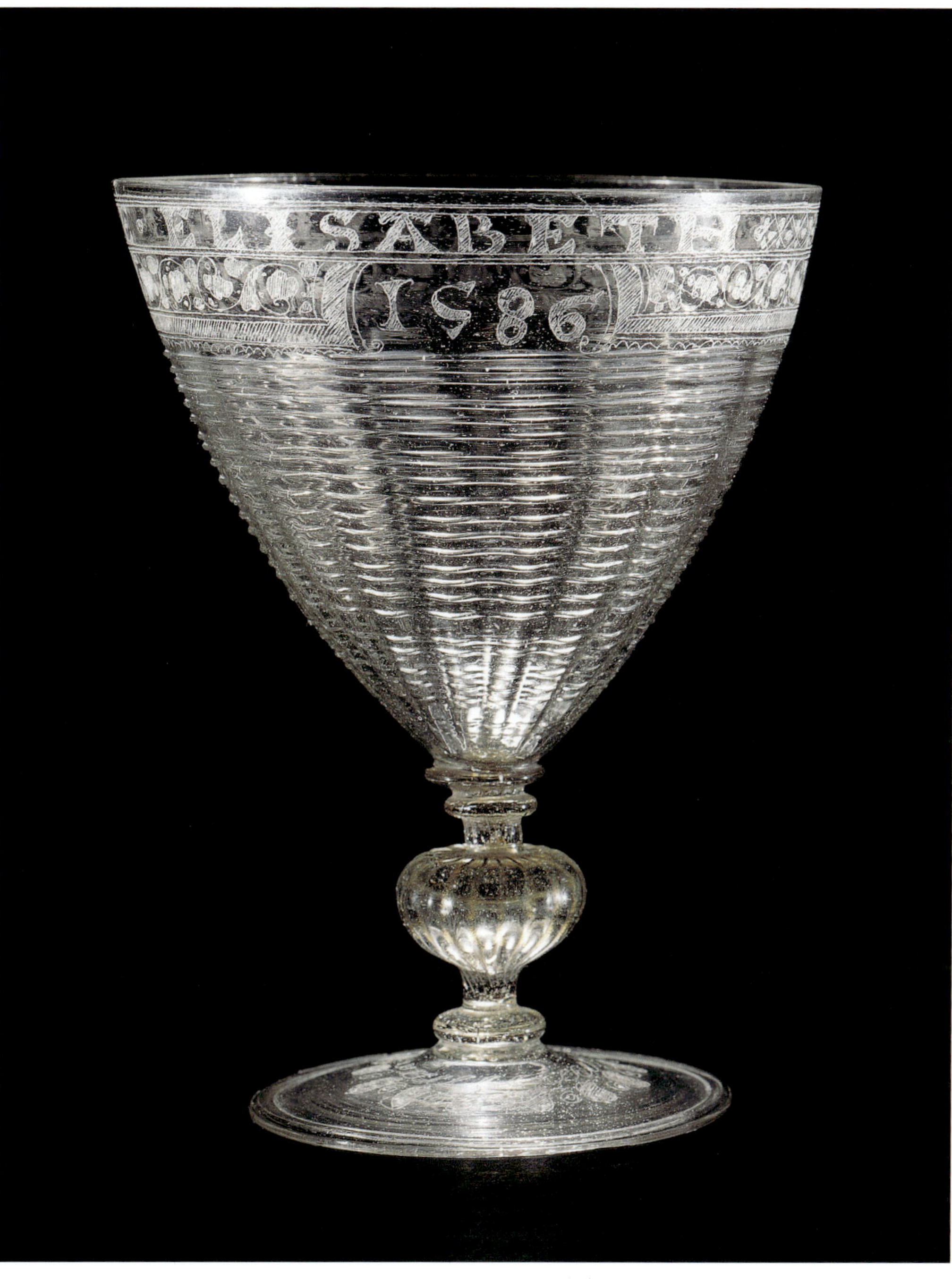

So in 1935 the powerful American financier Clarence Dillon came to the rescue. Today the estate, now rejoined to its prestigious neighbour La Mission Haut-Brion, is owned and overseen by his granddaughter Joan, Duchesse de Mouchy and her son, Prince Henry of Luxembourg.

While the great Bordeaux estates such as Haut-Brion were created for the glory and profit of their owners, those in Burgundy were created for the glory of God and the profit of the monasteries. The Cistercians controlled the finest vineyards in what was known as the Côte-d'Or, where vines on the lower hillsides produce the Grands and Premiers Crus so prized today. In 1241, the monks of the Priory of Saint Vivant, behind the hamlet of Vosne-Romanée, sold a tiny vineyard of 1.8 hectares called Cros des Clous, which they had received two centuries earlier from the dukes of Burgundy.

The Priory wine cellars still exist just below the vineyard of Romanée-St-Vivant, in which are found the barrels of the Domaine de la Romanée-Conti, Burgundy's most illustrious estate. The name comes from Louis François de Bourbon, Prince de Conti, first cousin of Louis XV, who outbid the king's mistress, the Marquise de Pompadour, for the estate, but who was to lose it 30 years later in the French

Revolution. Passing through several owners, always at astronomical prices, the estate was purchased in 1869 by Monsieur Duval-Blochet, whose direct descendant, Aubert de Villaine, manages it today. Known as 'the Domaine' or simply DRC, the estate owns only Grands Crus vineyards: La Romanée-Conti and La Tâche in their entirety, more than half of Romanée-St-Vivant, half of Richebourg, one third of

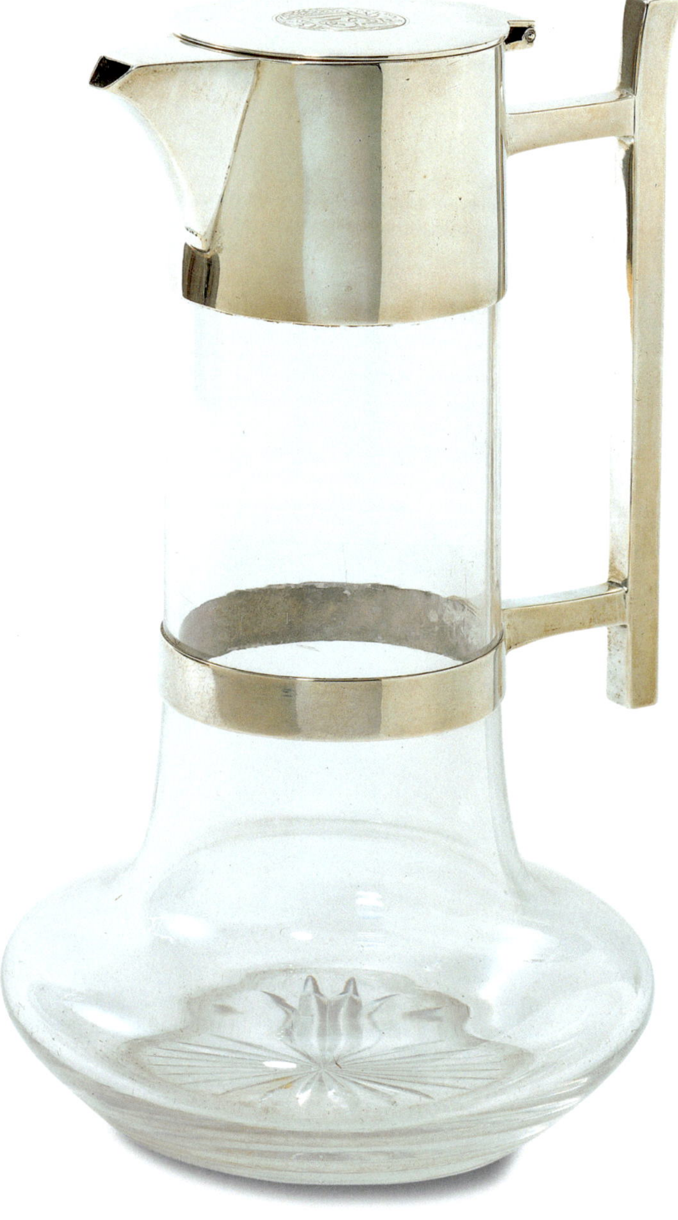

Claret jugs became essential for the polite table in Victorian Britain. The Duke of Wellington considered it bad manners to bring a labelled bottle onto the table.

Grands Echézeaux and one seventh of Echézeaux, all red wines, including a tiny parcel of the most famous white Burgundy, Le Montrachet. The wines, by which all other Burgundies are judged, are the pure reflection of their historic vineyards, never 'made' in the cellars. According to Aubert de Villaine, once the fully ripe grapes are transported to the winery, 'nothing is more difficult than to act simply: the ideal is to do nothing'.

Aubert de Villaine in person as a judge and Château Haut-Brion, in the shape of a bottle of its 1970, met in Paris on 24 May 1976 at an event known subsequently as the Judgement of Paris. This was a comparative tasting set up to show the French the fine wines emerging from California. The California wine producers' reds were made mostly with the Cabernet Sauvignon grape from Bordeaux and the whites from Burgundy's Chardonnay grape. At the tasting, which by happy coincidence was held 200 years after the American Declaration of Independence, they were set against Grand and Premier Cru white Burgundies from the finest estates, and First and Second Growth Bordeaux, among these Haut-Brion. The result, judged by nine of the most qualified palates of France, placed Château Montelena 1973, a Napa Valley Chardonnay made by an immigrant Croatian named Mike Grgich, and Stag's Leap Cabernet Sauvignon 1973, a wine from vines only four years old, first in both categories. The shock to French wine makers has now been forgotten and on 24 May 2006 a re-enactment was held at COPIA, the wine and food museum in Napa, California, more as a celebration than as a competition.

Stag's Leap Wine Cellars is owned by Warren Winiarski, whose Polish last name means 'son of a winemaker'. A professor in the liberal arts at the University of Chicago, Winiarski became fascinated with wine on visits to Italy, and in 1964 became a trainee winemaker in California. Five years later he bought a smallholding in the Stag's Leap district of Napa Valley and planted Cabernet Sauvignon grapes. His 1973, the first vintage to be bottled, so impressed the judges in Paris, that they thought it must be French and it was duly given high points. Sales, difficult for an emerging winery, took off and Stag's Leap Wine Cellars continues to produce exceptional wines. Winiarski's search to achieve the truest vineyard expression is a philosophy shared by Haut-Brion and Aubert de Villaine. Although French wines 'lost' in Paris, their quality has never ceased to gain converts.

In 1985 the Chilean Eduardo Chadwick was impressed by

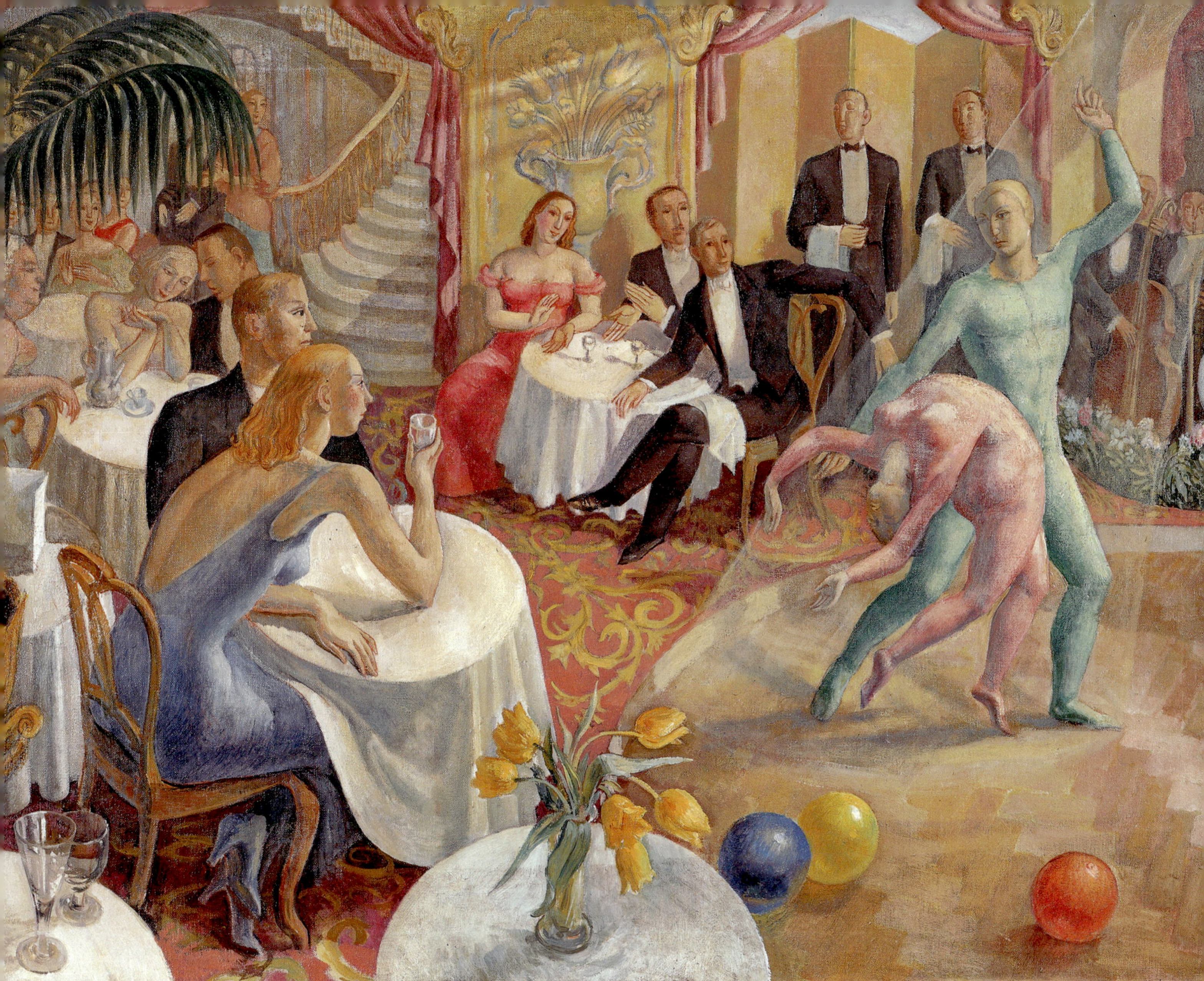

Café society in the 1930s: smart women were now free to drink in public and wine, spirits and cocktails were popular. However, the consumption of beer had dropped well below its pre-war level.

the French wines he tasted during a visit to Vinexpo, the biannual wine fair held in Bordeaux. Having joined Viña Errazuriz, the company founded by his great grandfather in 1870, he visited the world's great wine regions, returning home to incorporate new viticultural practices. By 2000, Chadwick was convinced that Chile, and Viña Errazuriz in particular, was producing world class wines; in 2004 a blind tasting in Berlin pitched his 2000 and 2001 'icon' wines Viñedo Chadwick, Seña (a joint-venture with Robert Mondavi, the doyen of California vintners) and Don Maximiano Founder's Reserve against the Médoc First

Growths Lafite-Rothschild, Latour and Margaux and cult wines Sassicaia, Solaia and Tignanello from Tuscany. In this Berlin tasting, Viñedo Chadwick 2000 and Seña 2001 were placed first and second respectively in a milestone for the Chilean wine industry.

Comparative tastings such as those held in Paris and Berlin have been derided by some critics as 'willingly controversial and attention seeking'. Yet how else is a wine to be judged, how else may vision, effort and persistence be rewarded? Vintners and vintages will come and go: the search for perfection from unique vineyard sites will remain.

Alcohol and Pictorial Advertising

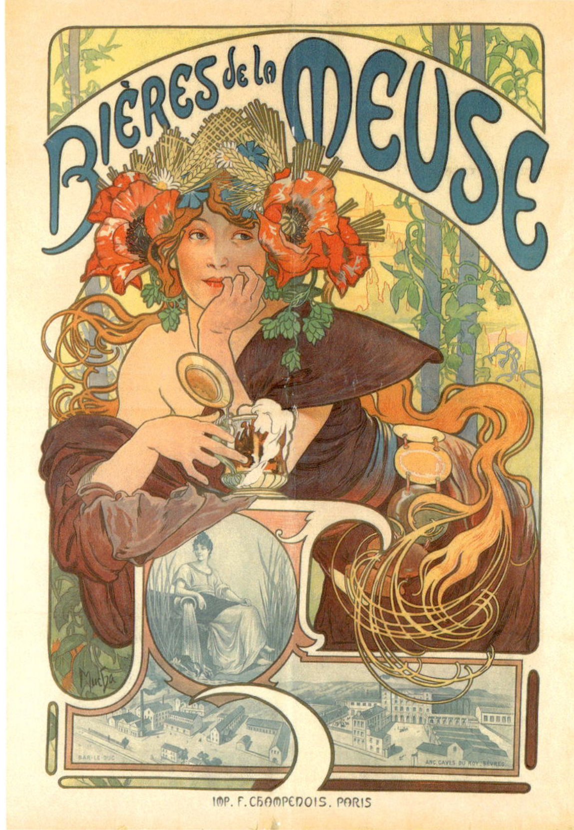

Left: A wreath of hops and barley, the trellis behind and the healthy girl with an overflowing lidded tankard unite the traditional associations of beer-drinking.

Opposite: Sophisticated café society and the delicious flamboyance of a Parisian showgirl are evoked in this frivolous advertisement for an aperitif.

Vivid and colourful, the evolution of alcohol advertisement is part of that enormous expansion of publicity that accompanied the great growth of manufacturing and an increasingly urban lifestyle from the late eighteenth century onwards. It flourished in the form of printed advertisements, trade labels and show-cards, bills and placards, wall advertisements and sign-boards. In Britain, the repeal of advertising duty in 1853 led to an explosion of advertising, and an increased demand for commercial imagery. This was further fuelled by the abolition of newspaper stamp duty in 1855, which created a new market for advertisement – including national campaigns – in the cheap mass-circulation newspapers and pictorial magazines that sprang up. The involvement of artists and professional designers to improve the commercial effectiveness of the advertisement was one of the factors that led to the emergence of a new art form in the 1880s – the pictorial poster. Wealthy drink manufacturers were eager to invest in this popular new medium.

Developments in colour lithography were crucial in attracting artists to the poster, and in France, Jules Chéret led

the way, demonstrating that art could be applied to new techniques of high-speed printing and commercial production. He was commissioned by numerous clients, among them *Vin Mariani / Popular French Tonic Wine* (1894), *Quinquina Dubonnet* (1895) and *Le Punch Grassot* (1896), in each of which an alluring girl is invitingly posed with the product. Lettering and imagery are combined in a harmonious composition. Among fine artists who essayed the medium was Pierre Bonnard, whose ebullient first poster for *France-Champagne* (1889) won him the immediate admiration of Toulouse-Lautrec. Alphonse Mucha, too, helped to establish the prestige of the French 'artistic' poster. Originally made famous through the patronage of Sarah Bernhardt, he was encouraged by his printer Champenois to design for a range of lucrative clients, including manufacturers of cigarette papers, infant food, biscuits and alcoholic drinks. For *Champagne Ruinart* (1896), he conceived an elegant muted design, whereas for *Bières de la Meuse* (1897) the design is less restrained, more florid, indicative of the popular appeal of beer. Controversy may have surrounded the social effects of absinthe, so notably captured by Degas in his painting *L'Absinthe* (1876), yet in an exquisite Art Nouveau composition the Belgian artist Privat Livemont glamorized the qualities of *Absinthe Robette* (1896).

Drink manufacturers were quick to associate themselves with other forms of 'artistic advertising' to gain commercial advantage. Following the lead set by A. & F. Pears who, from the 1880s, reproduced paintings by fine artists in order to sell their soap, John Dewar & Sons appropriated the epic image of Landseer's *Monarch of the Glen* (1851). Having bought the painting itself, they used the image as a trademark of their company, reproduced as an advertisement and on collectibles; Dewar's were keen to associate their whisky with high art, and with the qualities of majesty and Scottish heritage that the painting embodied.

A contrast to 'artistic advertising' was the revolutionary approach that characterized German design at the beginning of the twentieth century. In the new type of poster called *Sachplakat* (object poster), a realistically-rendered image of

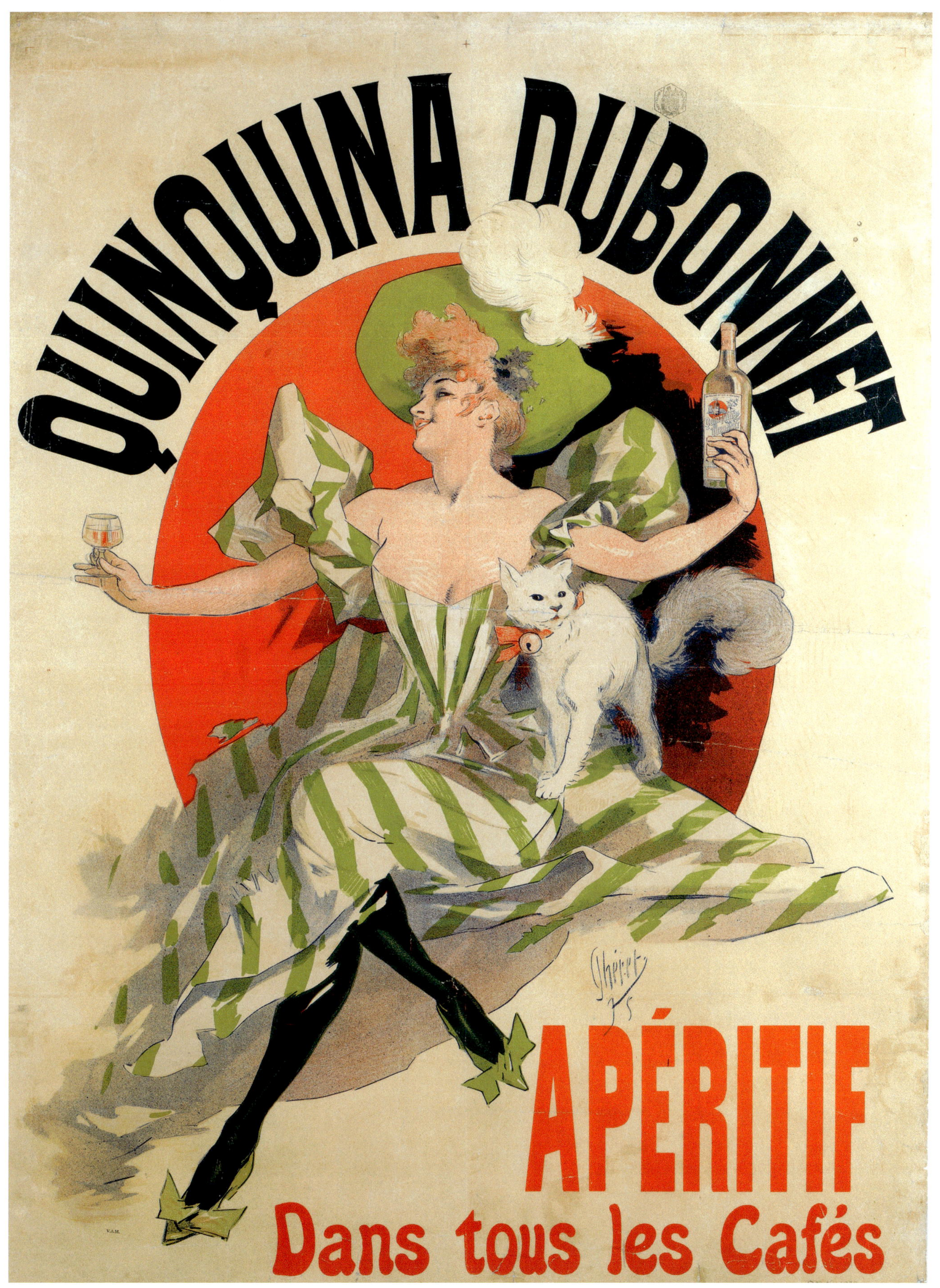
QUINQUINA DUBONNET
APÉRITIF
Dans tous les Cafés

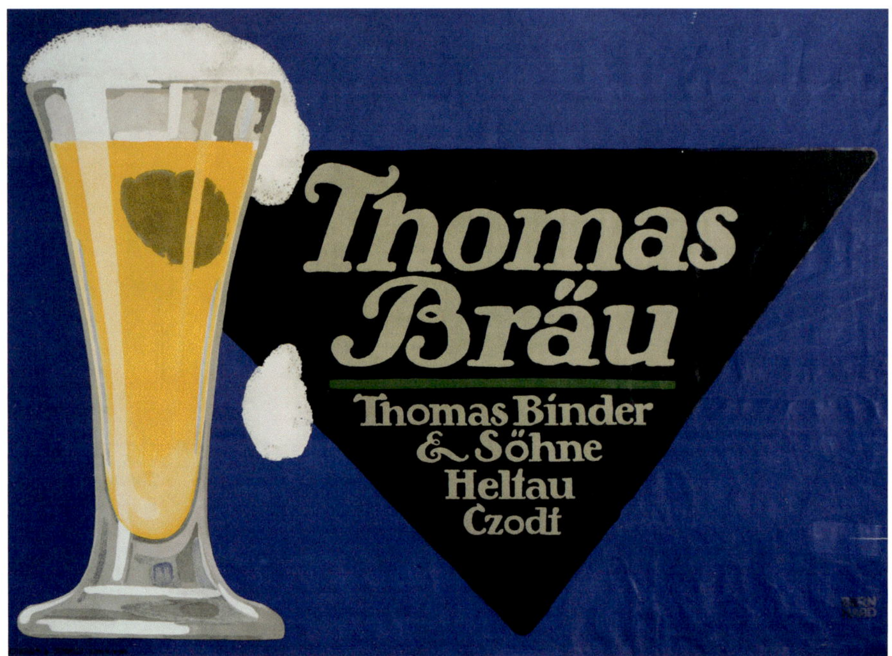

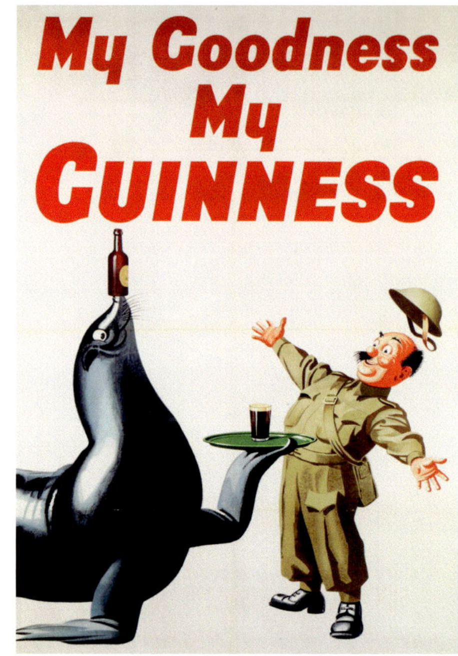

the product itself became the main subject of the composition. German advertising was led largely by manufacturers, and wealthy brewers and vintners were able to employ the leading printing firm in Berlin, Hollerbaum & Schmidt, who commissioned distinguished poster artists such as Lucian Bernhard, Julius Klinger and Julius Gipkens. Bernhard's poster for *Thomas Bräu* (1913) exemplifies the style – a masterpiece of graphic integrity, combining forms and lettering into a unified design that focuses attention on the product.

Elsewhere the early twentieth century marked a turning point in methods of advertising, spurred on by the rise of advertising agencies that pioneered the concept of campaign management from design, layout and copywriting to market research and distribution. S. H. Benson Ltd (1893–1971) was a leader in the field, responsible for some of Britain's best-loved campaigns, both in the press and on the hoardings. The Guinness campaign was launched on a national scale in 1929, and exemplified many of the features that remain crucial to advertising campaigns today: recognizable brand characters (the menagerie of animals drawn by poster artist John Gilroy), catchy slogans ('Guinness is Good For You', 'My Goodness My Guinness'), the use of humour, and association of the product with desirable qualities (strength and goodness).

Alcohol advertisement is responsible for many catch-phrases that have entered the language: 'Don't be vague, ask for Hague'; 'I'm only here … for the beer' (Double

Diamond); 'What We Want Is Watneys'; 'Heineken refreshes the parts other beers cannot reach' (1974); 'Well they said anything could happen' (Smirnoff Vodka); and 'Australians wouldn't give a Castlemaine XXXX for anything else' (1986). Commercial broadcasts on television, the ascendant medium from the 1950s, gave new scope to reiterative advertising: among memorable slogans were 'Looks good, tastes good and, by golly, it does you good' (Bernard Miles on Mackeson beer in an early 1958 commercial); and 'Suffused with herbs and spices from four continents' for the popular Cinzano ads (1978–83) starring Joan Collins and Leonard Rossiter.

Brand characters with whom the public learnt to identify include Tom Browne's early twentieth-century design for the striding figure 'Johnnie Walker, born 1820 – still going strong' (Johnnie Walker Scotch Whisky); Cassandre's angular little man for Dubonnet ('Dubo – Dubon – Dubonnet') who first appeared in the 1920s; the Sandeman Don, the mysterious cloaked figure in silhouette designed by George Massiot Brown in 1928 (Sandeman Port); and the little Red Grouse for The Famous Grouse Whisky, a witty recent television campaign. Advertisements showing the lone figure of the Osborne bull, a commercial trademark of sherry-producer Osborne (designed by Manolo Prieto in 1956), became so popular a presence in Spain that when roadside billboards were outlawed in the early 1990s, a great protest arose – the bulls had become a national symbol. The final

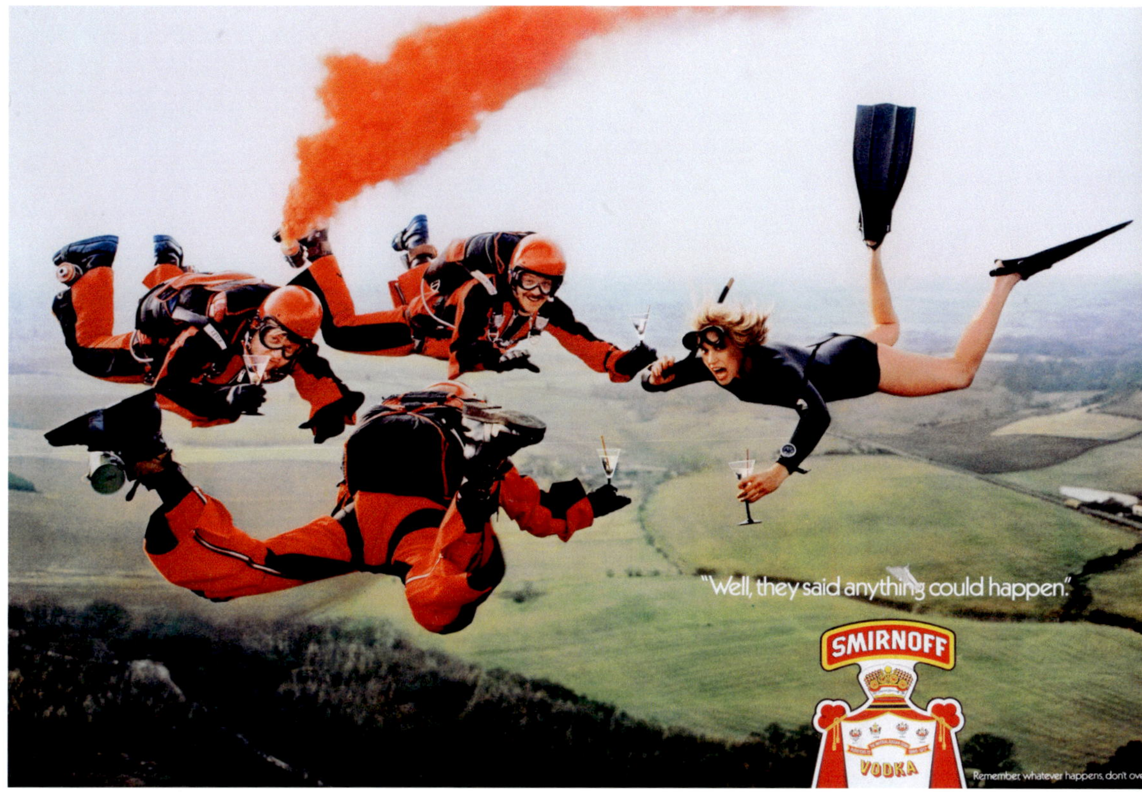

resolution was that they could remain, but with the company's name eliminated.

Now, in an age of global advertising of international brands, sophisticated research shows that consumers still respond best to marketing that is sensitive to distinctive cultural values and national sensibilities. Once that rapport is established, the audience is far more receptive to the message.

Anti-alcohol propaganda generated by temperance campaigns and state intervention, has taken many forms, from polemical tracts to television campaigns. In the late nineteenth and early twentieth centuries, temperance and prohibition were prominent issues, and posters with their popular appeal were recognized as a powerful medium for relaying messages of social and political concern; as with commercial posters, artists were commissioned to increase their impact. Walter Crane, for example, produced the poster *St. George against the Dragon Alcohol* (1911), promoting a campaign in Holland against the trade in alcoholic drink, and bad working conditions in the breweries. In a Swiss poster *Mein Vater trinkt keinen Schnaps*, issued in 1925 by the Nationaler Verband Gegen Die Schnapsgefahr (National Association Against the Danger of Schnaps), Zürich, the anonymous artist shows a child in the arms of his father rejecting the approach of a street-seller offering Schnaps.

Anti-alcohol campaigns also featured in posters by the Russian Constructivists in the post-Revolution period: a brightly-coloured simplified poster design of 1930 shows a bottle being targeted, with the exhortation 'Beat the Enemy of the Cultural Revolution!' Such a poster foreshadows the posters of *glasnost* and *perestroika* of the late 1980s and early 1990s, where issues of alcoholism and environmental pollution were among the issues raised in the new climate of openness.

In Britain after the Second World War, government policies on health, safety and security were relayed through posters to a population accustomed to receive information on the home front. One of the longest running campaigns has been that against drinking and driving: Mount/Evans poster *Don't Ask A Man To Drink And Drive* (1966) (see p.19), where the lettering lurches from side to side over the central line of a road, illustrated with stark significance the loss of control caused by alcohol consumption. More recent campaigns on the subject, whether through the medium of television, print or billboard, have been hard-hitting and deliberately shocking.

Currently, alcohol advertising in the UK is controlled by a combination of legislation and self-regulation, designed to ensure that drink brands are not portrayed as improving social or sexual success, and avoiding the encouragement of irresponsible behaviour such as drinking at work or while driving. The advertising of alcopops, sweet-tasting, brightly-coloured drinks aimed at younger people, has generated some of the most intense recent debates.

Nostalgia and the Old English Tavern

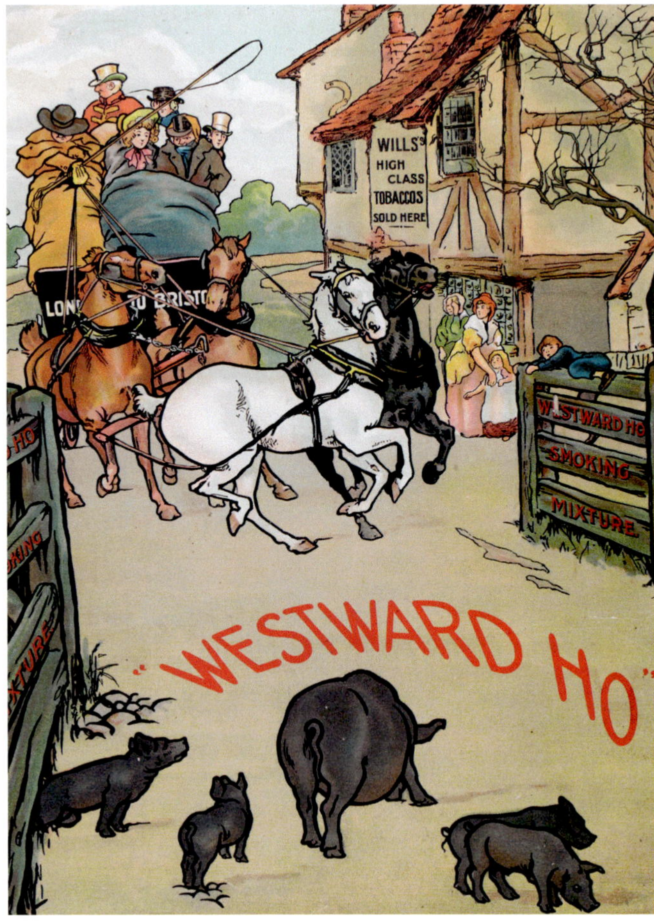

So powerful are the feelings of nostalgia that they often hopelessly glamorize the past, especially when we try to give these mental reconstructions physical form. However, the sentimental belief (or more politely, folk memory) that in the distant past there were cosy alehouses in every community, however small, does have a basis in reality. We know from medieval court records that alewives who brewed in their cottages (often illegally) provided a popular and welcome service at a reasonable price to their neighbours. Good fellowship, good cheer and unpredictable home-brewed ale were indeed available in these tiny and perhaps too cosy medieval front rooms. However, conditions were such that you might have had to step over a pig in the yard.

The same nostalgia clouds the later idea of the country hostelry or coaching inn. What cheer! How romantic! The gleaming mail coach appears before the inn, bringing its load of merry and thirsty travellers. However, the march of progress in the form of the passenger-carrying railways, burgeoning after 1829, killed many great coaching inns. Some

people, such as the novelist R.S. Surtees, were not sorry at the time. In *Ask Mama* (1858), looking back to the 1820s, he denounced the food and drink as poor and expensive:

Posterity will know nothing of the misery their forefathers underwent in the travelling way; and when we hear … unreasonable grumblings about the absence of trifling luxuries on railways, we are tempted to wish the parties consigned to a good long ride in an old stage coach…. Then the coach refreshments, or want of refreshments rather…. Anything was good enough for a coach passenger.

This idea is neatly enshrined in the Railway beer jug, designed in 1847 on which the driver of the stagecoach confronts the engine driver who is displacing him. One side of the design (labelled '1800') shows the exhausted passengers on foot, staggering up a hill too steep for a loaded stagecoach. They really need their all-too-short break at the inn for expensive refreshment. The other side (labelled '1847') shows the train gliding through a cutting in the hill.

Succeeding generations felt ambivalent, admiring progress yet nostalgically regretting its success. The railways encouraged the urbanization of Britain, and so the brewers grew more ambitious in their construction of modern public houses everywhere. These culminated in the huge, elaborately and eclectically decorated palaces that are frequently the best (and sometimes the only) surviving late Victorian buildings of any quality in many small towns. However, a deep nostalgia for the good old days of pre-industrial revolution Britain also encouraged the building of new old English taverns, and the best architects built them.

Most surviving country pubs of modest size used to be of the utmost plainness. They had survived in this state only because they were not worth the money to be spent on improving them. Even in the mid-twentieth century, many had the barrels arranged simply behind the bar and a minimum of long-lived sturdy furnishings. However, even this austerity has been imitated with the revived taste for real ale and the upsurge of microbreweries, and even primitive and remote rural pubs have reinvented themselves. A pub re-styling aimed at appealing to nostalgia can cut across the nostalgia of the regulars who prefer the tradition already built up by their own presence. Some do not want the imposition of a theoretical ambiance created on a drawing board instantly and expensively applied. Which do you prefer? Your favourite rough old boozer, complete with pig, or pub chain designer's olde-English-post-modern? Nostalgia is a contradictory thing.

Opposite: Nostalgic coaching scene poster advertising tobacco. In fact, coaches and coaching inns were uncomfortable, and very expensive.

Right: 'The Ashopton Inn', Derbyshire, about 1941. This watercolour by Kenneth Rowntree was part of the Recording Britain series, commissioned by the government. The plain and simple interior was typical of thousands of country pubs in the 1940s.

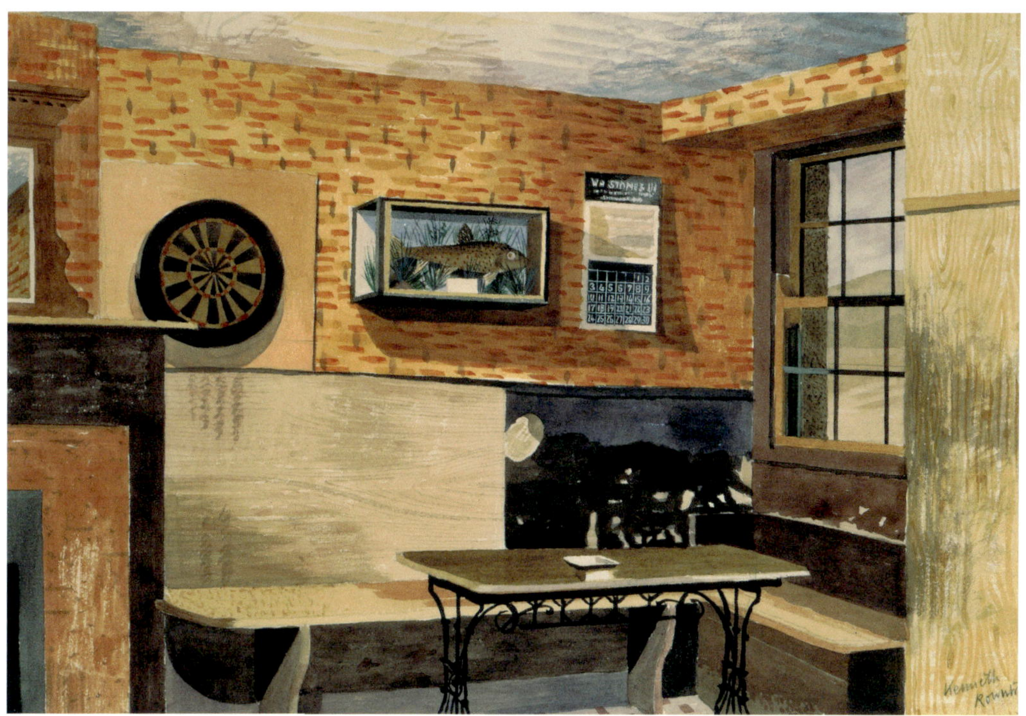

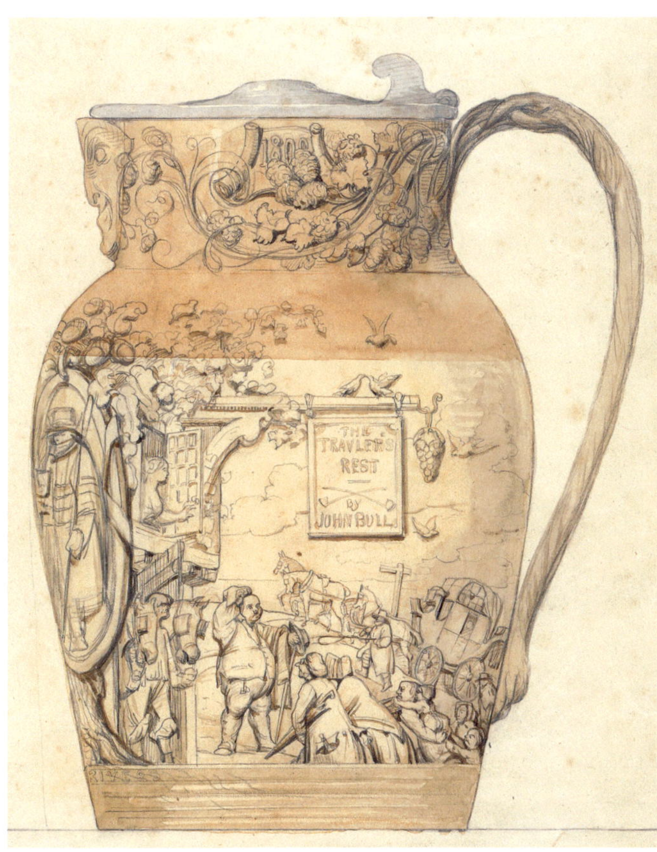

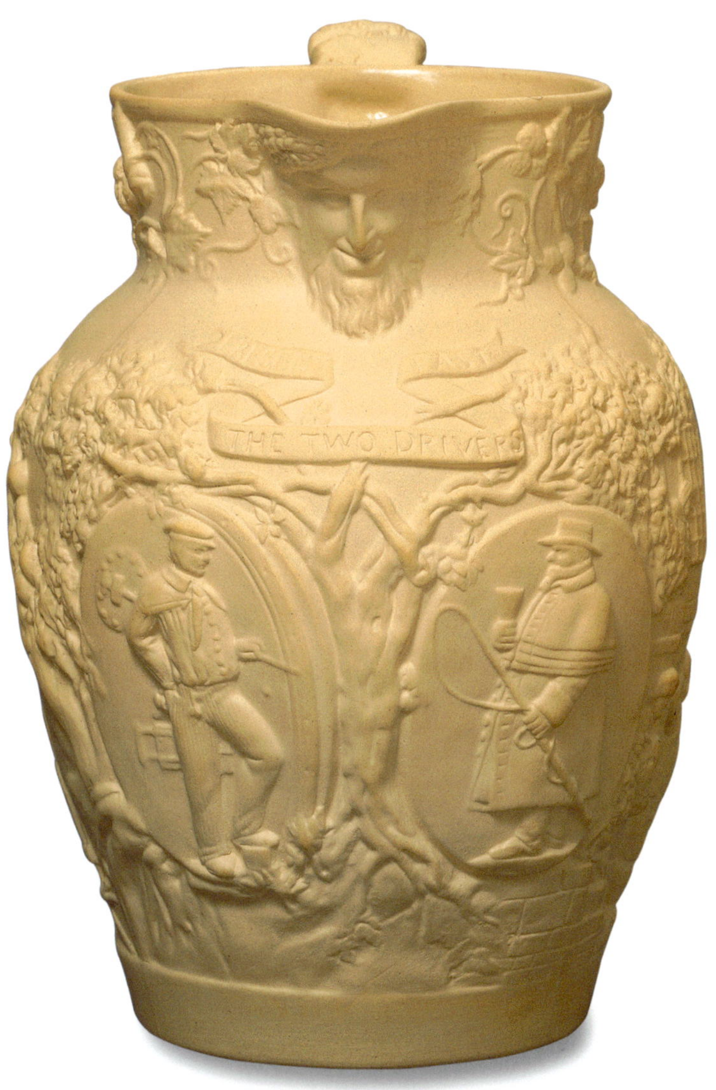

Above: This beer jug design shows the exhausted coach passengers staggering up a hill too steep for a loaded stagecoach.

Right: Detail of the Minton stoneware jug, designed by Henry James Townsend, showing the coach driver confronting the train driver.

Appreciating Wine

Its taste, colour and bouquet are delicious and it is consequently considered genuine by all. The wine is sweet, exceedingly full-bodied, slightly thick, and especially powerful and keeps well with a strong fragrance, its colour is clear and transparent. The wine spreads a good odour from the mouth. No wine is more highly prized than bastard wine with its noble sweetness.

OLAUS MAGNUS, SWEDISH ARCHBISHOP AND HISTORIAN, ON BASTARD, A SPICED WINE FROM THE IBERIAN PENINSULA, 1555

Colourful terms to capture the distinctive character of wine, devised for bottle labels, weekly newspaper columns and wine merchants' marketing brochures, are not a twentieth-century phenomenon. Discerning drinkers have enjoyed distinguishing between wine from different regions for centuries and good wine appeals to all the senses, as Olaus Magnus's comment shows. His is already the language of a connoisseur. Taste defined social superiority.

Transparent glass allowed wine drinkers to judge colour and clarity, as well as taste. Venetians appreciated clear glass for red wine, and Germans drank Rhenish white wine from green glass, dimpled for coolness, well before 1600. In the 1660s the English savoured red or white wine, cider or Champagne from distinctive glasses. A century later pattern books for the glass factory at Nøstetangen, Norway, show how widespread English wine glass forms were in the Northern market.

Sillery or still Champagne, fashionable at Charles II's court and enjoyed by Samuel Pepys on a summer picnic in Hyde Park, quickly became a treat. Etheridge in his play *Man of Mode* referred to 'sparkling champagne'; although it lacked the bubbles expected today, bottles were inclined to explode so private buyers often imported it in the barrel.

By 1700 it was considered the queen of wine, worthy of drinking from the finest fluted glasses (Augustus the Strong had his made of costly newly invented gold-ruby glass, engraved with his initials). This reputation was largely due to Dom Pérignon who arrived at the monastery of Hautvillers in 1668. His two great gifts were that he improved the vines and blended wines from different growths to create a consistent wine. He may have introduced the idea of tying corks down with packthread. Colour varied according to the pressing, producing varieties such as *œil-de-perdrix*, *gris*, *blanc* and *cerise*. Also the wine might be *non mousseux* or *pétillant* rather than the pale and bubbly drink that we'd expect today.

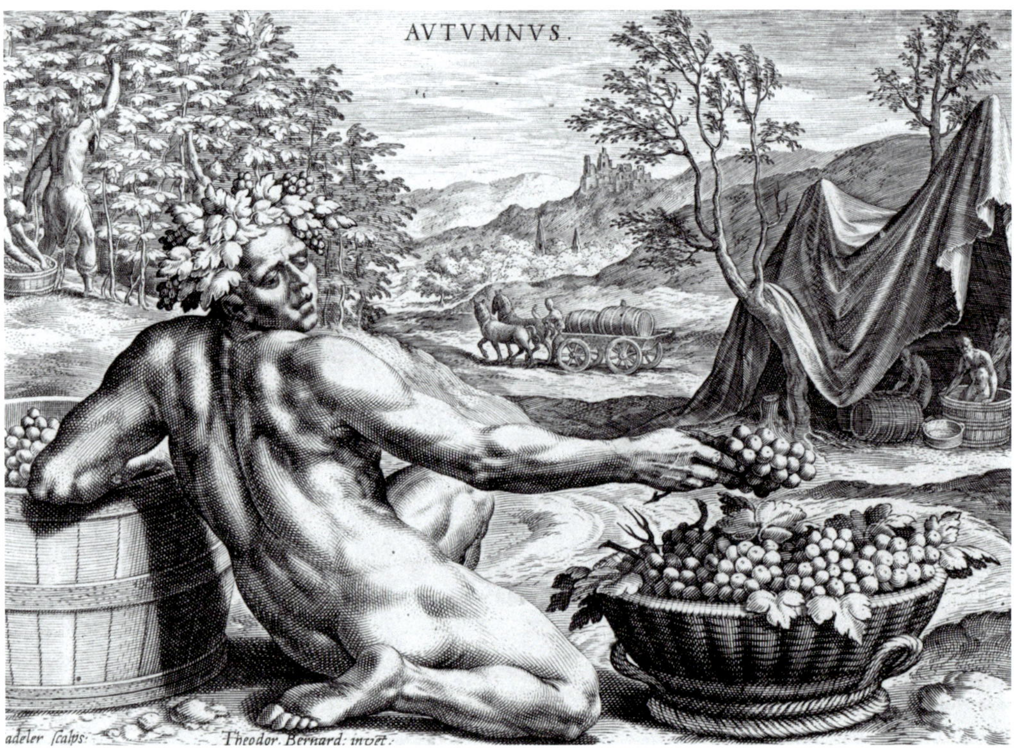

John Wilmot, Earl of Rochester, wrote a poem celebrating the joys of drink, 'Upon his Drinking a Bowl':

Cupid and Bacchus my saints are,
May drink and love still reign,
With wine I wash away my cares,
And then to love again.

By the 1730s a combination of improved techniques of
maturing, tight corks and stronger bottles had given
Champagne its proverbial character and made it the most
fashionable drink in aristocratic circles. William Pulteney,
suffering from over-indulgence, declined an invitation from
the Frenchified Duke of Richmond:

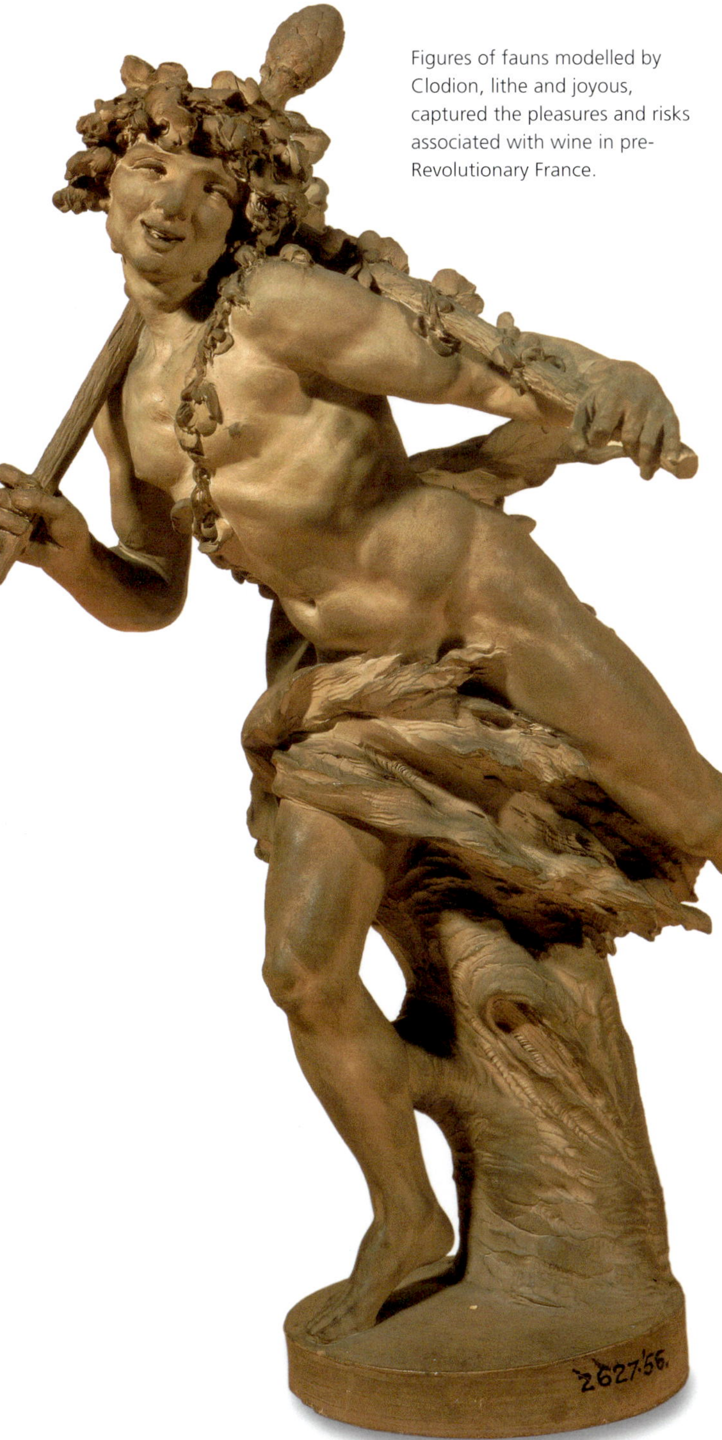

Figures of fauns modelled by
Clodion, lithe and joyous,
captured the pleasures and risks
associated with wine in pre-
Revolutionary France.

*Temperance and regularity are still necessary for me to
observe and at Goodwood I believe no one has ever heard of
either of them … I flatter myself I shall have the honour … of
swallowing many a bottle of popping Champaigne, but for
the present a little discretion is absolutely necessary.*

In Bath Champagne became an elegant accompaniment to
taking the waters, although it was still a very sweet drink
and for the British market became a dry wine only late in the
nineteenth century.

Champagne has its own bottle type. It is sold by the house
not by the château or vineyard, because of the five-year
fermentation period, and has retained associations of
celebration and superiority, a high price and a distinctive
glass shape.

English wine drinkers were often victims of adulteration,
as the Puritan Abraham de la Pryme noted in the 1690s. The
London trade was blamed; as Addison commented in *The
Tatler* in 1709: 'They can squeeze Bordeaux out of a sloe
and draw Champagne from an apple'. For almost two
centuries the English drinker was directly affected by
politics; a deliberate policy of excluding French wines was
finally reversed by Gladstone only in the 1860s. Customs
accounts show that in 1677 almost 10,000 tuns of French
wine were legally imported, far more than from Portugal or
Spain. In 1679, when Charles II declared war on France,
there was a huge drop, to half a tun, a barely credible figure.

Late Stuart politics divided drinkers, since supporters of
James II preferred claret, heavily taxed under William and
Mary and hard to find except in private houses, whereas
Portuguese wine became a symbol of the Whigs. In
1703/4 the Methuen Treaty required that the duty on
Portuguese wines should be two-thirds or less
the rate payable on French wines. William
III discouraged French wines at court,
apart from Champagne, supplied through
Holland by his friend the Duke of Zelle.
When out hunting, the king's party was supplied with 24
flasks of port, and 11 each of Rhenish and Canary wine and
sherry, plus Champagne.

So Anglo-Saxon consumers with no strong Continental
contacts were denied a reliable supply of French wine by
high duties and import restrictions. No bottles of wine could
be imported from 1729. Aristocrats could obtain French
wines in the barrel through friends, or by asking the British
ambassador in Paris, as Newcastle did in the 1730s, whereas

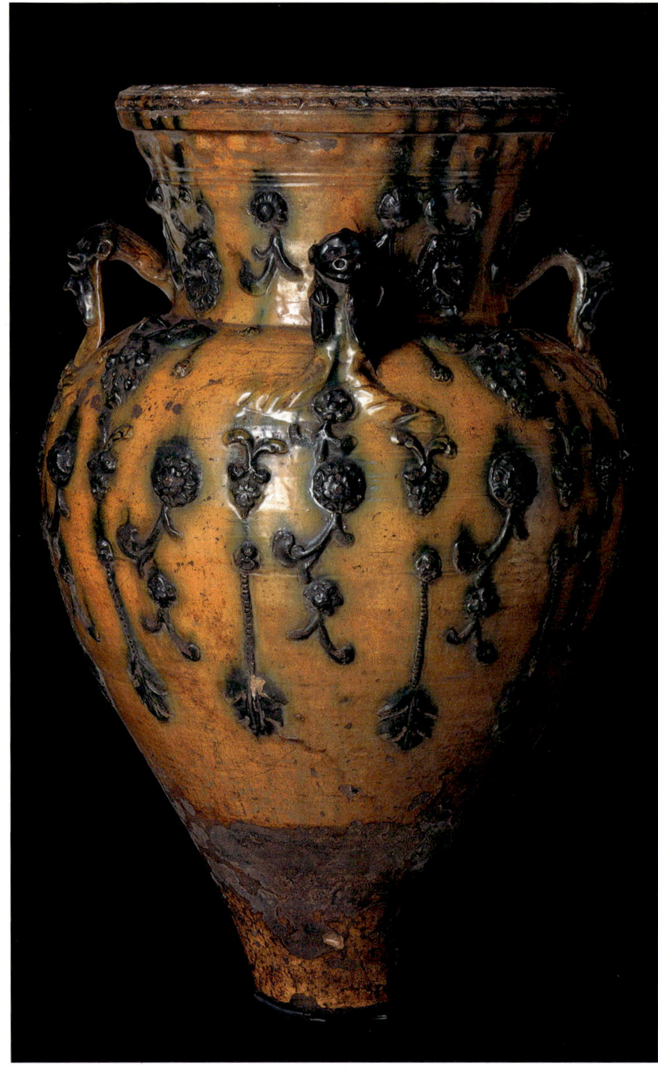

This large Spanish wine storage jar recalls a Roman amphora. Once filled, it was slung from ropes and set on a frame. In Spain, people stored young local wine for drinking and cooking.

Mary Wortley Montagu enjoyed the wine offered at court dinners; 'the variety and richness of their wines is what appears the most surprising. The constant way is to lay a list of their names upon the plates along with the napkins, and I have counted several times to the number of eighteen different sorts, all exquisite in their kinds'. Her choice probably included Tokay, 'the greatest wine in the world and only the masters of the world should drink it'. James Boswell, joining a law students' drinking club in 1764, was also delighted by Tokay, imported to Holland from a Hungarian estate. As early as the reign of James I, it was drunk with pleasure in London and a Hungarian count, Adam Batthyany, marrying an Italian countess from Friuli in 1632, gave her 300 Tokay vines as part of her dowry.

Although distinctive French wines were recognized in the seventeenth century, they barely travelled between regions because of transport difficulties and internal duties. The nobility often bought Spanish and Portuguese wines and those with vineyards shared their produce with friends. Savary's *Dictionnaire de Commerce* (1762) lists all the wines of France, headed by the regions of Champagne and Burgundy. When Thomas Jefferson visited French vineyards in February 1787, he had clear preferences; Chambertin was the best of the reds, followed by Vougeot and Vosne because 'they were the strongest and will bear transportation'. He had many bottles of Volnay, a cheaper red wine that was ready to drink after a year, shipped to America and admired Montrachet, the world's most expensive dry white wine.

Since many vineyards now bottle and label their own products, there is a strong incentive to ensure quality and maintain brand-consistency in a competitive international market. In France the legal strength for retail sale was 10 per cent, but the wine came from the vineyards at up to 15 per cent, so dilution was inevitable. A thriving internal market in wine became possible only when the railways arrived in France and wine expertise became a middle-class attribute and *Vin ordinaire*, a commercial product. When a municipal laboratory tested samples of wine from 300 wine outlets in Paris in 1880, 225 were badly adulterated, 50 were mildly adulterated and only 25 were uncontaminated. Too many outlets were trying to make profits.

others drank wines imported by London vintners. Customs figures give average imports for the decade 1717 to 1727: a modest 845 tuns of French wine, Portuguese 12,211, Spanish 8,467, Canary and Madeira each 380, Italian 237 tuns a year. Smuggling inflated consumption, but it was French brandy, rather than wine, which gave real profits.

Between 1720 and 1739, some 97 per cent of the wine consumed by the Barber-Surgeons' Company came from Spain and Portugal, 2.5 per cent from Germany and a tiny fraction from France. Meanwhile the Earl of Bristol, a wealthy courtier, drank as much wine from France as from the Peninsula, with a small amount of German wine.

The quality of German wines came as a pleasant surprise to one British aristocrat. Writing from Vienna in 1716, Lady

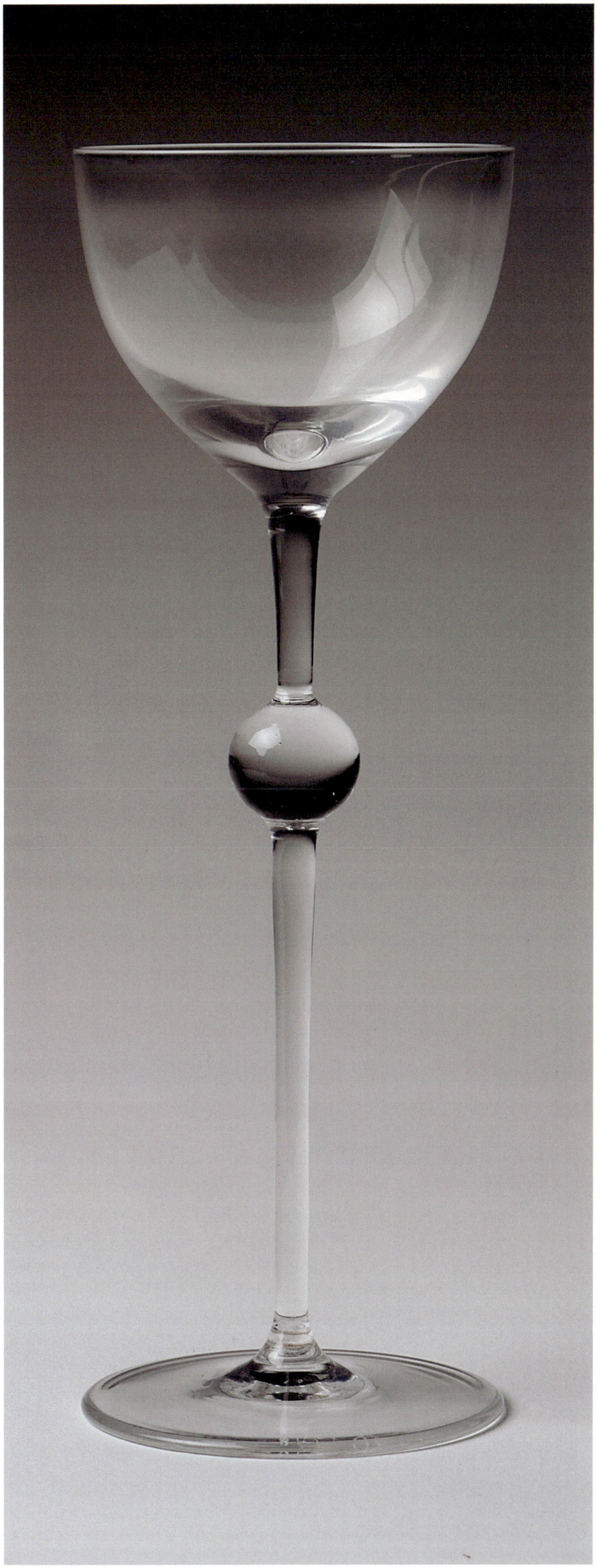

The science of tasting has been codified and tasters swill the wine around their taste-buds, almost chewing it; the Austrian master glassmaker Georg Riedel has developed goblets for connoisseurs, shaped for extracting the utmost flavour from specific grape varieties so that tongue and palate are in optimum contact.

Left: Clarity has always been admired in drinking glass. A guide to table etiquette (1788) recommended 'never give a second glass of wine in a glass that has been once used … the foot of the glass should be perfectly dry before it is given.'

Below: This table flask depicts the drunken Silenus and his entourage, a fit subject for a decorative container for luxury drinks, such as fruit-flavoured spirits or *eau de vie*.

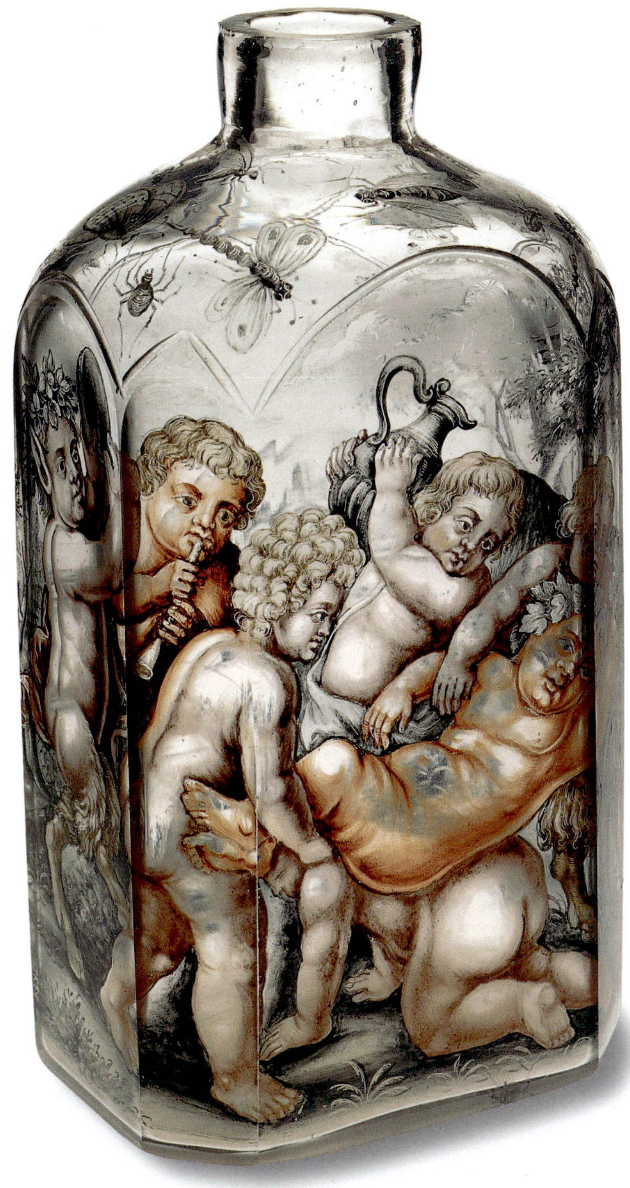

Bacchus and Ceres

ines and barley stalks, entwined around a ceramic jug, as background to a Victorian drink advertisement, or chased along the borders of a salver, have a long history as decoration in Europe, evoking the dual pleasures of wine and beer. Their symbolic meaning may now be merely of antiquarian interest, but they have had a powerful and persistent impact on the decorative arts.

The mysterious life that dwells in grape juice, causing such a miraculous transformation, has a magical effect on the mind, body and emotions of the drinker. The cult of the awe-inspiring and powerful wine god and hunter Dionysus, associated particularly with the wild excesses and flesh eating of his followers, can be explained physiologically in that wine-drinking releases endorphins into the bloodstream. These bring the drinker into a state of euphoria, a pleasurable state recalled in thousands of references throughout the visual culture of classical and post-classical Western art. A mural at Çatal Hüyük in Turkey, painted 8,000 years ago, is the oldest known depiction of this state; it shows a hunter, wrapped only in a pantherskin, celebrating with joyous energy.

For a post-medieval and masculine world steeped in classical imagery and educated from the classical authors as well as from the Bible, vines stood as emblems of fruitfulness, with a bunch of grapes often paired with a wheat ear, the one standing for Dionysus or Bacchus, the other for Demeter or Ceres, goddess of

agriculture. These life-sustaining essentials, bread, beer and wine, were evoked through a wealth of imagery. They could be alluded to visually in references to the more pacific Bacchus, the Roman deity particularly linked with viticulture, and the perpetual cycle of Nature's growth, harvest and decline. He also, as with Dionysus, hinted at the possibility of eternal life, so that the vine imagery could symbolize resurrection and continuity and the little putti tumbling about became reminders of rebirth.

Although vine imagery was claimed by the early Christians, these Greek and Roman associations ensured a long afterlife for all the pagan Bacchic imagery, such as Pan or Faunus, the ancient horned spirit of the woods, and the panthers and gambolling boys gesturing with cups and jugs of wine who also accompany his merry crew of Bacchantes. The putti adapted themselves into cherubs, small winged figures associated with immortality and appearing even on tombstones and wall monuments. As early as the fifth century BC, maenads or Bacchantes, female devotees of the cult of Bacchus, appear on Greek drinking cups. The Romans interpreted the graceful Greek god Dionysus as a fleshier creature and his annual festival, transformed into a three-day Bacchanalia, became so notorious for sexual excess that the Senate forbade it in 186 BC.

Italians considered that northern Europeans regarded getting drunk as a matter of prestige. In 1558 an Italian commented 'I thank God that, with other plagues which have come to us from beyond the mountains [Alps], this infamous one has not reached us yet: to get drunk not just for fun [*gioco*], but as an honour'.

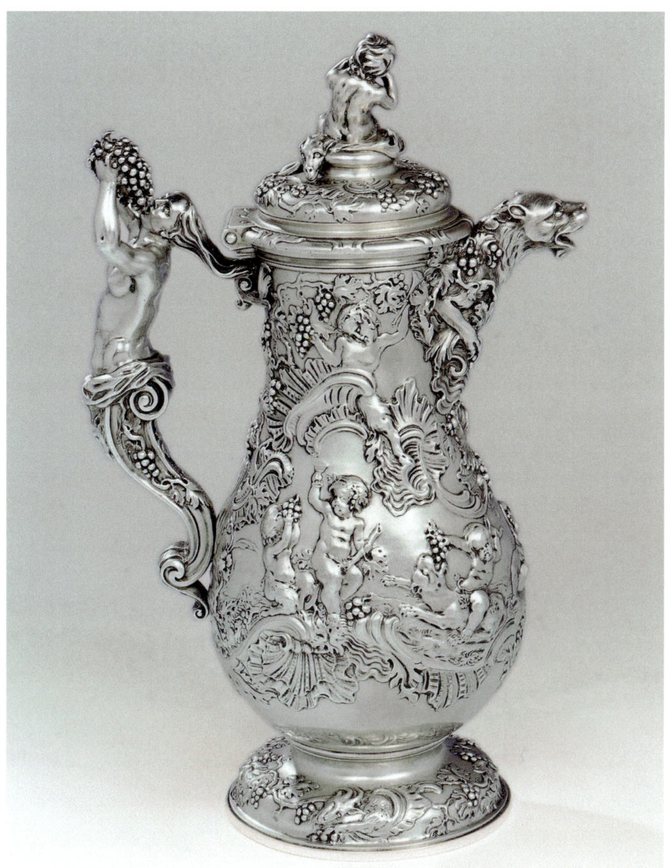

Engravers from Mantegna onwards played with the rich visual potential of lithe entwined half-naked male figures, their disordered curly hair crowned with vine-wreaths, tipping shallow bowls of wine in hospitable sharing gestures, which are recognizable as being on the cusp of drunken incapacity. Late medieval miniatures of the seasons and engravings of the months or the Zodiac depicted the bare vines in winter, the August or September vintage and sometimes the wine-pressing, as well as the pleasures of wine-drinking by the fire. About 1690 Nicholas de Larmessin issued a series of witty prints in which the figures are composed from the tools and attributes of their respective trades: a cooper, a *vigneron*, a cordial and liqueur seller and a *cabaretier*, whose thighs and arms are jugs, with glasses ornamenting his belt and shoe buckles.

In the decorative arts, particularly in the hands of the potter and the goldsmith, references to these classical antecedents abound in objects designed for convivial drinking. Their creations often brought humour to the table.

Above: Educated men would enjoy the Bacchic visual language of this beer jug. When the Prince of Brunswick came to London in 1764, the yacht was equipped with 12 dozen bottles of Burton Ale, brought to the table in jugs.

Right: Presented to the Chairman of the London & North West Railway Co., this cup has a vine stem and satyrs under the handles. A putto on the foot has a goatskin of wine, an archaic reference in the age of the claret jug.

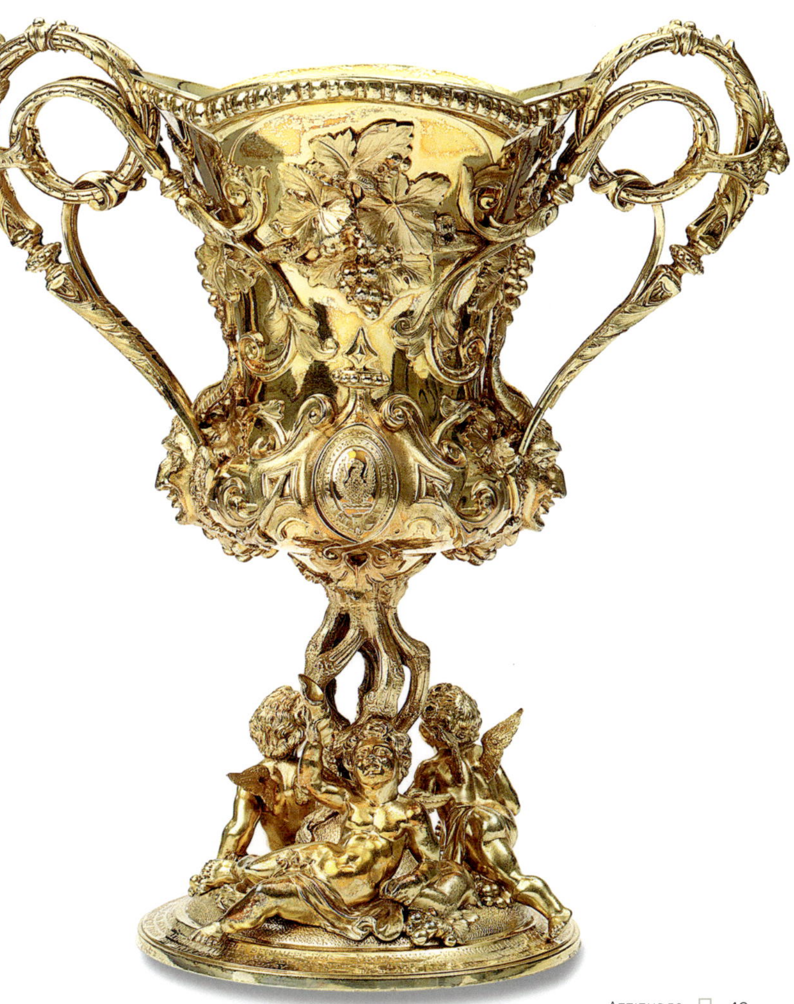

The graceful draped forms of Bacchantes, dancing in abandon, reappeared in the Renaissance, often paired with the Vestal Virgins, which could be read as embodiments of the female virtues. Light-hearted, joyful and sensuous, these subjects were popular in pre-Revolutionary France, and the French sculptor Clodion created lively groups in terracotta.

Painted around the Sala di Psiche in the Palazzo del Te, Mantua is a cycle of festive scenes celebrating the marriage of Psyche and Cupid; all the personalities associated with the pleasures of wine-drinking and fertility are present, appropriately for a wedding setting. Horned satyrs hand out grapes and bread, while one struggles to support the aged and drunken Silenus, tutor and foster-father of Dionysus, who is sliding off his mule. This is the most vivid expression of the Renaissance attitude to the classical deities, a set of attitudes memorably depicted by Poussin, Caravaggio and Rubens in many paintings alluding to the Feasts or Loves of the Gods.

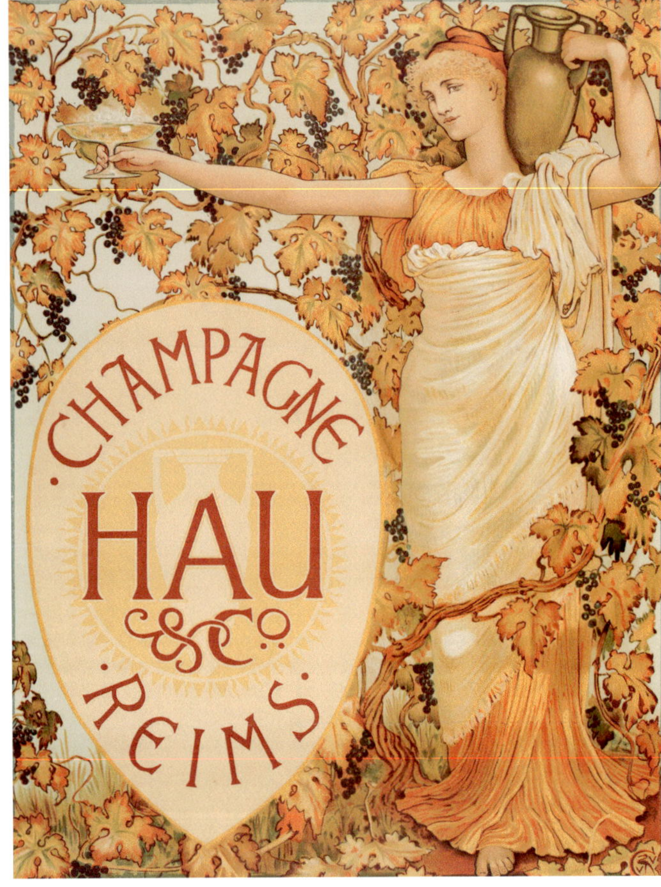

Left: Champagne attracted inventive designers. Richardsons of Stourbridge produced delicate 'Bubbles Bursting' glasses with putti cavorting among cascades of bubbling liquid.

Opposite: Costume design for the Spirit of Alcohol for the production of *The Vintage* at the Empire Theatre, London, 1903.

Bacchus may loll on top of a barrel, as in the tin-glazed Italian group shown here, or straddle a gilded silver wine wagon pulled by goats, as in a late sixteenth-century German table fountain in the Green Vaults, Dresden.

Increasingly from the late seventeenth century, as the demand for carved furniture and plasterwork, and tableware in ceramics and silver grew, designers disseminated reworked vine motifs across northern Europe. Shaped by training at drawing academies in Paris and Rome or derived from inexpensive suites of engravings, the visual repertoire of wine motifs became more sophisticated, less literal and more allusive.

But striking objects associated with wine still exploited this ancient imagery. In London in the 1730s, a close-knit group of sculptors and silversmiths modelled naturalistic vines, bunches of grapes, caterpillars and butterflies, to wrap around a massive wine cooler. Devised by an entrepreneur as an exhibition piece, this cooler, acquired by the tsarina in 1741 and now in the Hermitage, St Petersburg, was always admired more as sculpture than as a functioning object. It incorporates richly chased panels showing a Bacchic procession with putti and goats after Dieussart. An electrotype copy stands in the centre of the V&A's Whiteley Silver Galleries.

Formal drinking, the typical sociable gathering of the eighteenth century, when men (usually after dinner) met to share conversation and toasts, created a demand for richly decorated jugs and cups made of silver. In the 1730s, the prestige arising from a cup with Paul de Lamerie's mark, a salver by Archambo for a wine glass entwined with vines or a beer jug from Charles Frederick Kandler's workshop lay in the skill of the chaser, an artist who devised talking points all around these costly vessels. Motifs catch the candlelight and attract the eye of admiring drinkers, and subtly hint at the refreshing and intoxicating contents. Lions peer through foliage, snails perch on twisted vine branches and perky boys, virtually naked, play with Pan's pipes and tambourines, fondle the horns of goats or tumble in the grass with panthers.

The full ornamental value of twisting vine tendrils, spreading shaped leaves and fleshy bunches of grapes recur in many forms over 400 years. Learning to draw and to model such elements with a Bacchic theme became a standard part of an apprentice artisan's training in Paris, Rome and London from the early seventeenth until the late nineteenth century.

Sometimes it is hard for non-gardeners to distinguish between the grapes and bunches of hops, also with tendrils but with a differently shaped leaf, depicted on silver or in ceramic. Until relatively recently, both artists and consumers saw the two kinds of drink, grape- or grain-based, as coterminous. Beer, often home-brewed, found its place as an everyday table drink, alongside wine. It was only in the twentieth century that beer-drinking became a largely masculine and largely commercial activity, something that occurred as part of an outing and with particular connotations of class, so the two strands of imagery no longer read as partners and rarely appeared together.

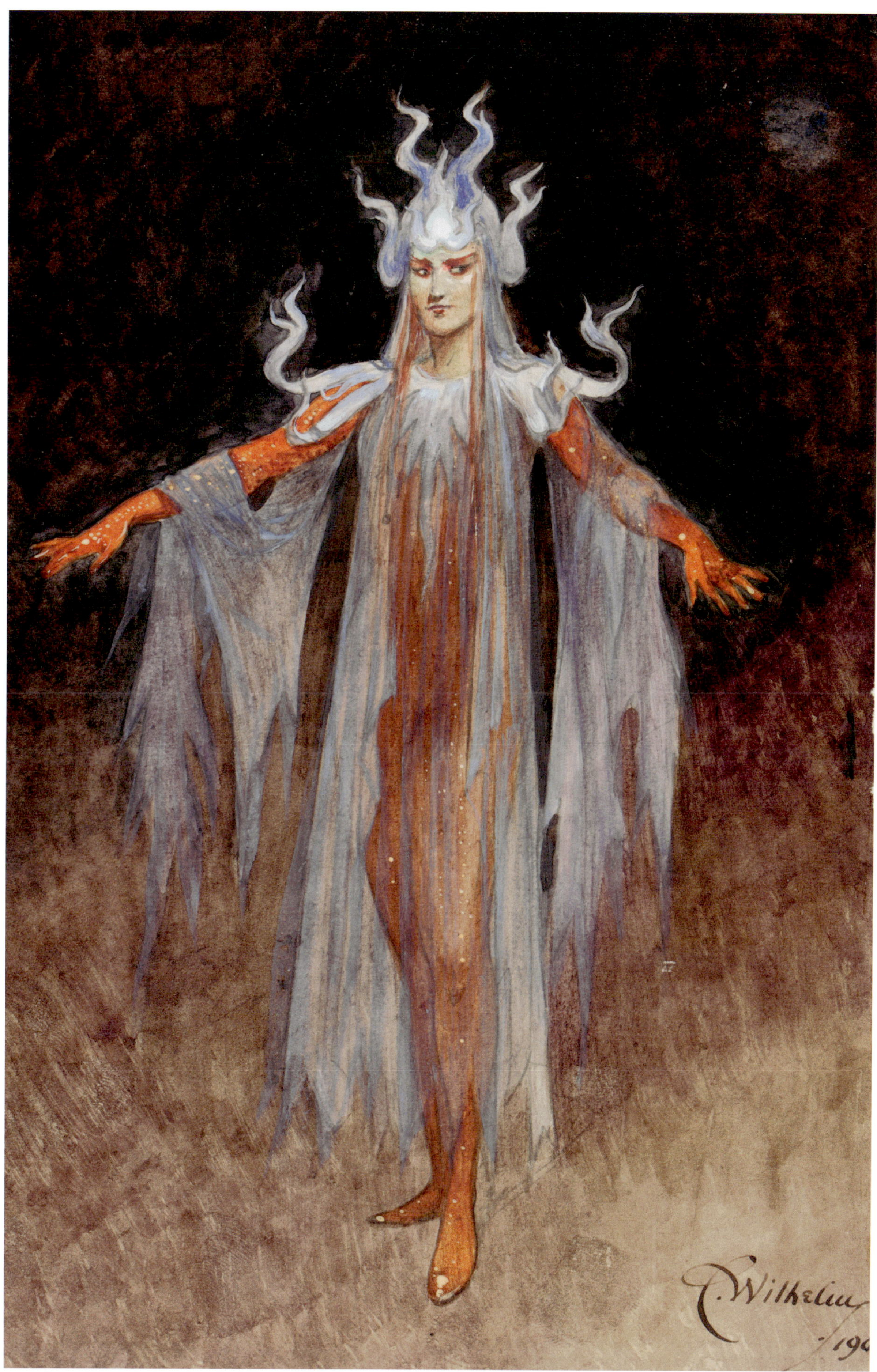
C. Wilhelm
/190

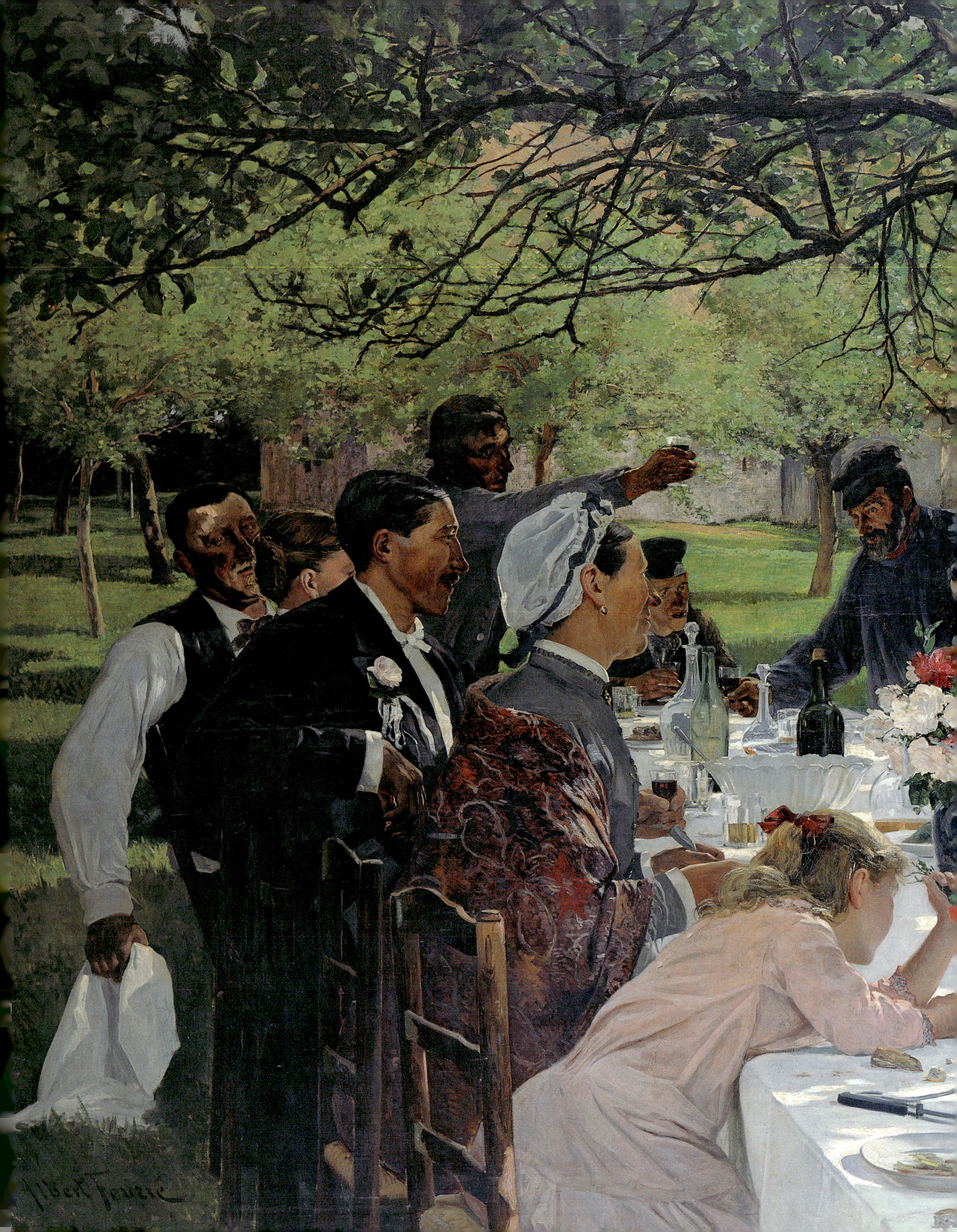

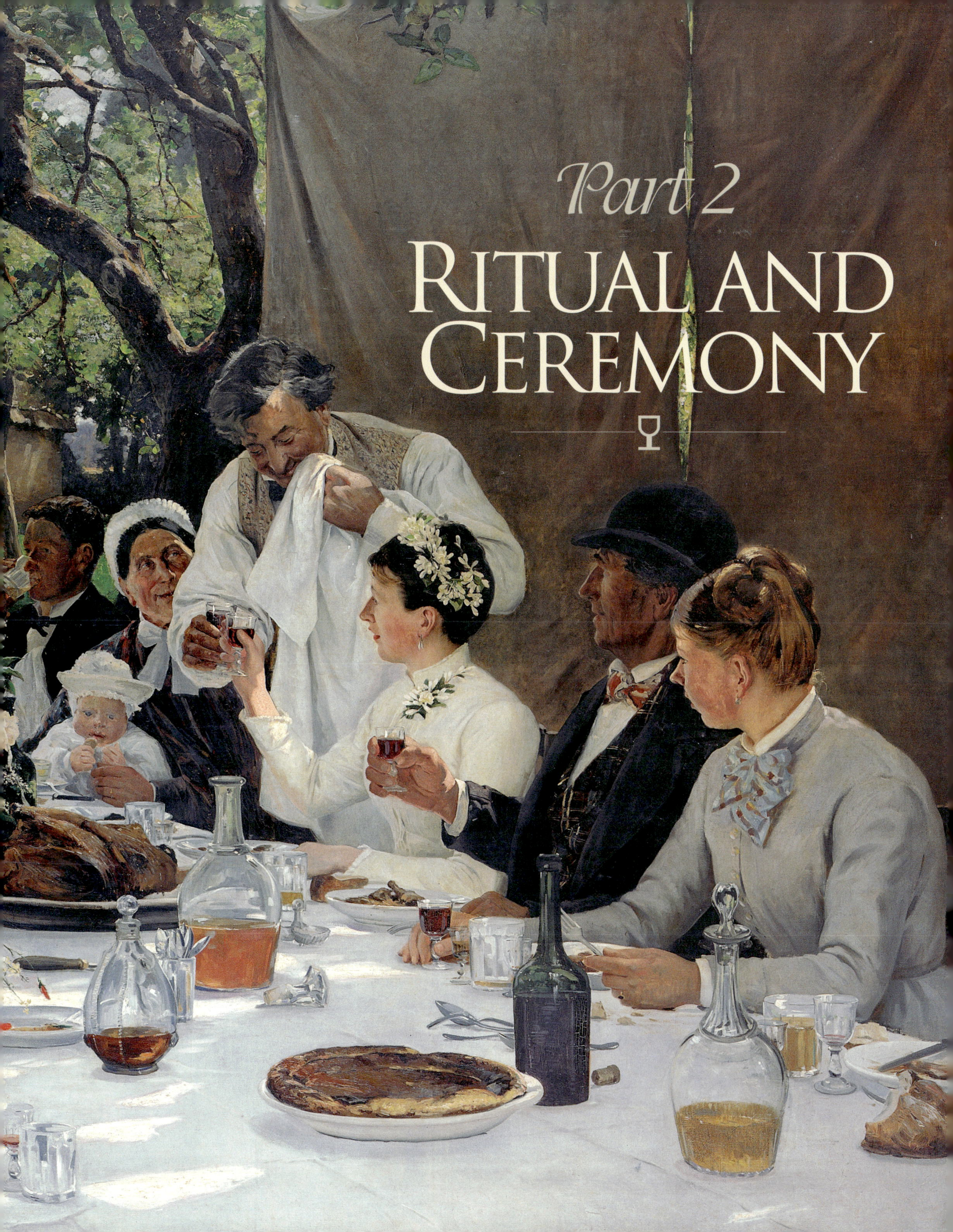
Part 2
RITUAL AND CEREMONY

Rites of Passage

Everything ends this way in France. Weddings, christenings, duels, burials, swindlings, affairs of state – everything is a pretext for a good dinner'.

So Monsieur Orlas sums up the universal relationship between the rituals of life and conviviality in Jean Anouilh's 1949 play *Cécile*. In Europe ceremonies involving alcohol have always been central to the rites of passage of family life.

A new addition to the family was a time of both risk and celebration. In seventeenth-century Holland news of a confinement was modestly conveyed by toasting the expectant mother with a wine cup called a *Hansje in de kelder*. After the birth, *kraam-bier* (crib ale) would be provided to herald the new arrival either on the day of his or her birth or a few days after.

Alcohol was also thought to strengthen women during pregnancy and to help the flow of milk. Obstetric manuals such as *Het Kleyn Vroetwyfs-boeck* (*The Little Book of Midwifery*), 1645, recommended that older mothers include wine in their diet to ensure a smooth pregnancy and safe delivery. Hot spiced wines or 'caudles' after the birth speeded up the mother's recovery. The *Batchelors Banquet*, a 1603 English version of the French text *Les Quinze Joyes de mariage* satirizes English childbirth customs, bemoaning how the husband must provide for the many 'gossips', brought into his household, 'to attend her and keep her, who must make for her warm broths and costly caudles'. As late as 1803, Lady Nugent, wife of the Governor of Jamaica, kept open house for ladies to admire her month-old son George; from 12 till 4, 'Cake, caudle, chocolate etc were devoured. The baby was shewn in his little cot'. Beer was routinely given for nourishment and a belief in its restorative powers for breastfeeding women persists into modern times. In 1887 William Gomm of the Beehive brewery, Brentford sold 'Double Stout (for nursing)' and Waltham Brothers' brewery in Stockwell proclaimed its SN stout 'particularly suited for invalids, ladies nursing or anyone requiring a good sound strengthening beverage'.

Baptism, welcoming a new Christian soul, was central to community life. Wealthier households hosted their own christening celebrations while alehouses or taverns were venues for the less well off. Costs could be substantial. In

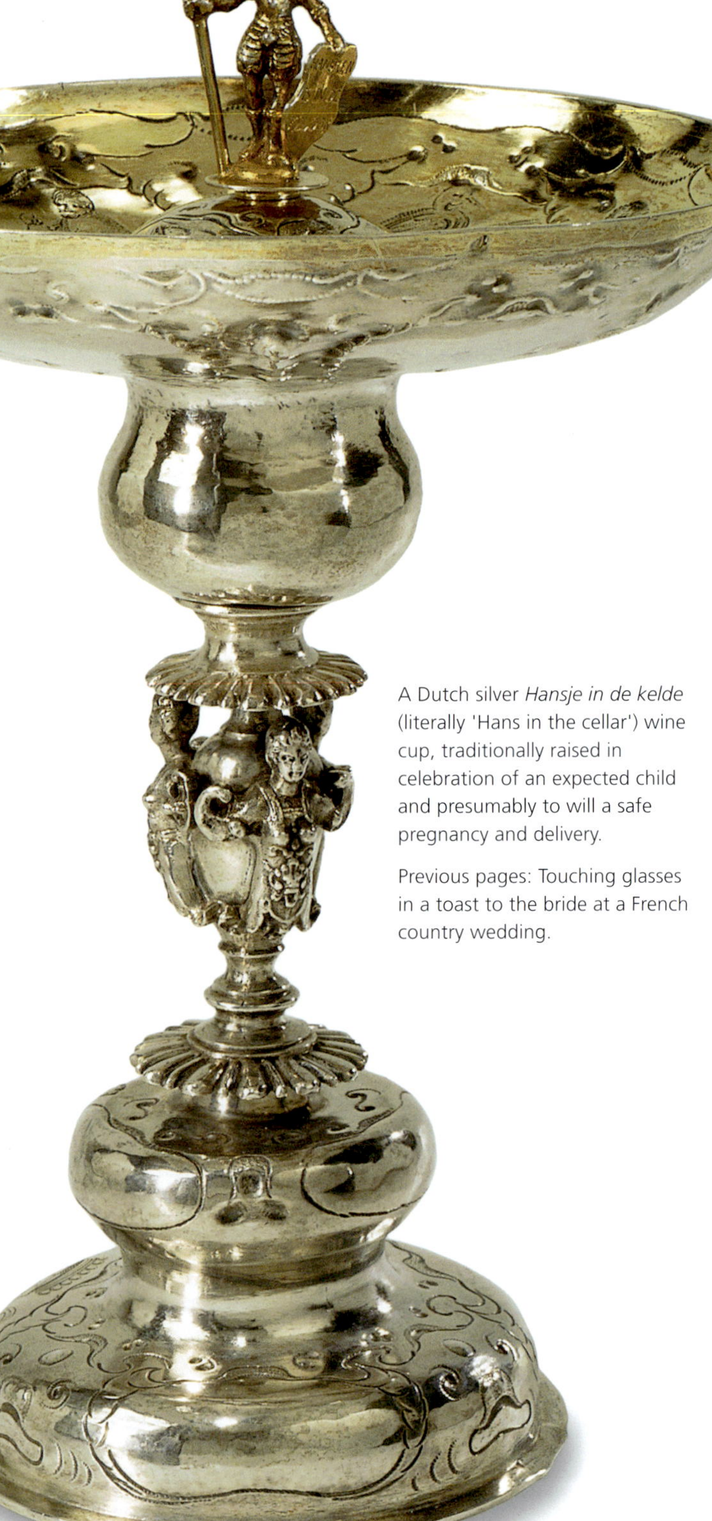

A Dutch silver *Hansje in de kelde* (literally 'Hans in the cellar') wine cup, traditionally raised in celebration of an expected child and presumably to will a safe pregnancy and delivery.

Previous pages: Touching glasses in a toast to the bride at a French country wedding.

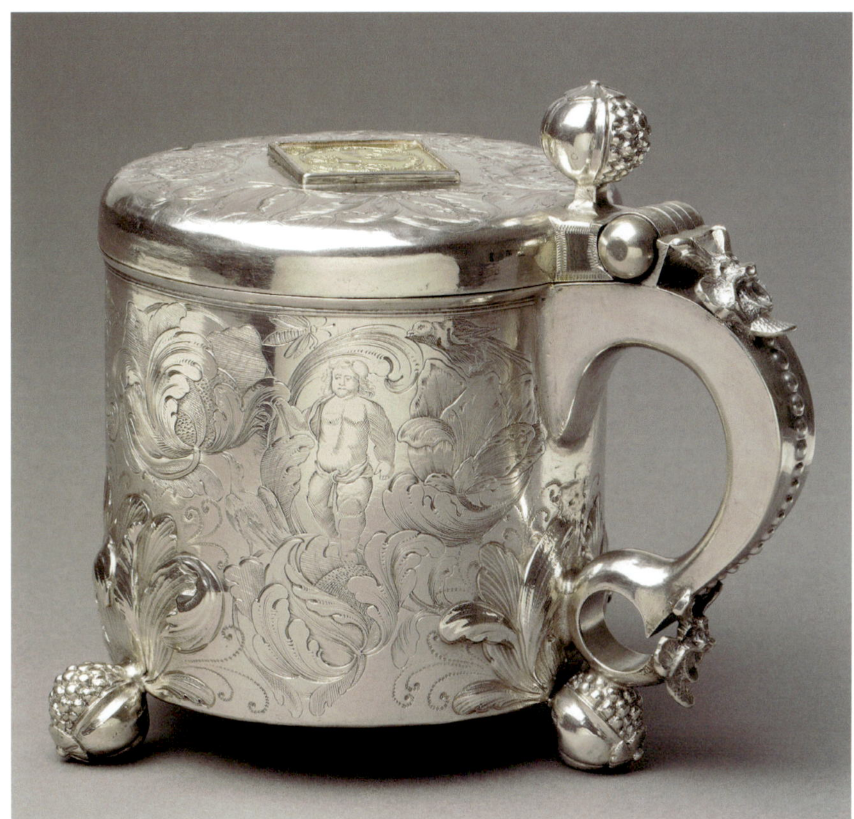

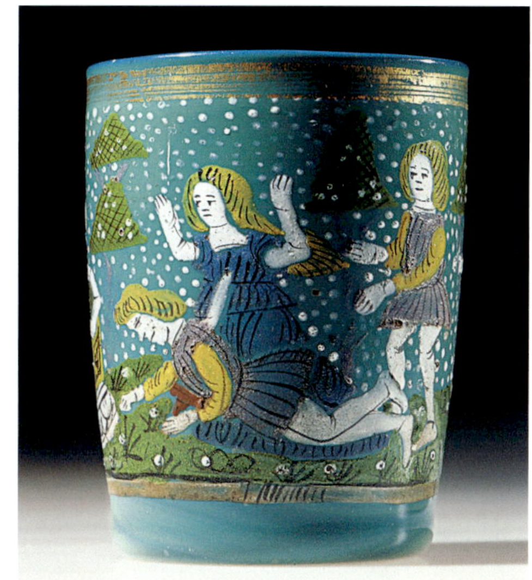

1642 an Essex clergyman Ralph Josselin spent over £6 for his first child's christening party, almost a tenth of his annual income. Special drinks were provided. Dutch customs transported to the American colonies in the seventeenth century included the passing round of a silver brandy bowl or a *berkemeyer*, a large glass goblet filled with sugared Rhenish wine. Another vessel, the *Kandeel pot* (cinnamon or caudle cup), was a tall drinking cup filled with wine in which a cinnamon stick, cut long if a boy, short if a girl, was used to stir in sugar. Festivities could be so extravagant that Dutch sumptuary laws restricted excessive eating and drinking, as in Dordrecht in 1618 and Amsterdam in 1655. The rowdy reputation of christening parties in England led to the seventeenth-century catchphrase 'as drunk … as women at a gossiping'. After the baptism of George, Lady Nugent's son at the Governor's house, 'all the gentlemen stood up to drink the new Christian's health, with three times three' and the entire household shared cakes and wine.

Christenings were times for gift giving as well as merrymaking. Silver tankards were common christening presents throughout northern Europe in the seventeenth and eighteenth centuries. Joyce Jeffries from Hereford made gifts of silver tankards to several of her 13 godchildren between 1638 and 1647. By the late nineteenth century the custom of a godparent laying down wine, notably port, for the male child to appreciate later was firmly established.

Birthdays and anniversaries offered customary opportunities for drinking. Mrs Beeton recommended a recipe for punch for children's parties of port diluted and warmed, sweetened and spiced. Coming of age festivities were the most important. Edward Phelips (1725–97) recalled his 21st birthday celebrations at Montacute House in Somerset on 12 April 1746: 'My mother was pleased to celebrate with unusual splendour and festivity … Eight hogsheads of cyder and four of strong beer were given the populace. One hogshead of wine and two of punch were Drank in the House, very little fuss or Disorder prevailed…'. The Bar Mitzvah ceremony for the coming of age of a Jewish boy at the age of 13 includes the ritual drinking of wine from a Kiddush cup, over which the rabbi has recited the Kiddush, a prayer of sanctification.

In medieval England weddings were usually a free-for-all, with 'church ale' brewed and passed around the congregation and an open invitation to the wedding feast, often held in a 'wedding house' belonging to the church. After reformers in the sixteenth century discouraged these practices, families unable to provide their own venue decamped to the local alehouse and guests were encouraged to contribute food, drink and entertainment to the feast, called a 'brideale'. Pieter Brueghel's painting *The Peasant Wedding* (*c*.1568) shows such a scene. Profits from the ale

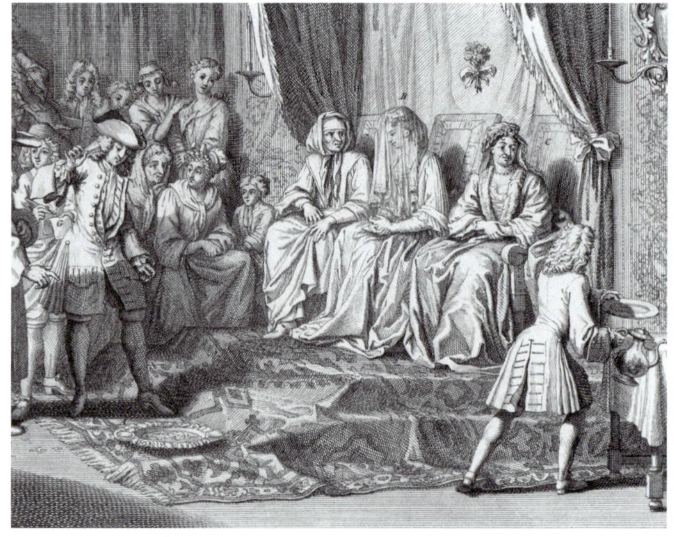

sold went to the bride and groom to help them set up house. In 1575 the Earl of Leicester's entertainments for Elizabeth I at Kenilworth Castle included a brideale celebration, possibly projecting his own marriage aspirations. In 1783, public houses in the parish of Stenderup in Holstein were granted licences to run their businesses, brew beer and distil brandy, as well as the exclusive right to supply alcohol for weddings, christenings and funerals.

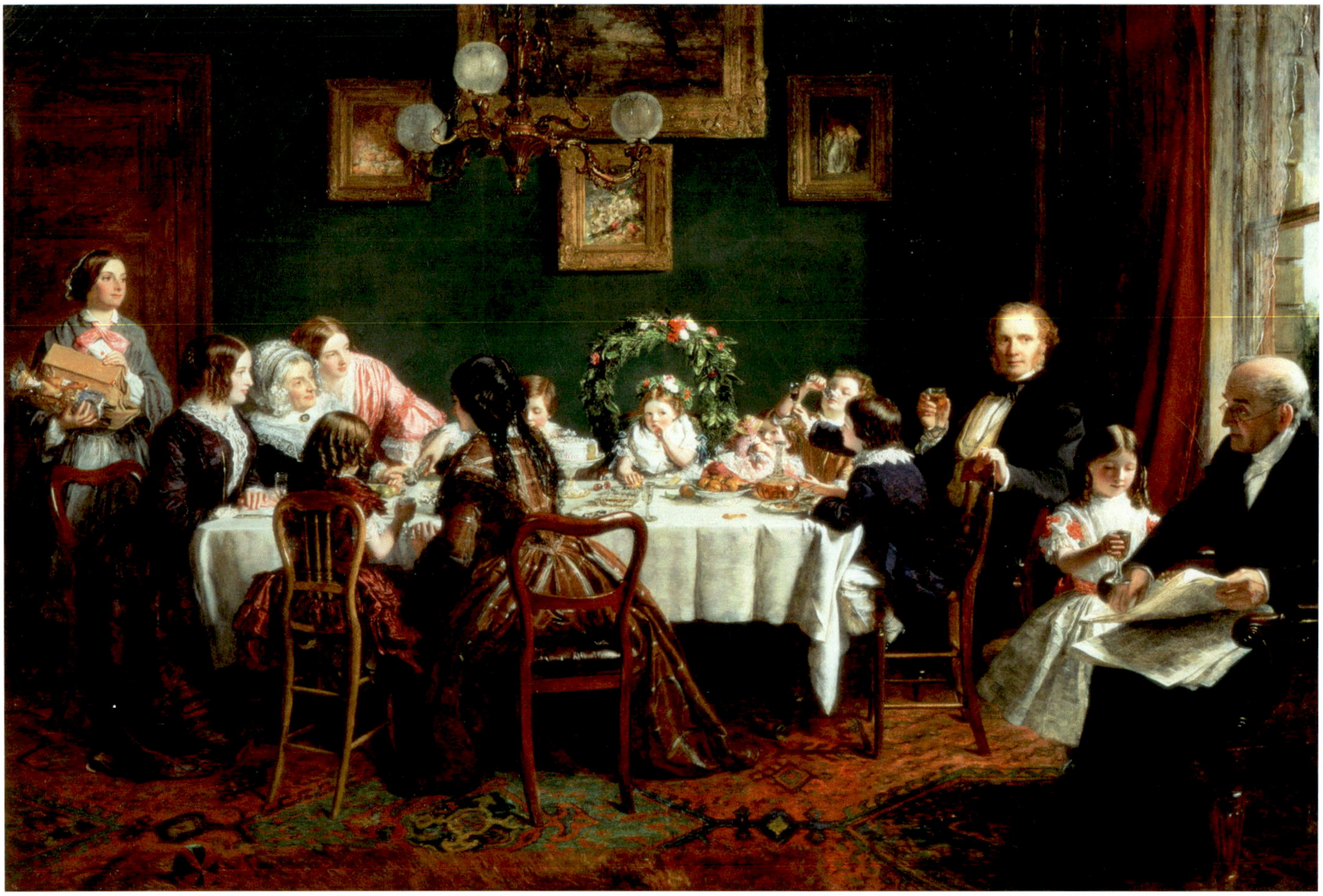

Private wedding feasts could be flagrantly excessive. At the wedding of Annibale Bentivoglio and Lucrezia d'Este in Bologna in 1487 the entire banquet, including precious sweet wines, was paraded through the piazza for universal acclaim. In *Madame Bovary* (1859), Flaubert's description of his heroine's wedding feast records what must be contemporary customs: 'At the corners [of the table] stood carafes of brandy, Bottles of cider were sending their thick froth up around the corks, and all the glasses had already been filled to the brim with wine'.

Wine cups were common betrothal or wedding presents; one was selected for a shared toast between man and wife, drunk with arms linked or from a single cup, as depicted in a mid-sixteenth century Urbino faience group. Delicate

Above: In William Powell Frith's 1856 painting of his daughter Alice's birthday party her brothers toast her health with alcohol. Children's parties were fuelled by negus, a light sweet warm punch made from port, water, sugar and spices.

Below: At Jewish weddings, a wine glass is broken by the groom after the rabbi has pronounced the couple man and wife. The couple would earlier drink the wine after the recital of the betrothal blessing (*kiddushin*) and the Seven Blessings (*Sheva Brachot*).

colourless glass goblets from Renaissance Venice, decorated with gilding and enamel, were probably made as wedding gifts, for display rather than use. At the feast guests would have drunk from less expensive coloured glass beakers. In sixteenth- and seventeenth-century Germany, gifts of silver-gilt double cups – one long-stemmed cup fitted upside down on top of an identical one to form a cover – symbolized the marriage, although engraved glass goblets later gained in popularity.

Champagne or sparkling wine became popularly associated with weddings only in the early nineteenth century. Earlier tipples included spiced hypocras, originally a digestive in the wealthiest medieval households, adopted for family celebrations when spices became more readily available in the sixteenth century. Recipes in manuals such as *The Booke of Kervinge and Serving* (1508) contained red wine (commonly claret) or white wine (Spanish or Portuguese) spiced with cinnamon, ginger, cardamom or pepper, sweetened with sugar or less expensive honey and sometimes scented with musk or ambergris. The wine was strained through a bag known as the sleeve of Hippocrates (after the Greek physician), which gave the drink its name. Appropriately for weddings, hypocras was also thought to act as an aphrodisiac.

Between the sixteenth and nineteenth centuries sack possets – or benediction possets – were traditionally drunk by the bridal couple from specially designed posset pots before they retired, presumably to relax them and smooth marital concord. Henri Misson, a French observer in London in the 1690s, noted the English custom of members of the bridal party preparing a

good posset, which is a kind of caudle, a potion made up of milk, wine [usually sack, a strong white Spanish wine], yolks of eggs, sugar, cinnamon, nutmeg, etc. This they present to the young couple, who swallows it down as fast as they can to get rid of so troublesome company.…

A ring was sometimes thrown into the posset pot and whoever fished it out would be the next to go to the altar. In Orkney a whisky and cream posset still survives, called bridal cog, which is passed around the wedding guests.

The solemnity of funerals did little to subdue hospitality. When Samuel Pepys' brother Tom died in March 1664 he gave all the mourners who escorted the body 'six biscuits a-piece and what they pleased of burnt [hot] claret'. Biscuits, popularly 'Naples' and 'Savoy' biscuits, and hot wine were traditional funeral food in England, although when a Mrs Howe was buried in 1782, the diarist Parson Woodforde recorded that the mourners were given an eclectic array of 'Cake and Wine and Chocolate and dried Toast'. At Farmer Keld of Whitby's funeral in 1760, mourners consumed 30 ankers of ale (one anker is about 40 litres) alongside vast quantities of meat, cheese and bread. The poor might also benefit from the funeral of a local worthy, through the distribution of dole in the form of food and beer.

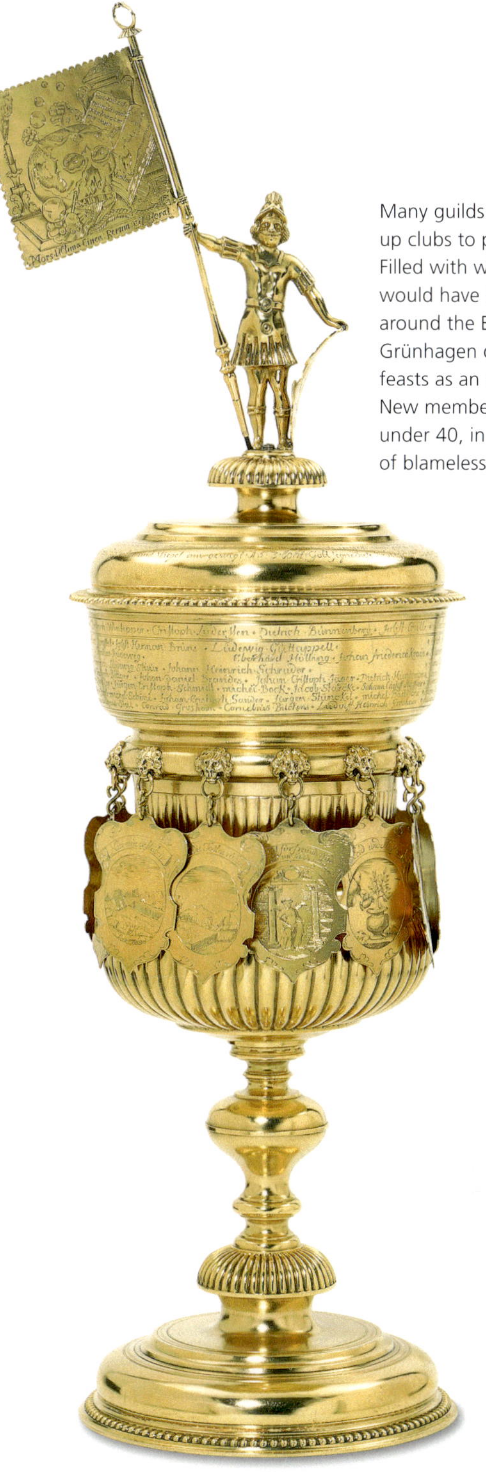

Many guilds and societies set up clubs to pay for funerals. Filled with wine, this cup would have been passed around the Brotherhood of Grünhagen during ceremonial feasts as an act of fellowship. New members had to be under 40, in good health and of blameless character.

Seasonal Celebrations

Although today we think of traditional celebrations as Victorian inventions or re-creations of a vanished rustic and pagan world, originally many were simply fundraising activities carefully controlled by the Roman Catholic Church, which even paid for morris dancers' costumes, the hiring of musicians and the erecting of the distinctly pagan May Pole. Ritualized plays performed at Christmas and Easter, with children elevated to king, queen or bishop, also presented both audience and players with an opportunity to invert and mock the social order that governed their lives – the feudal system and the church – as well as the chance to gather and drink. Not surprisingly, all such harmless fun was banned from church buildings by the Protestant reforms of Henry VIII and Edward VI, never quite dying out but palely reflected in our own times by pantomime, carol singing and the village fête.

Nonetheless the liturgical calendar, which began its New Year at Lady Day on 25 March, continued to play a significant part in punctuating the British year with opportunities for festive drinking together. For example, it was on 25 March 1724 that a huge stoneware jug (known as a *gotch*) was presented to the Hinderclay Society of Ringers in Suffolk with incised instructions 'to be fild with strong beer', while as late as the 1930s agricultural labourers were obliged to offer themselves for hire at the Lady Day fairs.

Christmas Day for Samuel Pepys in the 1660s, as throughout the eighteenth century, consisted of church followed by a feast of beef and mince pies; the giving of presents being confined to New Year when valued employees might expect some kind of bonus. For it was not until Prince Albert's marriage to Queen Victoria in 1840 that the romantic sentimental German Christmas, complete with Santa Claus, trees and stockings, made its appearance. But from at least 1600 alcohol, in the form of warm posset, served in a pot with handles and a sucking spout, was much consumed over the Christmas period by nocturnal revellers: My Lady How's recipe, for example, included sack (white wine or sherry), whisked eggs, sugar and cream, heated to make it 'thin(e) on the bottome, curd in ye middle and snow on ye top'. Just such a restorative sack-posset was enjoyed by Pepys and his friends at midnight on 16 January 1667 after an evening of dancing and singing. Related to posset, caudle also contained spices and might have a crust of toast on top. By the nineteenth century, and particularly in the north, rum or sherry, crumbled biscuits and Christmas cake were being added to these nutritious concoctions.

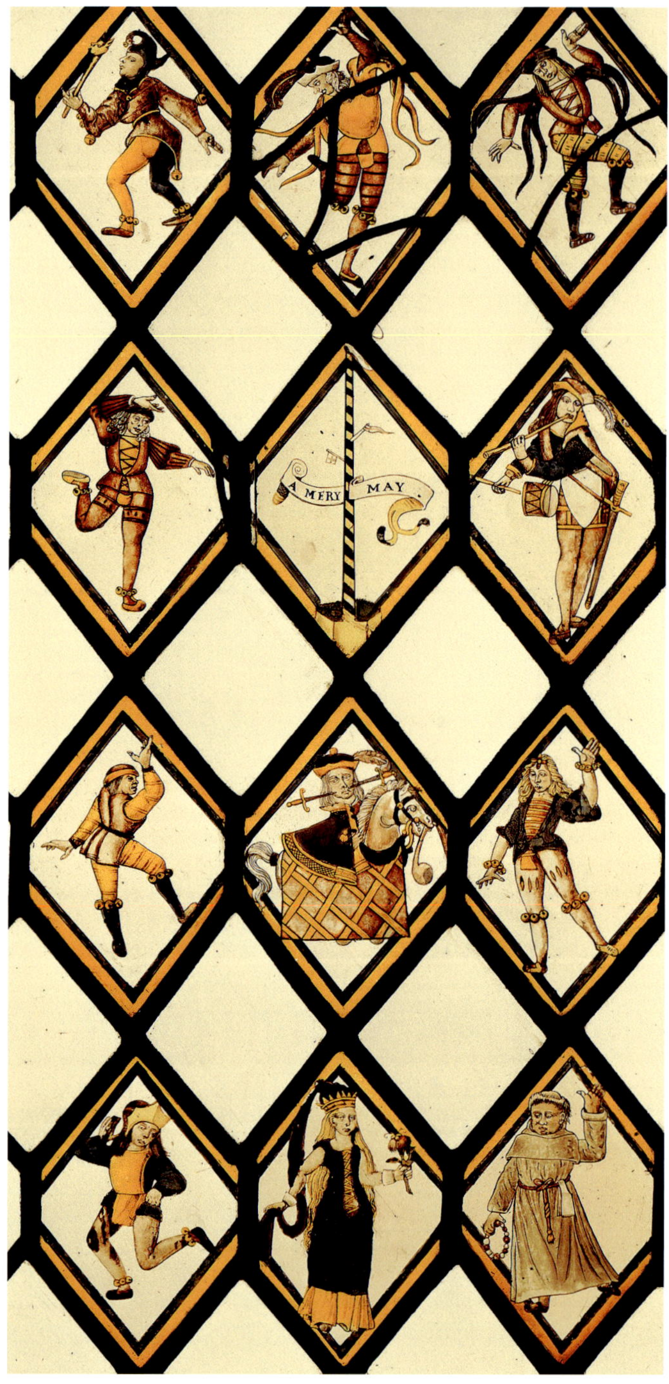

The Betley Window, 'A Mery May', showing the May-pole, Maid of the May, hobby-horse and morris dancers, traditional country pleasures fuelled by home brew. Criticised by Puritans from the 1580s, such pursuits survived until the late 18th century.

The continuity of such winter traditions can be traced better, not by the written word, but by the survival of the posset pots and loving cups themselves, mainly of robust lead-glazed earthenware or salt-glazed stoneware, but occasionally of more genteel glass. A large Staffordshire pottery example of about 1710 is inscribed 'Robart Pool mad

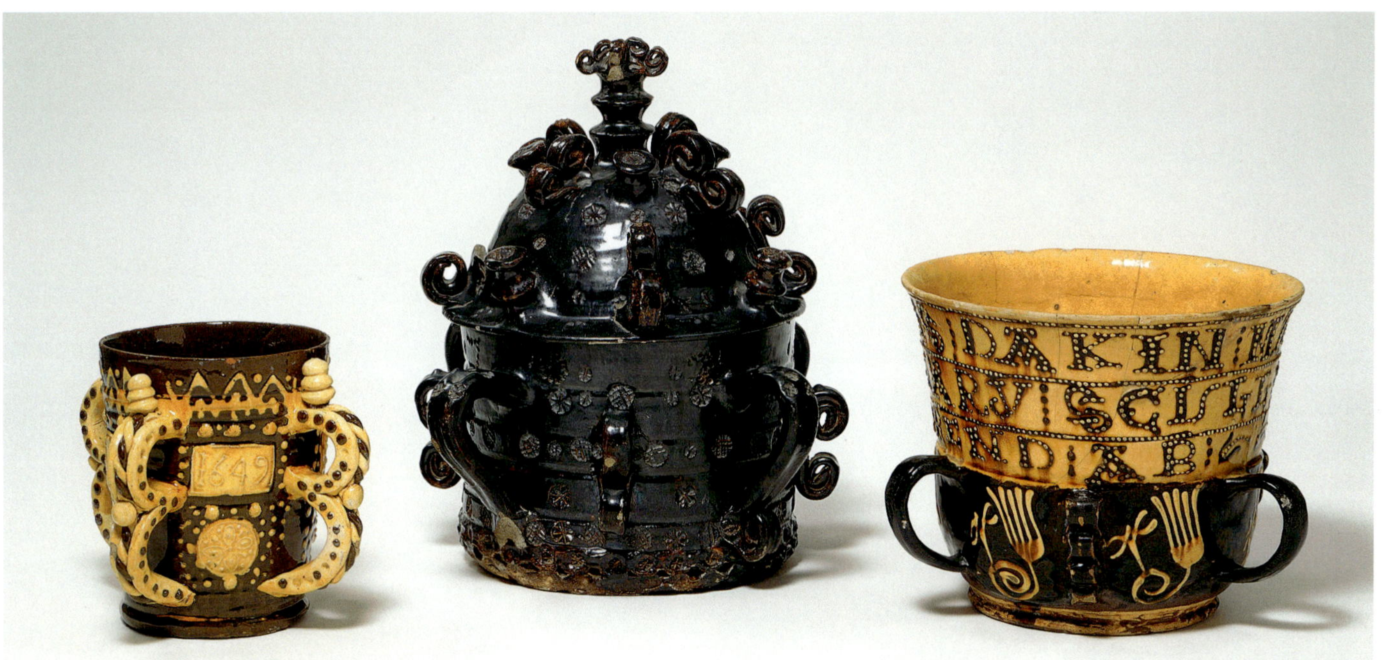

Cups for sharing: lamb's-wool, or ale mulled with sugar, spices and apple pulp, was drunk on Twelfth Night and taken out to bless the apple trees. Bragot, a spiced winter ale, was home-brewed from malt, sweetened with honey, liquorice root and spices, without the bitterness of hops.

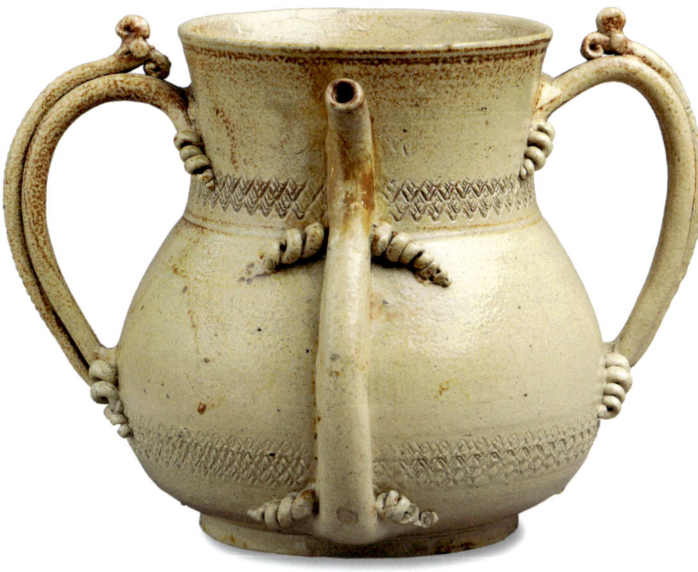

this cup and with a gud poset fil', from which we may suppose that the many Nottingham and Derbyshire stoneware 'loving cups' of similar form shared the same function: many nineteenth-century Derbyshire examples not only have dates around Christmas Day, but sometimes even carry incised carol verses. Of course such ceremonial drinking vessels may also have begun life as a christening or betrothal gift, but were always intended for communal drinking.

Wassailing, deriving from the toast '*Ves heill*' used by the Danish-speaking communities of England long before the Conquest, was later adopted in a spirit of patriotism by the English court specifically for the ritual greeting of the New Year, as the Household Ordinances of Henry VII for 31 December 1494 record. By 1548, at the court of Henry VIII, the entry of the wassail bowl had become the signal to 'brake up Christmas' on Twelfth Night – that is, 5 January or Old Christmas Eve. Pepys, on 4 January 1667, lavishly entertained his London friends, proudly displaying his new silver plate and making a triumphant ending to the evening when 'a flaggon of ale and apples, drunk out of a wood cupp as a Christmas draught, made all merry'. Though possibly of maple wood that had long been favoured for bowls and mazers, such a vessel is more likely to have been elaborately turned and burnished black lignum vitae, as made by skilled woodturners throughout the century. But whatever the material, inevitably the unhygienic custom of passing around a large bowl filled with warm spiced ale topped with the pulp of roasted crab apples (a drink known as 'lamb's-wool') began to move away from the metropolis and further down the social ladder.

It was in the cider-producing counties of Somerset, Wiltshire and Devon that unrestrained wassailing continued to flourish in the eighteenth and nineteenth centuries. There, its function also extended to ritual libation of the apple trees on Twelfth Night, perhaps using those splendidly massive vessels produced by local potteries. A final flowering of this ancient custom took place in Glamorgan towards the middle

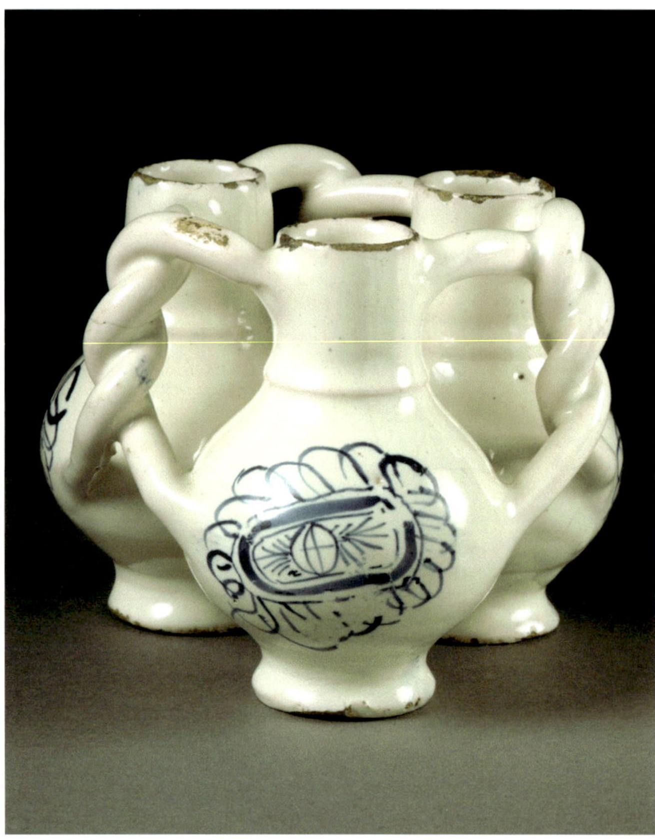

Stuart Londoners enjoyed the challenge of drinking from this 'jolly boys' or fuddling cup. Its twisted handles expressed conviviality and linked arms, but drinking without spilling required dexterity.

agricultural workers would have needed flasks or costrels with lugs for cord, either for clean water or, in winter, for something much stronger. So prized were these rough containers that fine decorated imitations were produced as love gifts. By contrast, the mass of moulded novelty gin flasks produced in Lambeth and Derbyshire from the 1830s were intended for hard use by the labouring poor, to be re-filled and carried in the pocket. For both the farmer's table and for his men in the fields, however, it was the great jug for ale, beer or cider that was an indispensable part of the harvest. Here the earthenware potteries in the great grain-producing counties of Wiltshire, Dorset, Somerset and North Devon, as well as the stoneware potteries of Bristol, provided all that was needed. Decorated slipware examples from Barnstaple or Bideford, with massive strap handles, frequently bear sgraffito variants on the rhyme 'Now I am come for to supply / Your workmen when in harvest dry / When they do labour hard and sweat / Good drink is better far than meat'. In the nineteenth century, the Fishleys at Fremington ably continued this tradition of wide-bellied harvest jugs.

Some idea of the thirst generated by harvesting in midsummer may be gleaned from a contemporary landowner's brewing account: in July and August 1843 Col. Robert Hildyard of Stokesley Manor in North Yorkshire supplied 61 gallons of beer 'to the Haymakers', and the following year another 52 gallons to the haymakers and stackers. In the hotter south-west of England, the consumption must surely have been greater still.

If the harvesters of the West Country worked long and hard, never wasting a minute of fine weather and indulging in mild competition with those in neighbouring fields as to who could reap and hold up the last sheaf of wheat with the triumphant cry 'A neck, a neck!', they were also encouraged to play hard afterwards. The great annual task of completing the Barley Mow united all in one great effort, after which farmers were

of the nineteenth century, with the difference that instead of offering a sip of their lukewarm soggy brew in exchange for a monetary contribution, the wassailers carried an empty wooden bowl and sang their verses in expectation of strong liquor. Specifically for domestic use, elaborate 18-handled covered slipware bowls of archaic design were made by the Ewenny potteries from about 1825 until the end of the century. In the West Country, wassailing never actually became extinct so that, after a little 're-booting' in the twentieth century, the tradition is still very much alive.

In London, the drabness of the Lent period in February and March might be relieved by a Frost Fair when the Thames froze thickly enough for booths to be set up for refreshment and for souvenirs. Then came the celebrations of Easter and Whitsun, and finally May Day, which heralded the beginning of summer. Though firmly rooted in the country village with its maypole and dancing, in the London of Pepys's time it encompassed all social classes who put on their finery, decked their hair, their horses and even their carriages with ribbons and paraded in Hyde Park. Then for Londoners the next great event, at the end of August, was St Bartholomew's Fair, which for several days sprawled over Smithfield.

But if Londoners remained oblivious to the harvesting, in the country the months of July and August were the busiest and thirstiest of the year. Of course, all year round,

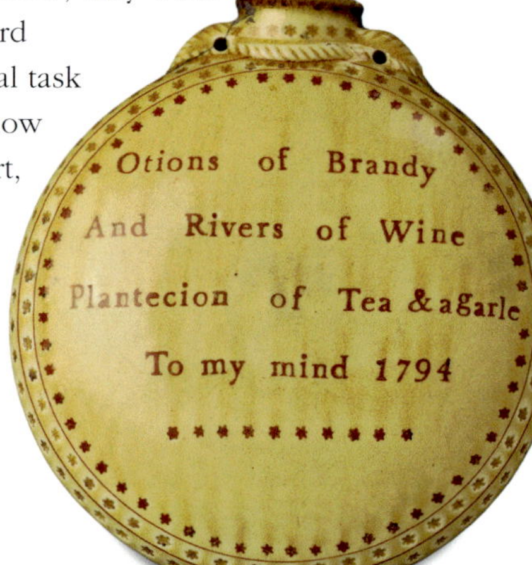

'Otions of Brandy And Rivers of Wine…'. The promise of good things is inscribed on this handy pocket flask, a present to be filled, plugged with a cork and strung with decorative cord as a modest gift.

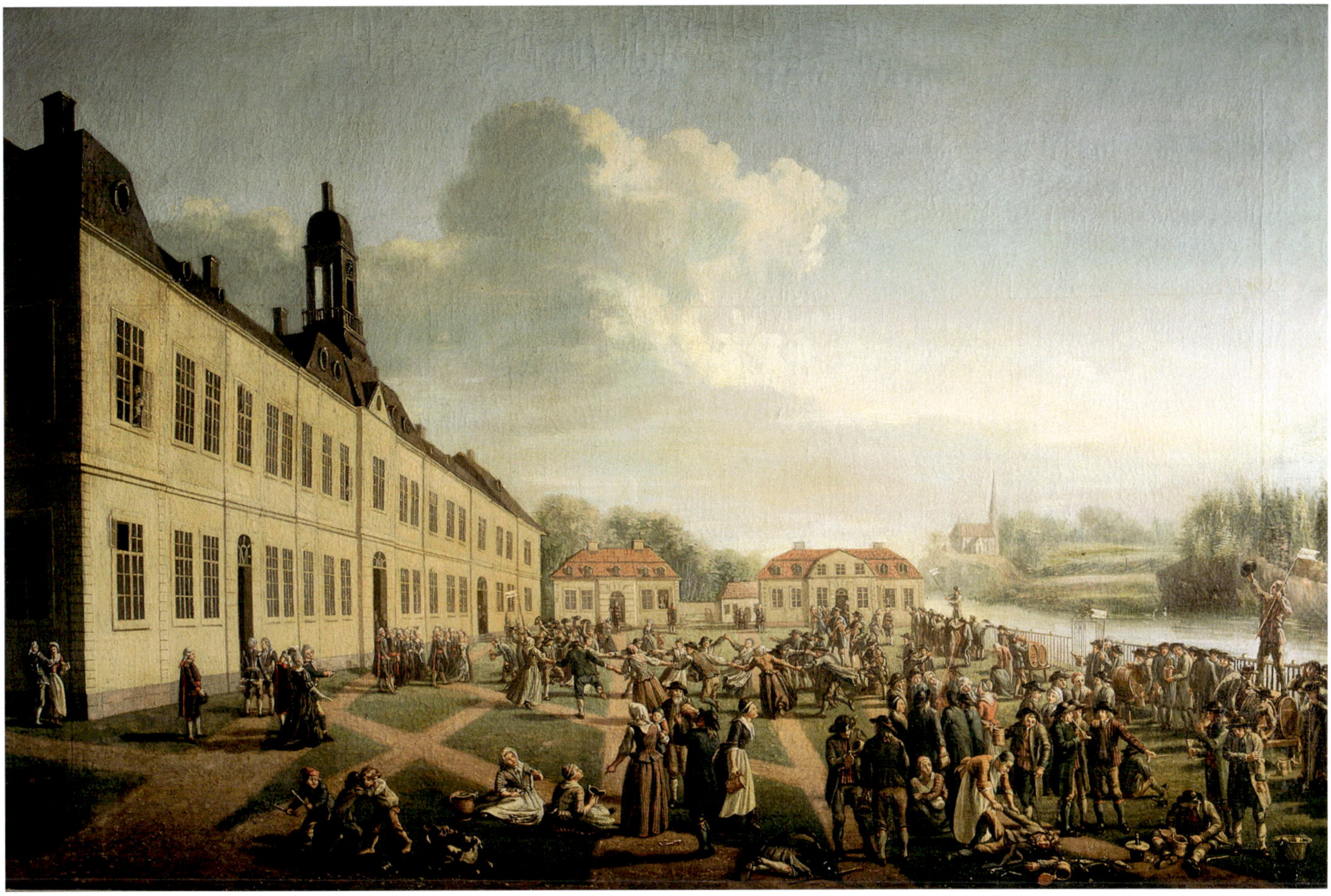

more than happy to entertain their haymakers with a great spread of food, ale or cider and perhaps a fiddler. Such communal activities barely survived the upheavals of the First World War and, in the 1920s, the introduction of the American tractor with mechanized reaper and binder, which spelled the end for the scythe, hay rake and pitchfork.

As autumn turned to winter, while the middle classes in the big cities returned to their theatres and clubs, and while the manufacturing towns returned to their smoky existence after the obligatory holiday known as Wakes Week, much of the country turned its attention to the hunting season. Here the cycle of furious and dangerous activity rewarded by alcoholic orgy was regularly enacted; for when the hip-flask and the stirrup cup were empty, there were abundant hunt suppers where the restraint of etiquette and moderation was unknown. Here the punch bowl remained traditional, along with a copious supply of beer for which large stoneware hunting jugs, such as those made around 1800 by the Kisheres at Mortlake, were certainly used: indeed, anything less robust would not have survived.

Above: Dancing in a ring: jollity and celebratory beer-drinking on the terrace are traditional rewards for the shared effort of bringing in the harvest on a Swedish estate.

Below: On this late 19th-century Christmas card, the sentimental message evokes a lost vaguely medieval world, as chilled musicians are given drink by a maid servant: 'May Christmas bring thee peace & plenty, barns and cellars never empty. A horn of honest wholesome beer will warm the heart – the spirits cheer.'

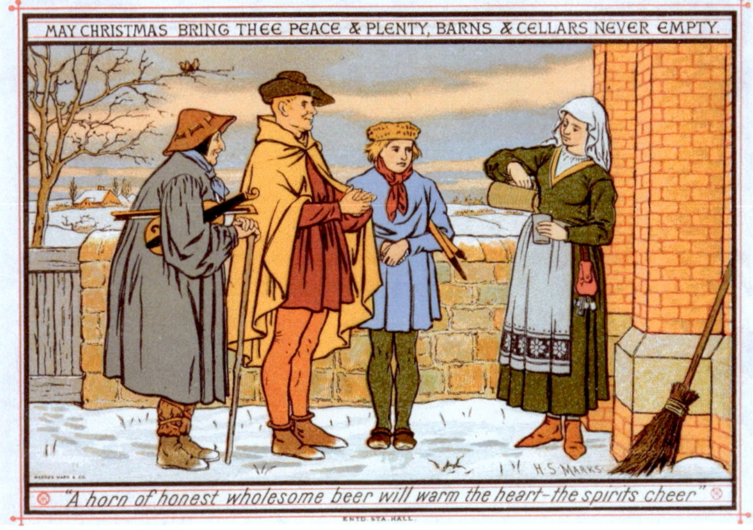

Assays and Liveries: drink at court

Four marvellous great pots or crudences as they call them of gold and silver, I think they are a good yard and a half high. By the cupboordes stood two gentlemen with napkins on their shoulders and in their hands each of them had a cup of gold set with pearls and precious stones, which were the Duke's own drinking cups....

RICHARD CHANCELLOR ON A SPECTACULAR KREMLIN PLATE DISPLAY, 1553

*A*t all courts, from Moscow to Windsor, ceremonious drinking was governed by strict protocols and toasts were heralded by trumpets and drum rolls, as at Anne of Brittany's coronation feast. High status visitors, as when Christian IV visited James I, exchanged precious drinking vessels in public after a mutual toast. Coronations and public receptions for ambassadors are well recorded, although it is harder to glimpse the private drinking habits of royalty.

A tiered buffet displaying plate, dominated by treasured drinking vessels, was central to all court occasions. Tiers were dictated by rank; within the Habsburg domains, five steps for the emperor, four for a duke, three for a count and two for a baron. At the electoral dinners in Frankfurt, each elector had his personal buffet, dressed with his own treasured objects, but these were for show, not for use. Drink was served from a separate table. Although western European monarchs abandoned these displays as they retreated into private dining, buffets rising from massive wine coolers, past pilgrim bottles and jugs up to covered cups remained a feature of state rooms in Moscow, Berlin, Dresden and St Petersburg.

A ceremonious tasting, or assay, of all drink before the monarch consumed it required small assay cups, or the taster might drink from the lid of the king's cup. At the Danish court around 1520, 'The taster shall take my Lords glass or tankard, whichever is my Lords drinking vessel, and its bowl, shall bear it to my Lords side and shall remain there as long as my Lords meal lasts'. Tasting remained the custom at the Hanoverian court. Lady Bristol in January 1728/9 complained of her sufferings in bending her knees when as lady-in-waiting she played the part of Cupbearer: 'giving the Queen drink and tasting upon the knee, which at this time I am not able to bow to a superior being'.

Ale, beer and wine were supplied on contract to princely tables. Hundreds of royal servants and officials received a daily allowance, or livery, of beer or wine. Four massive silver livery pots, made in Hamburg in the 1640s and 1650s for the Duke of Brunswick, weighing 31kg, offer a rare glimpse of the scale of this customary provision, in a room of silver from the Hanover Court now on show at the Museum of Fine Arts Boston. In December 1729, when the Purveyor of Claret Mr Rawlinson delivered a hogshead of wine, the officers of the Board of Green Cloth in St James's Palace were to taste a bottle and to be informed of 'the name, growth and quality of every bottle so brought in'. Although few details of royal consumption are recorded, unlike the daily menus listed in the Lord Stewards' ledgers, special orders were noted. The Prince of Wales's Christmas shooting days in December 1729 called for claret, Champagne and Burgundy, Rhenish wine and the inevitable bottles of arrack (palm spirit) for punch. For Queen Charlotte's supper party in January 1764, Edward Marjoribanks delivered two bottles of Champagne to her kitchen, probably to poach fish, plus Rhenish and Madeira, claret, port and Burgundy for the table.

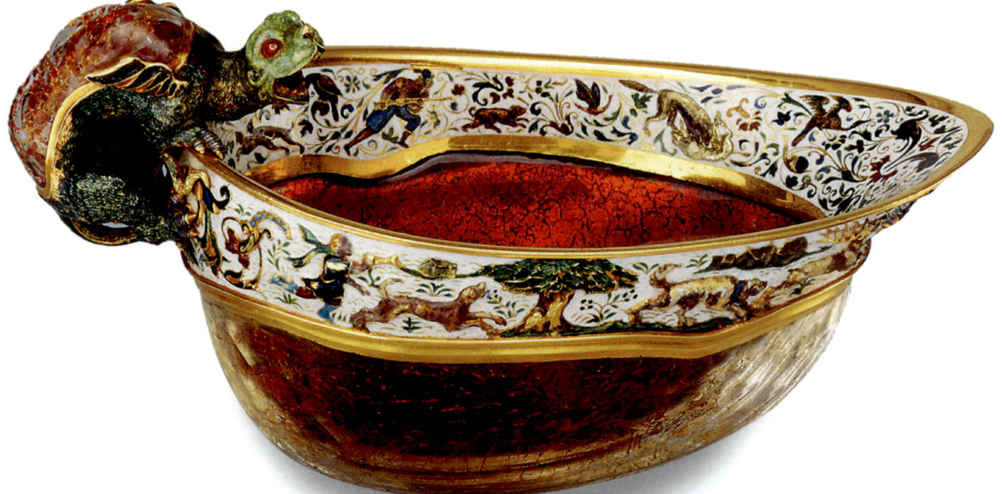

This delicately enamelled cup fits into the palm of the hand. Perhaps made for a Habsburg and Wittelsbach marriage toast, its dragon breathes fire over the cupful of brandy or fruit spirit. Hunting was a metaphor for courtship.

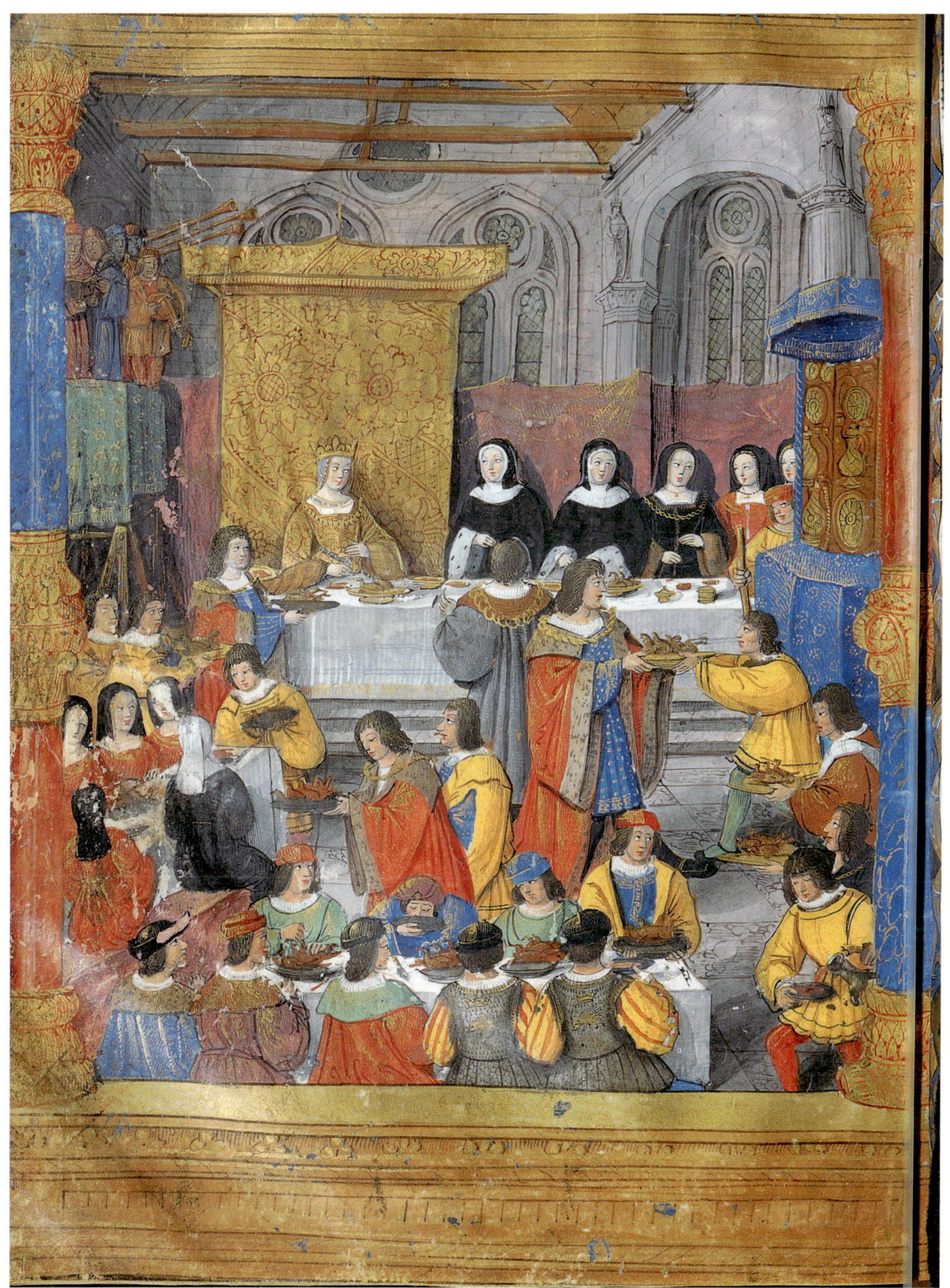

Trumpets blew when the monarch drank, a custom recalled in the thumping of glasses on the table at a toast. Anne of Brittany's coronation feast shows toasting, and the ranking of vessels, from tall cups to shallow bowls.

Drink has always been driven as much by personal taste as custom. In the 1590s James VI of Scotland asked his London agent, James Hudson, to ship 12 tuns of London beer up to Edinburgh for his queen, Anne of Denmark, 'because it is nocht unknown to you that our dairest spouse is daylie accustomed to drinke the same'. Distinctive beers depended on the brewer's skill in obtaining pure and fresh ingredients as much as the local water; Charles II enjoyed Eton College beer, which had so high a reputation that he had it sent up to Windsor Castle when in residence; his father had drunk it in captivity at Windsor before his trial. When the Dowager Queen Catherine of Braganza left England for her native Portugal in July 1687, her ship was stocked with her favourite mead, cider, rum, three kinds of beer, wine from Cahors, Bordeaux and Hermitage, as well as Canary, sherry and Rhenish. Ten dozen bottles of 'Spaw Water' cost 6s. 6d. a dozen plus hampers to carry them, almost as much as the

Assays and Liveries: drink at court

Below: Charles II being given drink on bended knee at a Garter Feast. Taking assays (samples) of the king's drink in public continued even when he was absent.

Right: This tin-glazed bottle, painted with Charles II, was a decorative commission, filled with sack, or sherry, a strong sweet wine from Jerez in Spain.

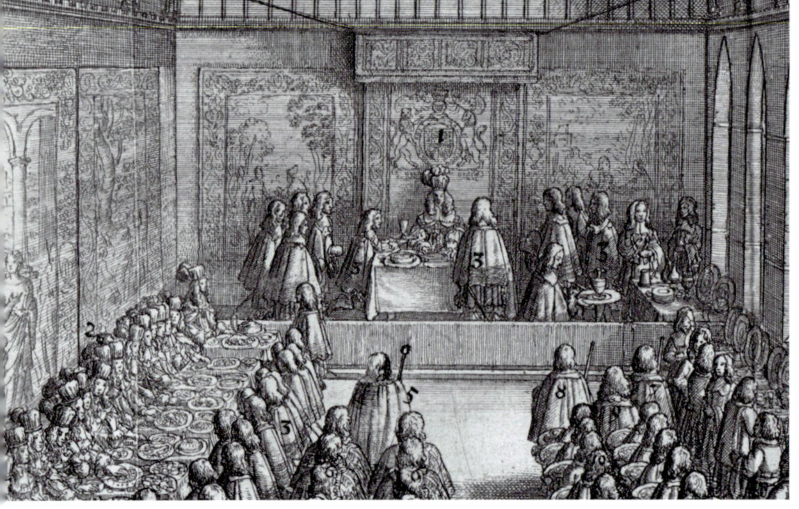

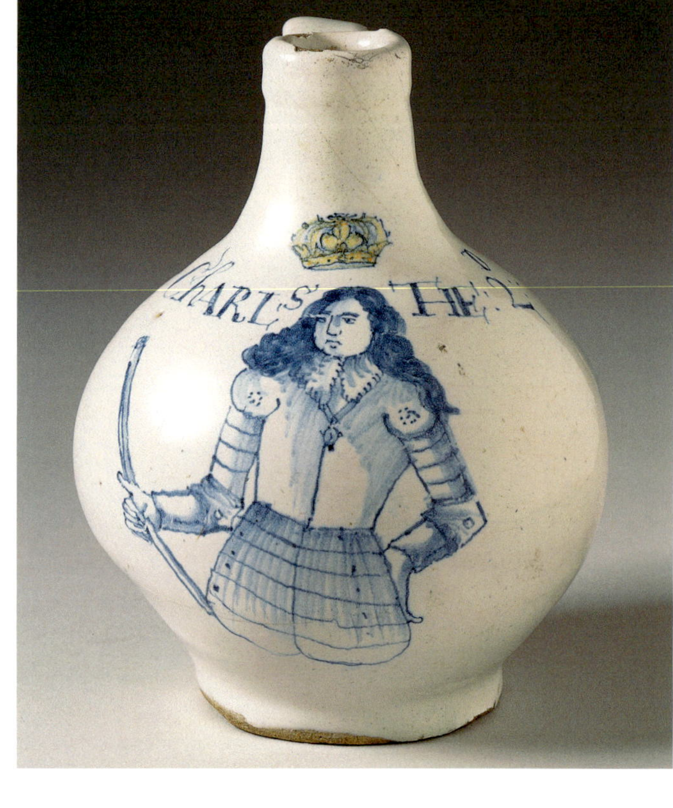

Nottingham ale. Louis XIV always drank French wine, diluted with water from the Versailles springs, which had been judged the purest by his doctors. His last drink at night was a glass of water slightly tinted with wine.

Rhenish wine remained a Hanoverian favourite; for Victoria, 'a little drop of hock keeps away the doc' and her visit with Prince Albert to the Hochheim vineyards in 1845 led to their being renamed Königen Victoria in homage. As a young woman she sampled in 1843 the grog supplied to sailors on the royal yacht, a curiosity about drinks she demonstrated again on a French visit in August 1887, trying a glass of Chartreuse.

In July 1822, when George IV sailed from Greenwich to Scotland, his yacht the *Royal George* was stocked with his preferred drinks, from cherry brandy to Bristol water, plus ice from Carlton House, bottled beer, brown stout, ale, cider, liqueurs and ginger beer. His Edinburgh visit, orchestrated by Sir Walter Scott, included his public consumption of illicit Highland whisky. This gesture triggered the following year Parliament's decision to reduce drastically the huge fee for a Scottish distilling licence (over £100). Over 12,000 distilleries opened within 12 months, a rare instance of royal approval bearing directly on the drinks trade

At coronations and royal entries, open-handed hospitality was expected. At the Field of the Cloth of Gold a wine

conduit, with liveried servants drinking from bowls, is visible outside Henry VIII's temporary palace, serving his train of almost 5,000 courtiers and servants. In 1579, Frederick II of Denmark ordered 4,000 banded glasses, 6 barrels of wooden beakers, 150 large and small stone tankards, all for use at the coronation; the glasses were then broken after drinking his health. Charles I, who revived formality at court and was concerned to reduce intemperate behaviour, ordered that guests at Whitehall be served 'not with niggardliness nor yet with prodigality and waste'. As Charles II passed through the City in 1660, London's water conduits were temporarily adapted to run with wine. In 1766 the city of Philadelphia celebrated George III's birthday with 'an Elegant Entertainment', an ox roasting, punch and 40 rounds of toasts. When Napoleon 'liberated' Utrecht in 1812, a dozen tuns of beer were ordered by the city for a citizens' celebratory 'beer run'.

This tradition of free-flowing drink, and its risks, is vividly described by Lady Anne Conolly, newly married daughter of Thomas Wentworth, earl of Strafford. The duke of Dorset's extravagant entertainment at Dublin Castle in October 1733, when appointed Governor General in Ireland

was finer than anything I ever saw in my life. I don't well know how to discribe it but the super was mighty well disposed and lighted and in the middle there was a large basan with three sea horses and three sea serpents that was twisted around a pilar and at the top out of their mouths play'd a fountain of Burgandy which lasted three hours.… I was glad to come home quietly for the Duke gave 16 hogsheads of wine to the guards and servants, by which as they grow drunk there was mischief done. My servants were very sober… one of the soldiers run my Lord Mountcashels footman thro' the body with his Bayonet just by my chair and he was taken up for dead.… The wound is mortal and the soldier will be hanged.…

An unusually rigid etiquette at the Portuguese and Spanish courts demanded that male attendants, other than the cupbearer, could not serve the queen. In the 1660s, Anne Fanshawe, wife of a diplomat, was entertained solely by the queen and her female attendants. At the Sublime Porte, this custom of high status women, such as the Valide Sultan (the Sultan's mother) or his senior wife, entertaining wives of envoys was well established, as Lady Mary Wortley Montagu recorded. The London Levant Company, concerned to maintain the standing of its representatives in the late

A porcelain flask of saffron spirit, presented by the Shogun (effective ruler) of Japan to Queen Victoria. The spirit was considered a great delicacy.

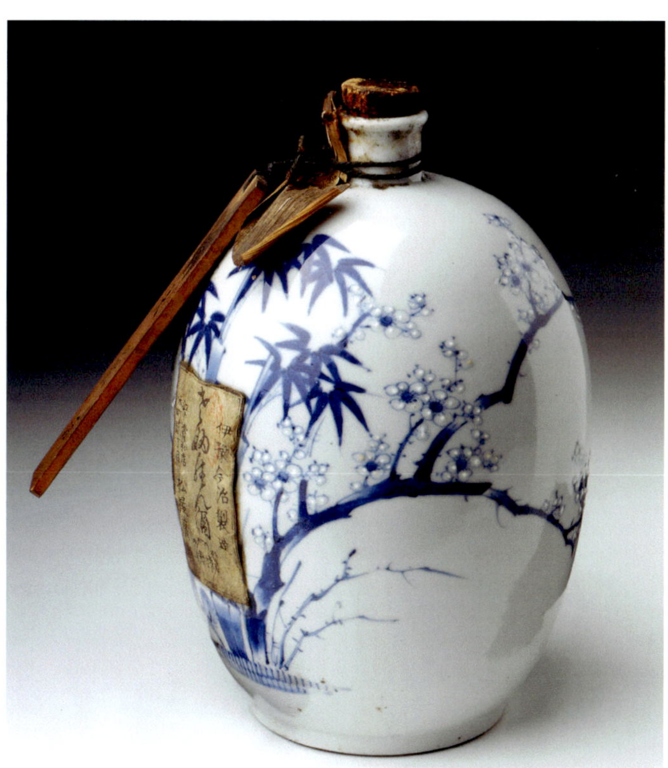

The ancient concept of a cup as a symbol of royal favour is expressed in Thomas Coventry's Cup, made from the Great Seal of James I and still set out on the table at festive occasions 350 years later.

seventeenth century, presented the wives of envoys to Constantinople with lidded gold beakers, to ensure that they could compete in splendour and honour.

Novel or distinctive local drinks could serve as diplomatic gifts; in February 1614 the cargo of the East India ship *New Year's Gift,* taking goods to Emperor Jahangir, was loaded with velvet-covered bottles 'filled with strong waters'. These London-distilled spirits cost almost as much as 30 paintings, also a gift for the Indian emperor, bought in Rouen.

Dealings with non-European monarchs were tricky, since some did not appreciate alcohol. Elizabeth I sent tall standing cups to the Sultan, a precious and impressive gift that could be melted for other vessels, but Louis XV chose a more appropriate sherbet set of gold and crystal in the 1750s. When the Shogun decided to send a symbolic gift to Queen Victoria in 1865, the porcelain flask contained saffron spirit. For the Habsburgs as kings of Hungary, with a right to the precious Tokay essence and dessert wines, these made an appropriate gift for the king of France in 1744 (who reciprocated with Champagne). Champagne's reputation made it a suitably princely gift from Louis XV to Frederick, king of Denmark too. Rulers of Denmark and the Netherlands benefited in the mid-nineteenth century, reciprocating with Cape wines and Madeira.

Prizes and Presentations

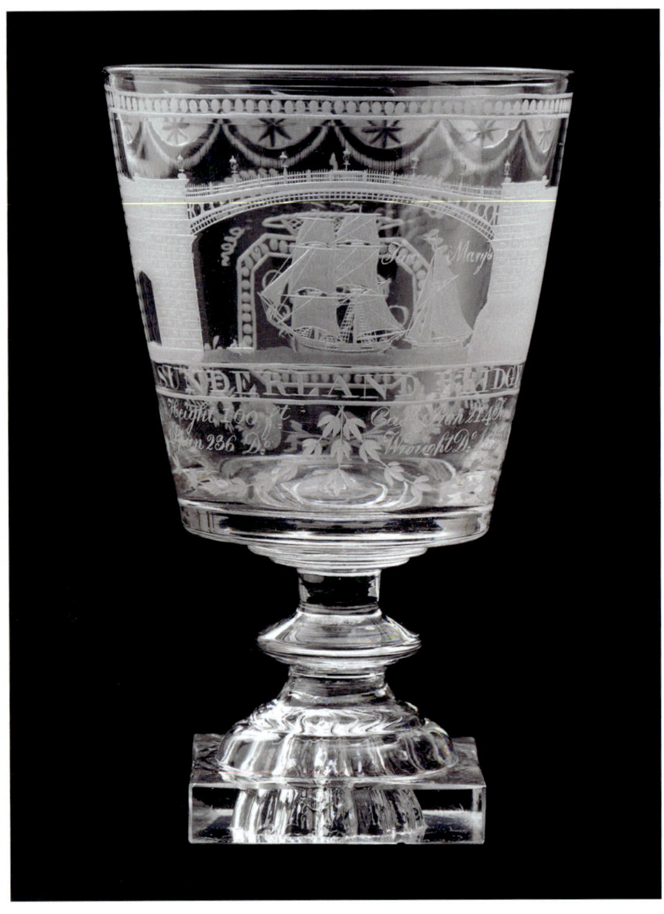

Horse Plates For a silver dish … for a silver tankard…. Paid for drinking in the plate … for 2 dozen of French wine and 6 bottles of sack to entertain the gentlemen at Stamford…. Paid for cider to fill the plate at Preston … for punch to fill the plate 15s.… For 4 bottles of sack in the painthouse [penthouse] where they deliver the plate … paid at the delivery of the plate…. Paid for several times filling the plate in the field to treat the gentlemen … paid at Barlow Moore in wine to fill the plate….

Payments for race prizes and drinks, accounts of the Earl of Rutland, 1668–95

Striking objects in silver, glass and porcelain offer mute testimony to past ceremonies, centred on drink from a costly and eye-catching vessel. Curators treasure such objects, notably British race prizes, tall engraved Saxon glasses and Germanic welcome cups, without exploring their true purpose or long-evaporated contents. Once archives and objects are brought together again, these showy objects reclaim their histories.

Left: This large glass rummer or goblet, engraved with the new Sunderland Bridge, combines pride in industrial achievement and glass cutting. Statistics of the weight of iron and the height of the bridge commemorate its 1796 opening.

Below: Beaker made from the horn of a prize ox. The astonishingly large ox, which Sir Richard Osbaldeston killed on 28 January 1707, weighed nearly a tonne. Its tallow alone, valuable for candles, weighed 21 stone.

Drinking before witnesses could bring specific benefits. At York Minster the drinker draining four quarts from the Cordwainers' Bowl, a 96-cm (38-in) wide wooden vessel, given to the Corpus Christi Guild around 1400, was promised a pardon for 40 days of sins. In 1525 the Howard Grace Cup, an ancient ivory cup associated with England's favourite saint, Thomas à Becket, a famous drunkard before he reformed, was given new silver-gilt mounts. Mottoes invoking his blessing, *Estote Sobrii* and *Vinum Tuum Bibe Cum Gaudio* were prominently engraved. A similar purpose lies behind the donors of silver wine flagons to celebrate communion in the Protestant liturgy. Their names, engraved large enough to be visible when the plate was set out on the altar, linked them to the act of worship in perpetuity.

Race prizes offer the most familiar English secular link between activity and reward. Silver prizes go back to the mid-sixteenth century, if not before. In 1540 at Chester Town Races, silver horse-bells were awarded, for stamina rather than speed, since the race on the Roodee took all day. But from 1609, the race was moved to St George's Day in April and the prizes became silver cups. Almost every English town set up its race meeting; 308 meetings are recorded by 1771.

Horse-breeding and training were encouraged by government policy. Competition among the élite inevitably attracted the silver vessels and fashionable drinks that lubricated their lives. Race meetings were social events for merchants, gentry and aristocracy, with plates or races graded by the value of the prize cups of silver, silver-gilt or even gold, from £5 to a 100 guineas or more. 'Horse Plates' (a silver dish and tankard) for prizes cost

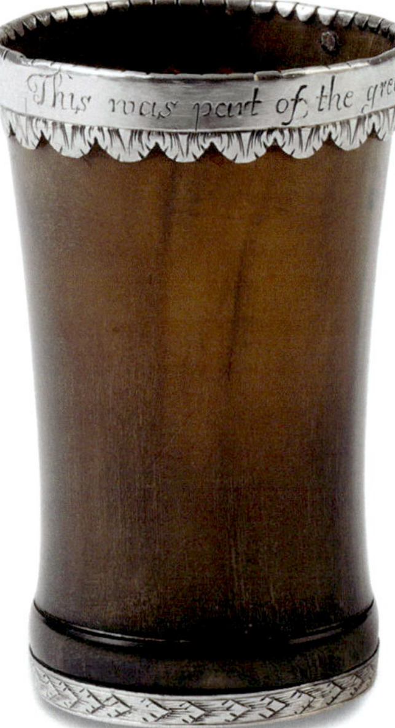

the Duke of Rutland £22 in 1667, but his share from a large
bet on Bo-Peep, a winner at Newmarket, was 200 guineas.
At York in 1720 prizes ran from the King's Golden Cup of
100 guineas, a silver soup dish of £30 value (the Ladies' Gold
Cup) to a £20 cup. Paid for by the corporation, by tradesmen
and by landowners who were subscribers, these civic
occasions were marked by music, bell ringing and
processions. Cock fights, a fair, assemblies and balls added
to the jollity.

For this competitive masculine world, drink at the winning
post was a relatively small but essential expense. Displayed
on the winning post, silver monteiths and two-handled cups
would reflect the sunlight and dazzle onlookers, until the
winning owner and friends drained their contents. Described
in December 1683 as a 'bason notched at the brims to let
drinking glasses hang there by the foot' so that they could
chill in iced water, the monteith had been invented only the
previous summer. Punch, drunk in England from the 1650s,

was the perfect high-status, high-impact and convivial drink
for celebrating a victory, and a monteith's detachable rim
meant it could double as a punchbowl. So the Basingstoke
Monteith, won by Edward Chute at the Basingstoke Races on
2 October 1688, weighing a hefty 86 troy ounces (almost
3kg) and chased with fashionable *Japon* motifs, had a triple
purpose. The expense of racing contributed to the ruin of
William Hogarth's Rake; his jockey, a winner on Silly Tom at
Newmarket, clutches an extravagantly large prize monteith
while the trainer pesters the Rake with a long bill of costs.

In Chester the Grosvenor family, major landowners from
the late seventeenth century, gave gold tumbler cups as
prizes between 1741 and 1800, although only three survive
from this period. What was drunk from these is not recorded

This early race prize, a massive silver monteith, combines fashionable
'Chinese' or *Japon* motifs with the wining jockey. It was won by the
Hampshire owner Edward Chute of the Vyne in October 1688.

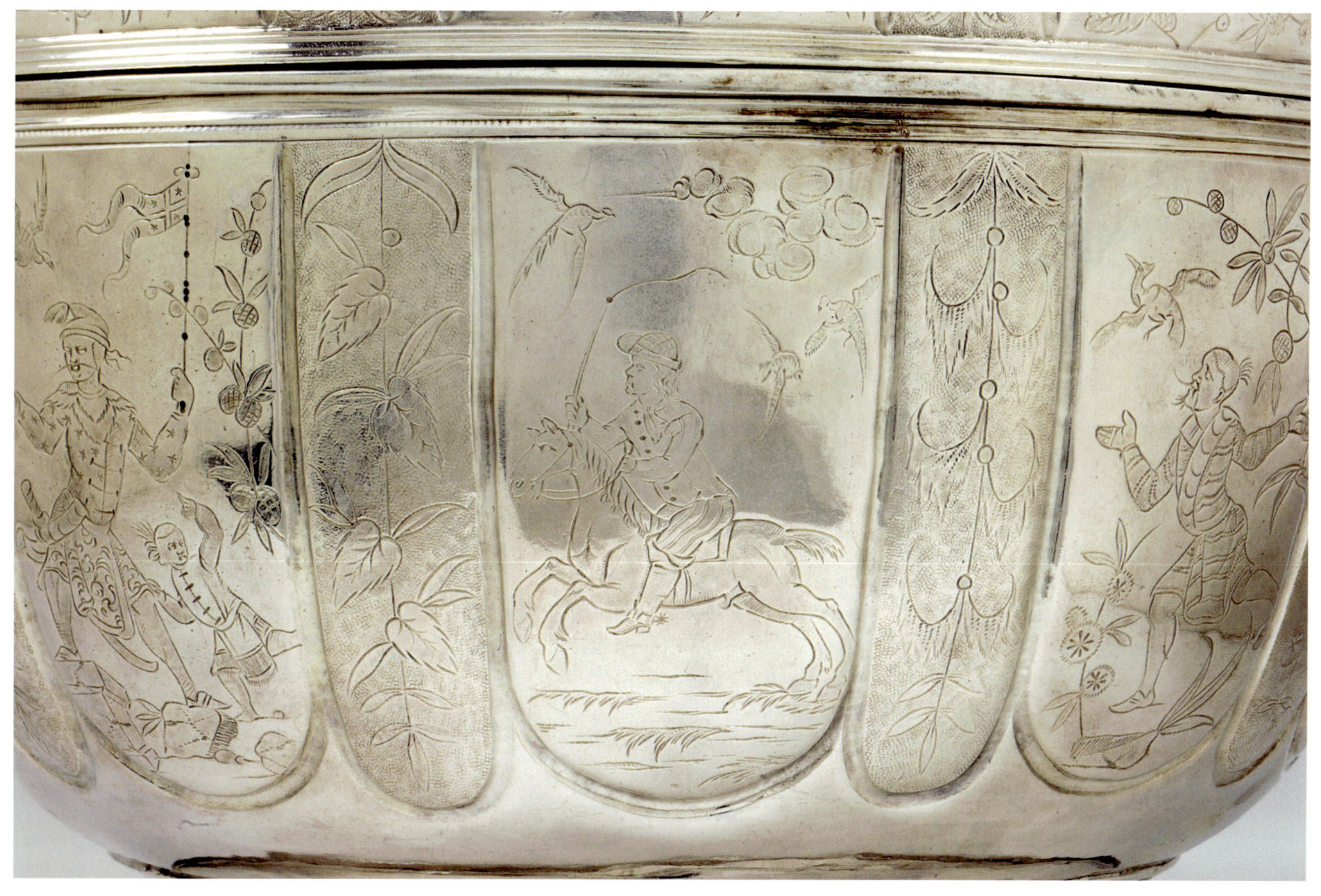

– presumably punch. A two-handled cup supplied by the London manufacturers Emes & Barnard for the Doncaster Gold Cup in 1828 contained 6 quarts, plenty for a shared toast; this firm supplied all the Doncaster cups during the 1820s, several inspired by the Buckingham Vase, an antique marble wreathed with vines at Stowe House, Buckinghamshire. But the Municipal Corporations Act 1834 made it illegal for public funds to be spent on racing, so the last race cup was given by Chester in 1835. Today, race meetings and sporting competitions continue the celebratory tradition in scenes of cricketers or footballers spraying themselves with Champagne, while flourishing a large silver cup before a crowd of spectators.

Although the custom of welcome cups is rarely recorded in England, some towns observed this gracious greeting. In Sheffield for instance, the Duke of Norfolk, the local lord, was greeted by the Town Trustees with the customary six

dozen bottles of wine, which in 1693 cost £6.10*s*. Some towns substituted a sugar loaf, as at Gloucester, for a welcome, but the principle was the same, the ancient wish to honour a good lord (it was hoped) with a symbol of homage.

Corporations that presented drink to a visiting dignitary gradually chose the politer and more commercially appropriate form of local manufactures associated with drinking. A splendid suite of locally made glasses and decanters was presented to George, Prince of Wales, when he visited Liverpool in 1806, speaking in support of the slave owners. Liverpool's merchants in gratitude ordered this huge suite from Perrin & Geddes.

North of the Alps certain free cities observed the welcome cup tradition, greeting an arrival with drink, in a special vessel and with ceremony, at a distance outside the town gates determined by the visitor's rank. It was also an opportunity to check his credentials and intentions. German

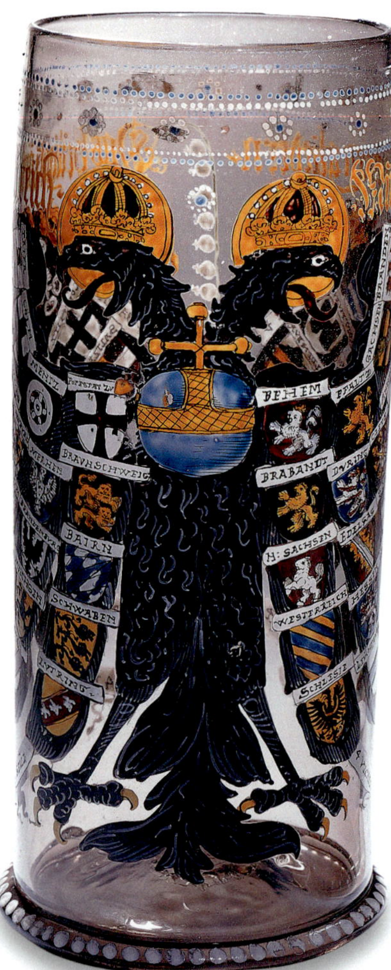
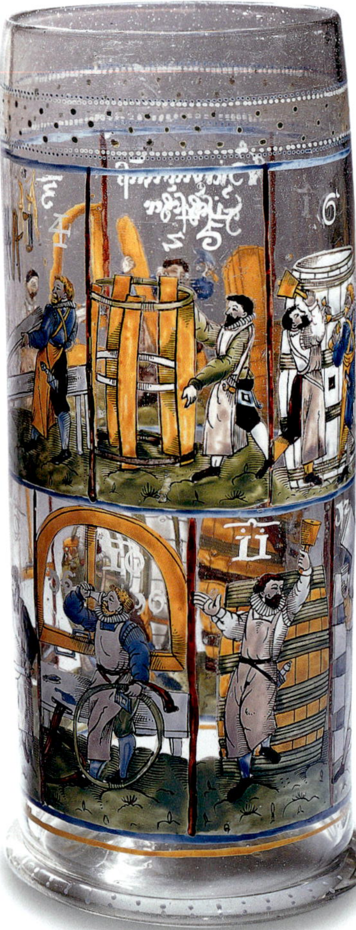

Left: These tall *Humpen*, one for a Coopers gild, enamelled with scenes of barrel-making, were for shared drinking. Tall 4-litre glasses for beer were admired around the Baltic and displayed on buffets at the Brandenburg and Danish courts.

Opposite: Racing drivers drinking champagne in 1906. Media images have established a popular convention of sporting champions treating themselves at the moment of victory, a custom with deep historical roots.

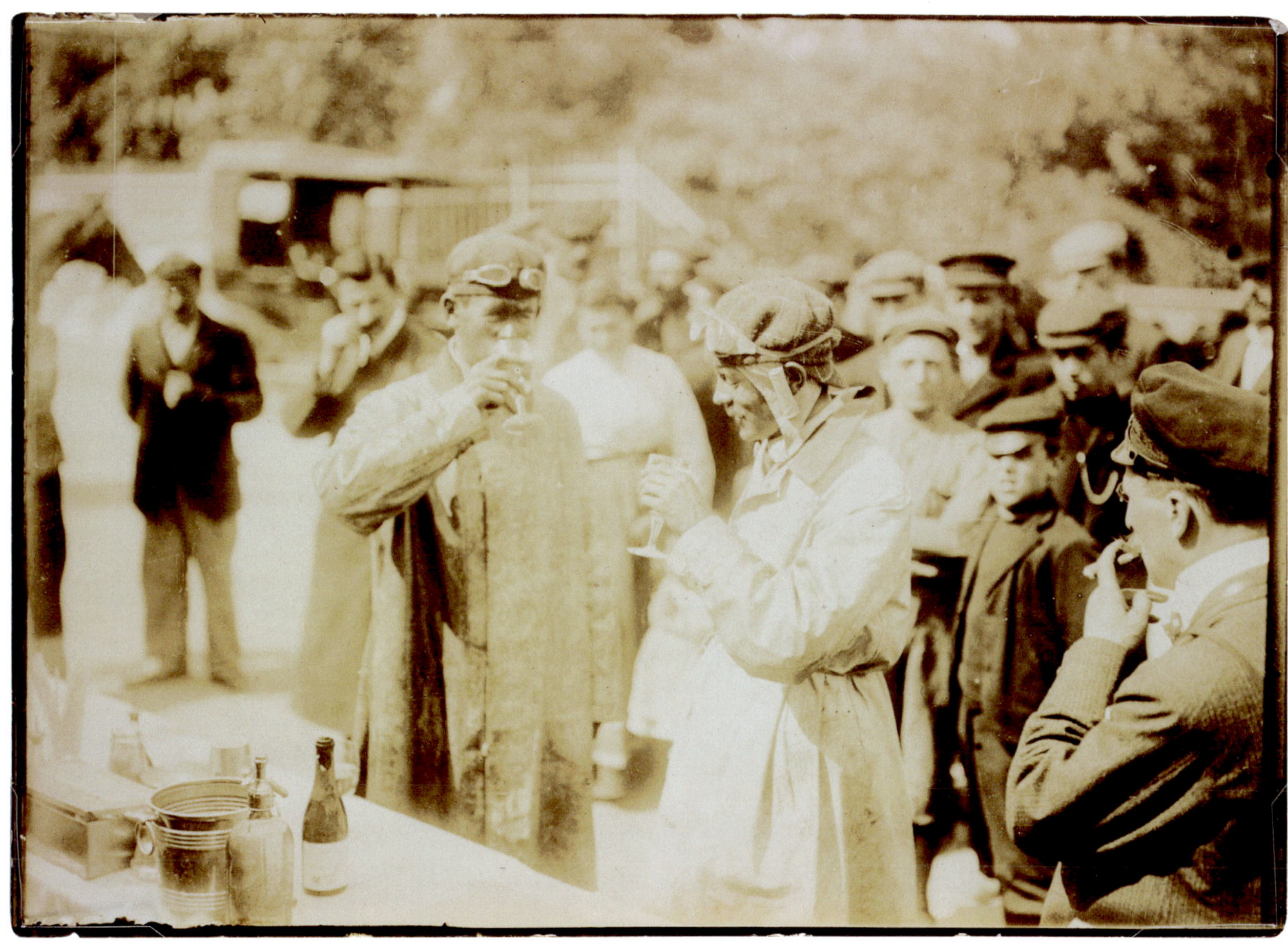

princes embraced the welcome cup tradition, devised to demonstrate mutual bonds of obligation, hospitality from the prince and loyalty from the distinguished visitor. An envoy had to be briefed about the niceties of drinking rituals and show grace in draining a full cup while representing his princely master. In Dresden, the Saxon electors had a highly developed sense of theatre, creating two extraordinary buffets of silver ore, mountains crowned with 20 silver-gilt drinking vessels. An automaton, a rider in Roman costume, rode on a track through the mountain to bring a *Willkomm* goblet to the stupefied visitor. Sadly this astonishing expression of princely magnificence was destroyed in the Second World War, but Dresden's newly restored Green Vault retains several splendid welcome cups, one a massive coconut cup for the Stable Court, decorated with three horse's heads in reference to the tournaments held there, another chased with scenes of silver mining, and made for the elector when visiting his foundry in Saxony.

Strikingly tall glass goblets, perhaps used for a single toast, were treasured and costly items; they became part of the ceremonial buffet display at the Prussian court around 1700. Single examples have survived, treasured by their recipients as a souvenir of royal office and the honour of a mutual toast with the ruler. The Danish royal collection is rich in such documented glass, including anointing or installation glasses up to 50 cm (20 in) tall. Some were engraved for military victories and marriages; at the coming of age at 14 in 1713 of Crown Prince Christian, a suite of specially engraved decanters and glasses was ordered by the Royal Wine Cellar from Saxony.

Glass has a long history at the Danish court. Some 20,000 table glasses were ordered for Christian IV's coronation in 1596, all broken in the toasting finale of throwing them out of the window. A single 3-pot glass (almost 3 litres) is preserved from his father, Frederik II of Denmark. Made at Hall in the Tyrol in 1558, it was passed around for a toast, but preserved because of the names engraved on it. These included William the Silent, one of Frederik's allies, who attended a key secret meeting with the king at Kronborg Castle, planning the Nordic Seven Years War in 1568.

Drinking in Fellowship: colleges, guilds and societies

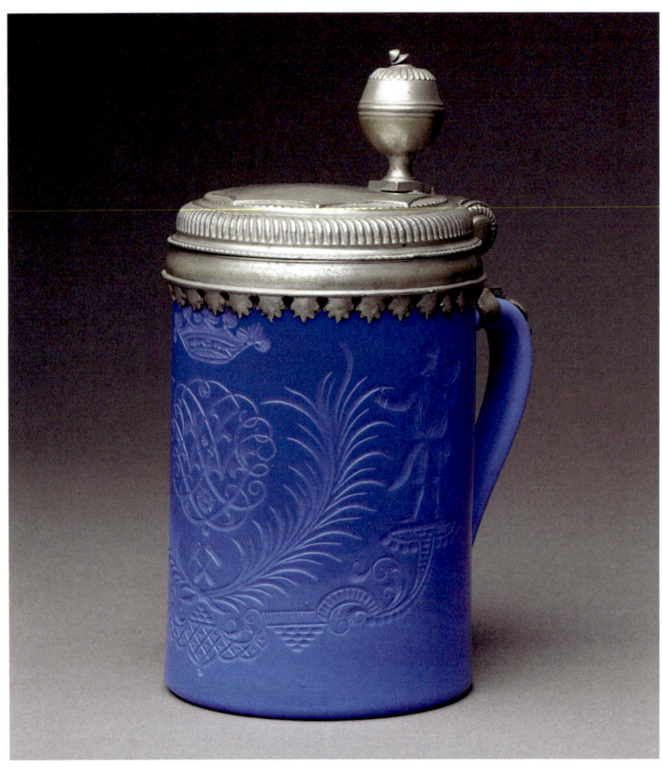

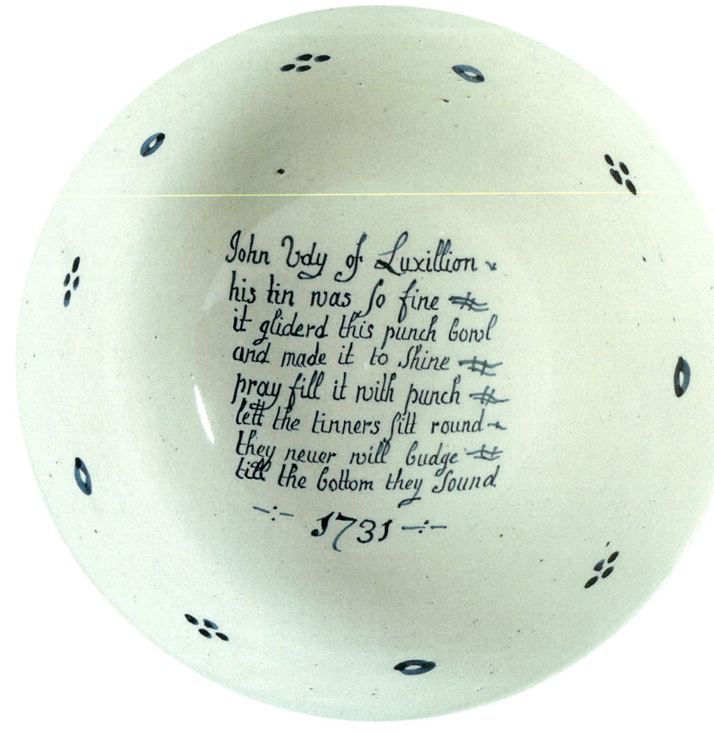

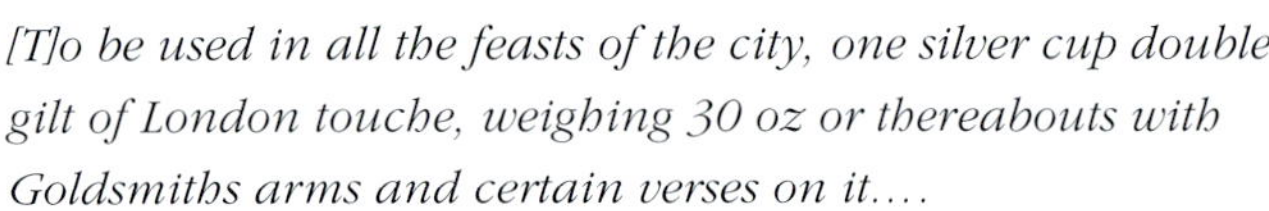

[T]o be used in all the feasts of the city, one silver cup double gilt of London touche, weighing 30 oz or thereabouts with Goldsmiths arms and certain verses on it....

NORWICH GOLDSMITH PETER PETERSON, BEQUEST, 1603

For groups that share fellowship and common purposes, drinking together has always been a central act of affirmation, with wider purposes and effects than mere conviviality. Peterson was in a long and still flourishing tradition. In 1768 Boston patriot Paul Revere (1735–1818) made a silver punch bowl to commemorate the Massachusetts Representatives' campaign against British tax policies. Revere's bowl, now in the Museum of Fine Arts, Boston featured at a dinner; 45 toasts were drunk and a new song was heard 'In Freedom we're born, and in Freedom we'll live.' This typifies alcohol's enduring role in Western society, combining a cause, a ritual celebration and a commemorative object.

Rites of passage such as moving up in the craft, appointment to a post, completing a term in office, were all occasions for drinking, as temperance enthusiast John Dunlop pointed out in *The Philosophy of Artificial and Compulsory Drinking Usage in GB and Ireland* (1839). His infinitely detailed list of rites of passage in the trades, when money was required for 'a drinking', criticized the tyranny of custom, the expense and the time lost.

For all trade and craft guilds in northern Europe, at the annual election of officers and to commemorate benefactors, ceremonious drinking followed a prescribed ritual, all present wearing their livery gowns and hoods, after a dinner or a light repast of fruit and sweetmeats.

Hypocras, an ancient spiced wine mixture, was the traditional city drink. Samuel Pepys decided at the Guildhall Lord Mayor's dinner in 1663 that sharing this compound mixture did not break his vow to avoid strong drink, as it was 'only a mixed compound drink and not any wine – if I

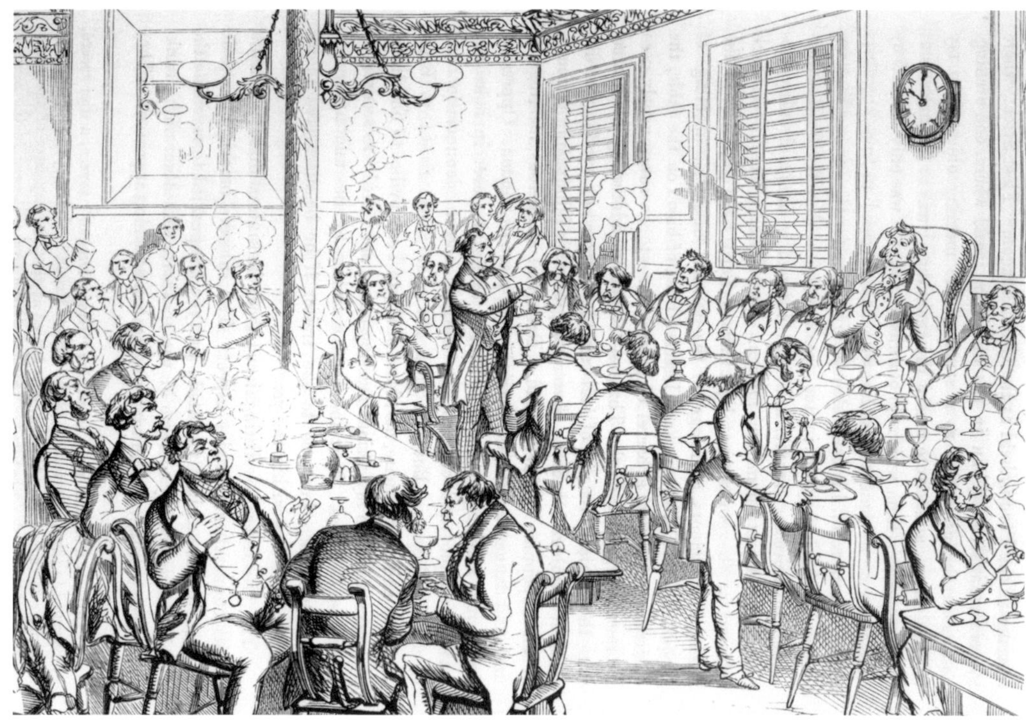

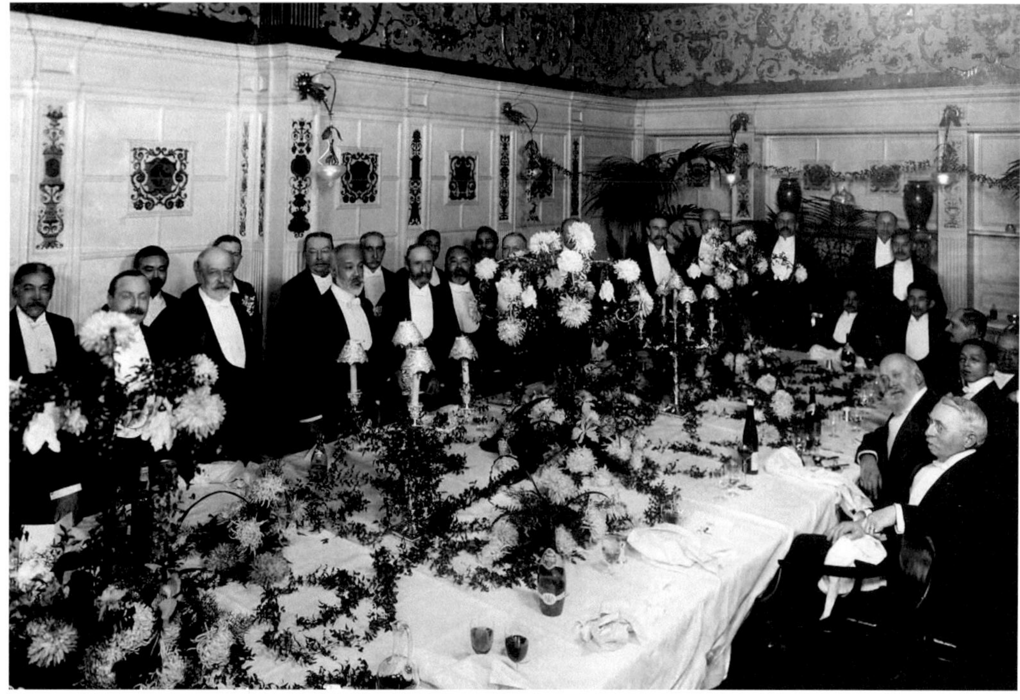

am mistaken God forgive me'. At the ancient universities,
lamb's-wool, a mixture of warm ale, apples and spices,
marked winter festivals. Specific cups were dedicated for the
final collective toast or grace, a tradition continued from
before the Reformation. Several Cambridge colleges retain
pre-Reformation drinking horns, although only for very
occasional ritual drinking today. That enamelled medieval
rarity, the so-called King John's Cup at Kings Lynn, was

formally drunk from 'by the fraternity' with a ceremonious
turning of the foot on its stem. It has been restored several
times over the past 600 years, as a symbol of the
Corporation's former status.

Such characteristics – the unusual vessel, its striking colour
and age, and its royal associations – single out certain
drinking vessels treasured by societies for ceremonial toasts.
Napoleon's military campaigns in the Low Countries and in

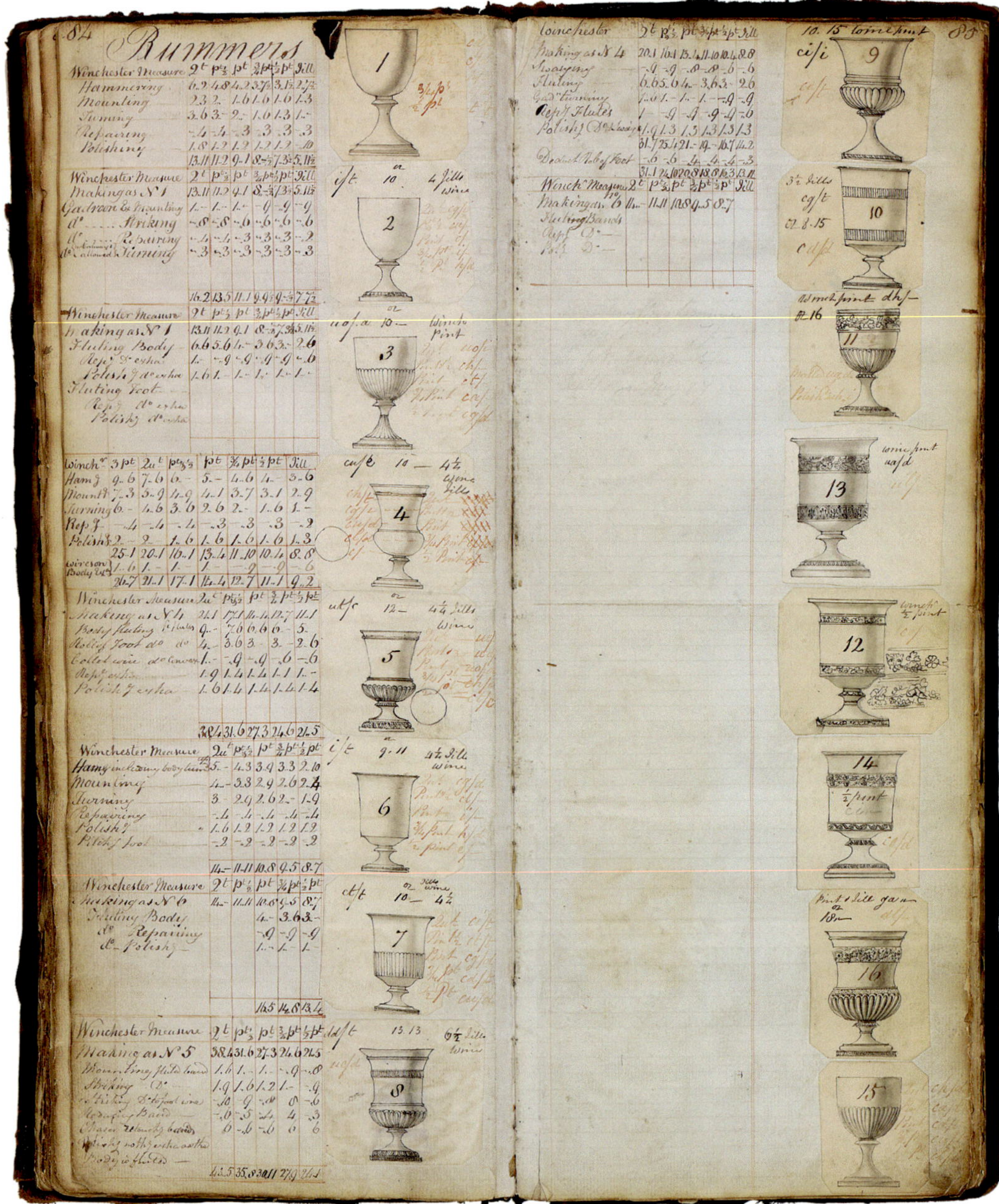

Designs for rummers in white or gilded silver made by Edward Barnards, a Regency firm that became the largest silver manufacturer in the world. Many variations of these neoclassical goblets, sold for punch or wine, were exported to the Americas and India.

Opposite: Skull and crossed bones toasting cup. This expresses the deep-rooted German tradition of ritualized drinking, with a hint of the Vanitas tradition, 'drink with pleasure but remember all will come to dust'.

the German states disrupted fraternities and corporate bodies, broke their traditions and dispersed their treasures. In Berne, for example, William III of Orange presented to the *Aussere Stand* (young burghers) the Lion Cup, a gold and enamelled vessel weighing 100 ounces and shaped as a crowned rampant lion, 'as recompense for the zeal so splendidly displayed towards him'. The young soldiers had paraded on his birthday and might have joined his alliance against Louis XIV. In the event, they kept the handsome cup and did not fight. After 1798, when the *Stand* was abolished, the cup passed to the Archers of Berne, who in 1801 'drank from the wonderful cup that King William gave … singing and clanging of cups resounded till midnight … Governor

Graf emptied the King William nine times.' In Britain, the Reform Acts of the 1830s attacked civic customs and much historic silver was sold off. Ironically this coincided with the flourishing of antiquarian attitudes and several livery companies introduced the supposedly medieval ritual of the shared loving cup, passed from hand to hand and drunk standing back to back.

Drinking in northern Europe often involved challenges, in which a favoured male guest might be expected to drain a couple of litres of beer from a large cup, while retaining his dignity and setting the vessel down with grace. Today some rugby clubs retain a memory of this exercise in self-discipline, combining competitive drinking with a prohibition

on leaving the table. In 1654 an English diplomat, Sir Edward
Walker, described with relief his entertainment in Berlin
when delivering the Garter to Elector Frederick William,
Marquess of Brandenburg:

*Table excellently Served … At Second Course Elector
proposed Charles's health, and a while after the Duke of
Zeemerne, the Queen of Bohemia's and these were all the
healths past as to the Table so as to my great Joy, instead of
drinking after the German mode, I rose from the Table thirsty.*

Challenges developed to stretch the skills and self-control
of military men; Prussian officers played a drinking game,
draining off a toast from the muzzle of a loaded pistol, then
discharging a volley, and reloading and re-priming for the
next toast.

Until 20 years ago a version of this competitive drinking
flourished at the colleges of Oxford and Cambridge in the
custom of sconcing. A student had to down a hornful of cider
or beer, or a tankard of wine or sherry, as penalty for some
small infringement of undergraduate etiquette such as
mentioning work or a woman. The challenger paid the cost,
if the drink went down in one long swallow. This custom has
declined (but is maintained in private sporting and military
groups) as women joined the men's colleges and as attitudes
to alcohol-fuelled behaviour became more judgmental.

Amenities within the Oxford and Cambridge colleges were
bare and monastic. Small, tight-knit masculine societies, they
not only developed distinctive forms such as the ox-eye cup,
and rituals for drinking, but also retained individual silver
drinking vessels well into the twentieth century. Glasses and
stoneware tankards might be hired by the dean and canons
of Westminster College for their audit dinners, as was
happening in the 1770s, but these were not practical
materials for high-spirited youths and the wear and tear of
daily meals in hall.

By custom, which hardened into a rule, each new entrant
brought a drinking vessel of silver, depending on his rank,
for his daily beer, which then became buttery property on
his graduation. By the mid 1640s, Exeter College Oxford had
25 wine bowls (probably on stems), 27 'Ear'd Pots' (ox-eye
cups with two handles), 40 tankards and 11 beakers.
Schoolboys took their personal tankards to Eton and
Westminster with them, pouring their small beer from sturdy
leather blackjacks, and Roman Catholic boys crossing the
Channel to attend college at Douai in Flanders ordered
distinctive cups from London goldsmiths.

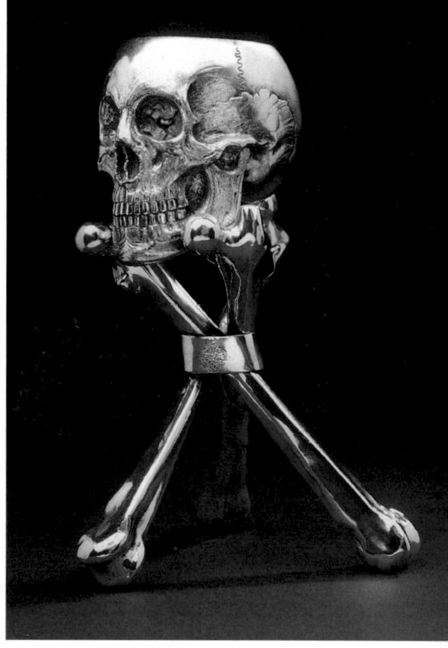

Undergraduates, schoolboys
rather than young men, were
forbidden to visit alehouses,
taverns and wine vaults; since
they shared accommodation
with their tutors, communal
spaces in college were
needed. The first senior
common room appeared at
Merton, in 1661. Wine vaults,
such as the 'Tippling
Philosopher' in the London
Road, Oxford, sold 'what
passes for wine at high prices
and small measure'. Dons
wanted something better, as
the maturing of wine in
cellars evolved. In 1749, the
bursar of All Souls, William Blackstone, bought wine
wholesale in London, and purchased 4,200 bottles from a
Stourbridge glasshouse. In 1761 fellows could buy port for
6*s.* a gallon, and Mountain (a strong red wine like Madeira)
for 7*s.* 6*d.* including bottling.

Many sets of silver cups and bowls for shared drinking
appear in the inventories of Henry VIII and were offered as
prizes in the 1567 London lottery. Small stacking beakers,
only one with a cover, for toasts, called *Monatsbecher* were
popular in Germany, each engraved with a month of the
year. Sets of gilded goblets for serving punch survive from
the mid-eighteenth century. Ten glasses and two decanters
engraved with Jacobite motifs are now back at Chastleton
House, home of the ardent Jacobite Henry Jones. Masonic,
trade and craft societies commissioned drinking vessels,
punch bowls and wine impedimenta; in the early nineteenth
century Britain's largest manufacturing silversmiths, Edward
Barnards, offered a great range of designs and sizes in
goblets for shared toasts.

Today London livery companies retain the custom of a
master's gift, a piece of silver to recall the donor and to
enhance corporate life. At the Goldsmiths' Company each
new entrant to the court commissions a cup, which is placed
at his or her seat at lunches and dinners, to stimulate, it is
hoped, admiration for contemporary design in the precious
metals as well as reinforcing a sense of membership as he or
she drinks (see Frontispiece).

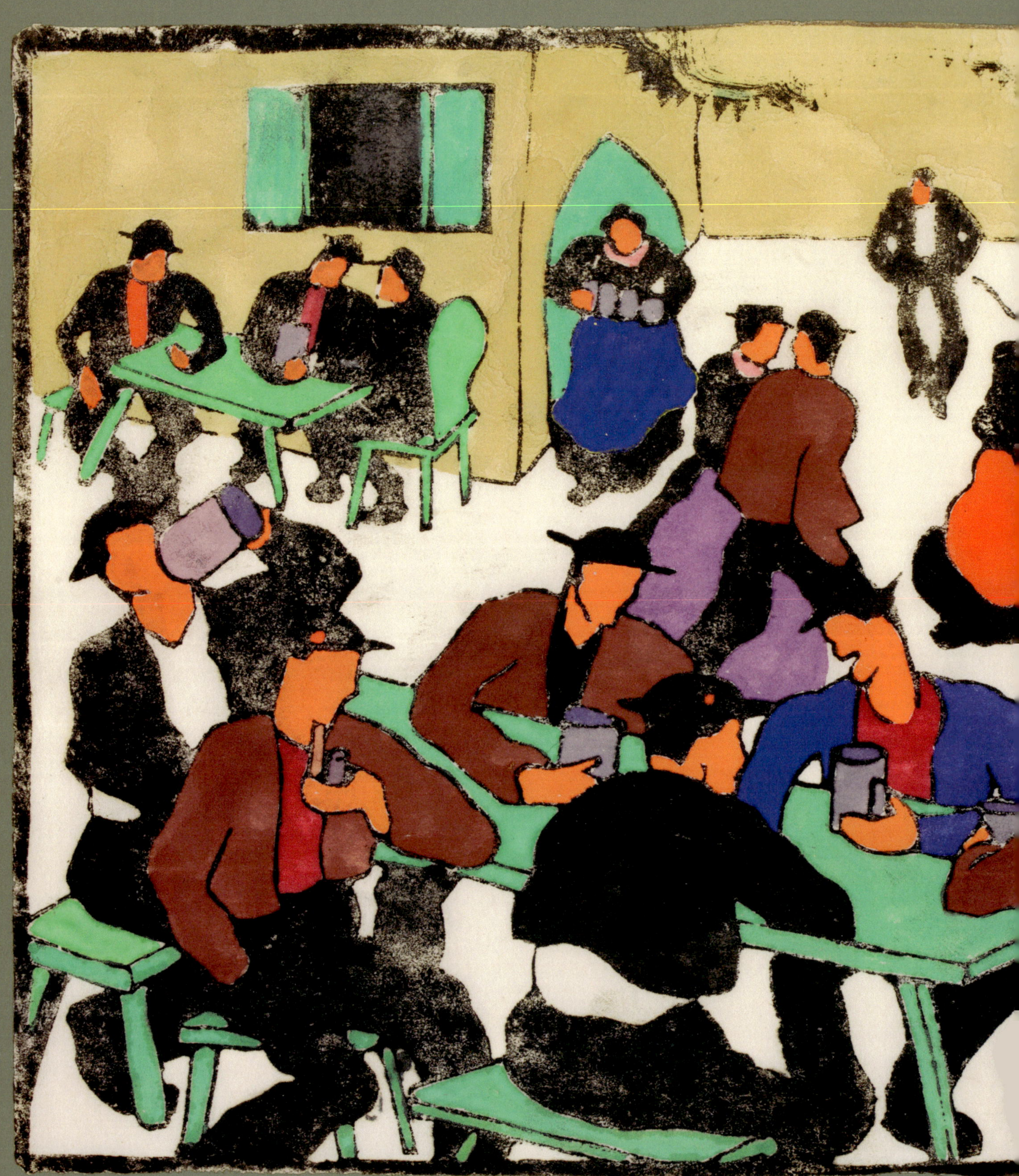

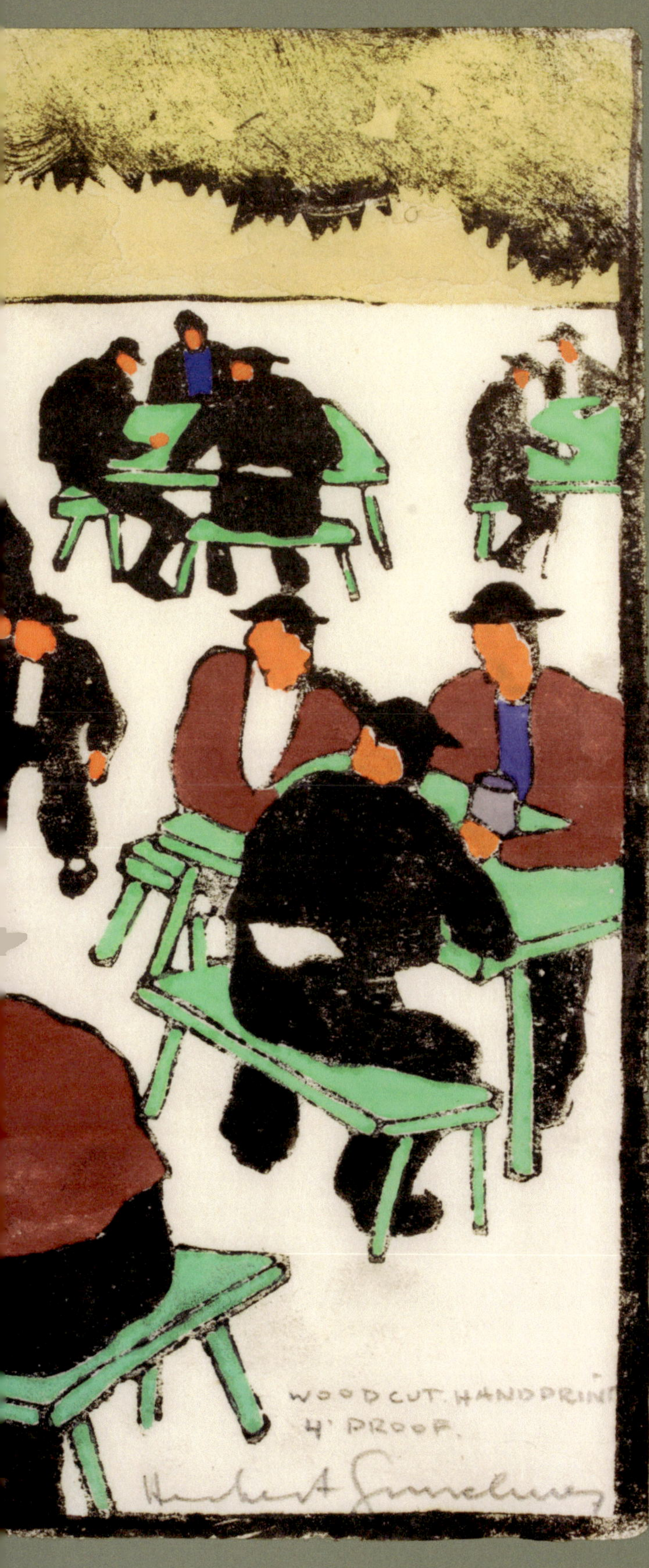

Part 3

CONTEXTS AND SETTINGS

Alehouses, Taverns and Inns

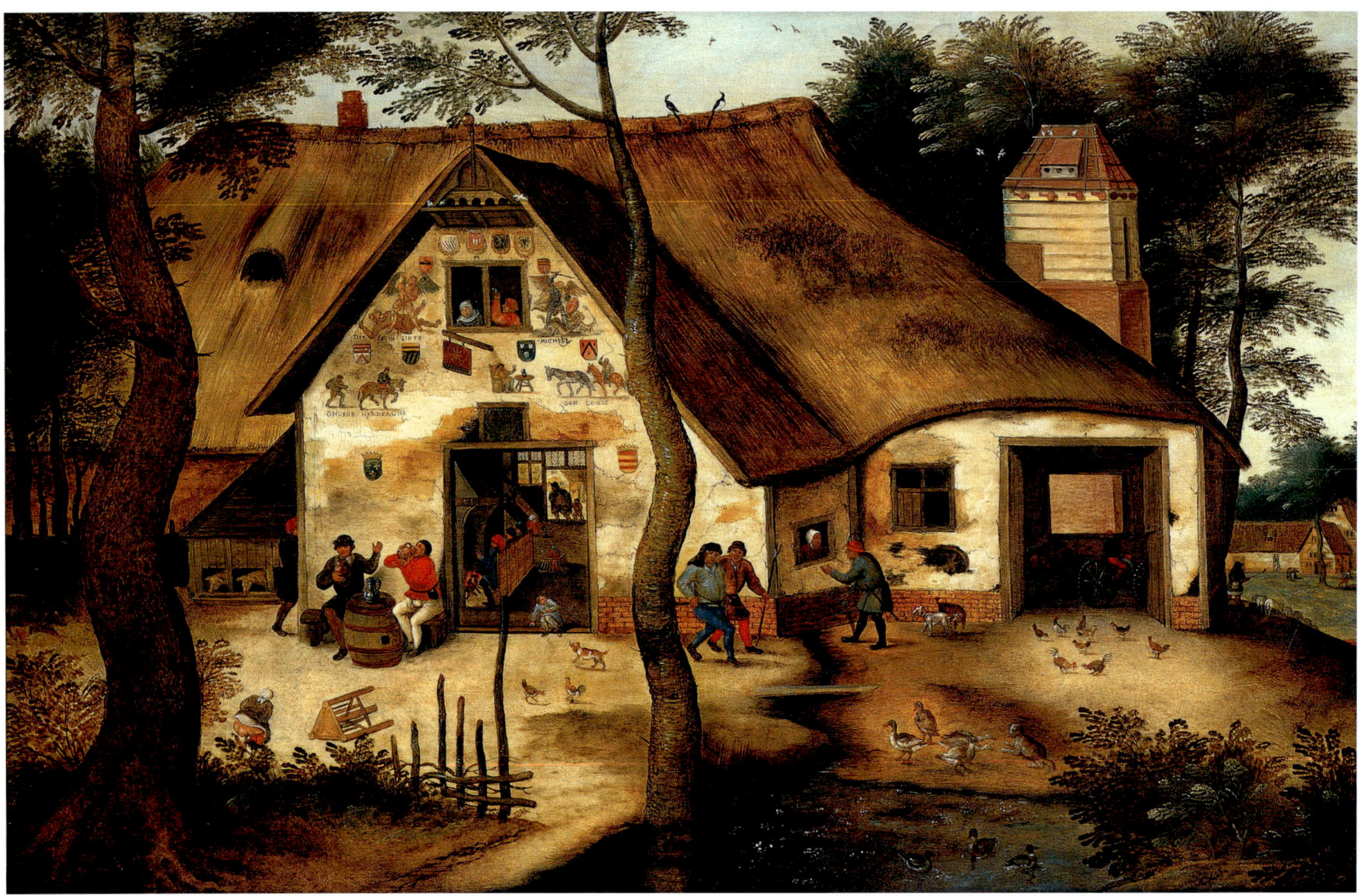

Services to travellers are advertised on the façade of this inn, typical of Breughel's socially observant paintings.

Upon all the new settlements the Spaniards made, the first thing they do is to build a church, the first thing the Dutch do … is to build them a fort, but the first thing the English do, be it in the most remote parts of the world, or among the most barbarous Indians, is to set up a tavern or drinking house.

BARBADOS PLANTER T. WALDUCK, *LETTER FROM BARBADOS* (1710)

Our perspective on the pre-industrial tavern has been coloured by German and Dutch depictions. *Kermis* or festival prints of 1530s Nuremberg by Hans Sebald Beham portray taverns as symbols of anti-culture. A dilapidated tavern, identified by its alestake, plays host to a riotous assemblage of gamblers, fornicators and vomiters while in the background an orderly congregation queues stoically outside the well-maintained church. Crumbling buildings with broken wine vessels and overturned furniture feature in Dutch tavern scenes by Jan Steen and Adrian van Ostade as symbols of moral and physical decay.

In fact, the tavern was central to community life in early modern Europe. A shared meal washed down with a nourishing ale; a lively conversation with music, dancing, skittles and dice: alehouses and taverns offered an escape from cramped homes, a venue for a business deal, catering for a parish festival and a place to gather news. Quart tankards of pewter or stoneware, passed around, encouraged conviviality, which was seen as a social good. Recent research by university psychologists in Pittsburgh and Canterbury on male drinking in small groups, reported in the *Journal of Studies of Alcohol* (2006), has confirmed the social benefits of drinking together rather than alone, as 'an informal means of regulating and moderating' risky behaviour.

In the German-speaking territories, Flanders and the Netherlands, distinctive drinking establishments served guild members, rural workers, journeymen and wealthy farmers. Bavarian taverns graded the tables by social rank, so reinforcing, rather than threatening, the social order. Licensing, locally organized, was often strict. Nuremberg council issued regular mandates against excessive drinking, gambling, dancing and political discussion: in the 1530s playing cards, knives and drums were all forbidden. In Berne frequent surveys listed alehouses to be closed for keeping poor order. Heavy fines were levied on the owners of unlicensed *winkelkrüge* or corner-houses. In Holstein in the 1630s inspectors were promised 20 per cent of any fines levied.

Among English drinking establishments, alehouses and their nineteenth-century successors, modest beer houses set up under the 1830 Beer House Act have been subjected to the most legislation. *An Act for Keepers of Ale-houses to be bound by Recognisances* in 1552 introduced formal licensing to combat the 'intolerable hurts and troubles to the common wealth of this realm' blamed on alehouses. All alehouse keepers were to apply to two justices at the Quarter Sessions for a licence to sell, which doubled as a bond to maintain good order and prevent unlawful games.

Bleak economic conditions in England in the 1590s and concern about moral discipline led to more Parliamentary bills aimed at alehouses: 25 were introduced in the sessions of 1601, 1604 and 1606. A petition in Bradford-on-Avon in 1628 blamed alehouses for enticing poor labourers away from their domestic obligations.

Inns, normally in market towns and ports or on post roads, had a different role. Their 'ancient true and proper use', according to an English act of 1606, was 'for the receipt, relief and lodging of wayfaring people'. Networks of inns spaced about a day's coach travel apart covered England between the 1660s and the arrival of rail travel in the 1830s. Supplying stabling (the inns of Horsham in Sussex in 1686 offered 83 beds and stabling for 365 horses) and meals, they served a higher social class, with few restrictions on opening hours and the right to serve spirits and wine. At Stamford, Lincolnshire, the sprawling 'George and Angel', one of four inns, contrasted with 10 smaller inns and 30 taverns, modest local drinking dens, licensed to sell just ale and beer 'by sealed quarts'. So in 1656 the 2,000 inhabitants had one drinking den for every 50 inhabitants.

Equipment in many drinking establishments offered comfort in a neighbourly shared setting, including porcelain punch bowls, prints and mirrors on the walls and upholstered chairs. In 1695/6 an Act to encourage the

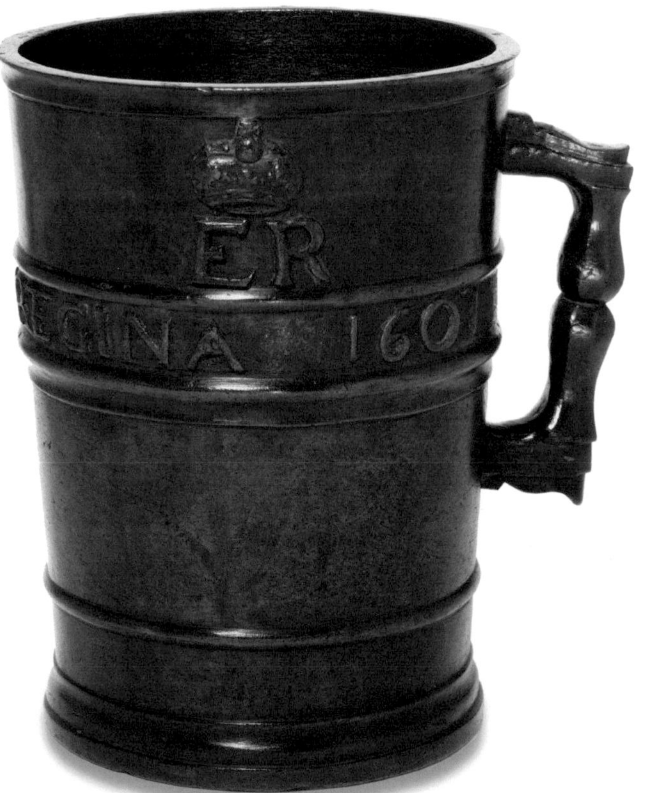

Left: A standard gallon measure of 1601, with crowned 'I' stamp showing its capacity had been verified. This essential equipment for local inspectors of tavern measures was made in bronze so it could not be easily tampered with.

Right: Tavern keepers kept sturdy drinking vessels in silver, pewter or stoneware. Engraved with the tavern keeper's name and the Bell, this half-pint beer mug is stamped with an 'AR' capacity mark.

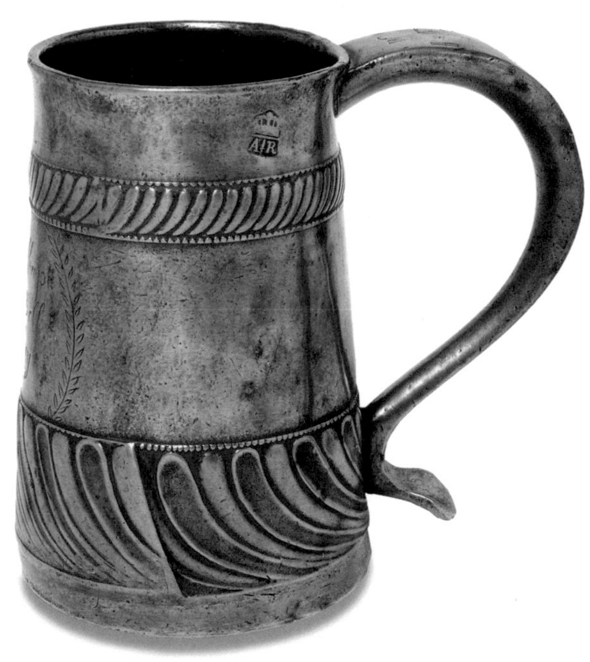

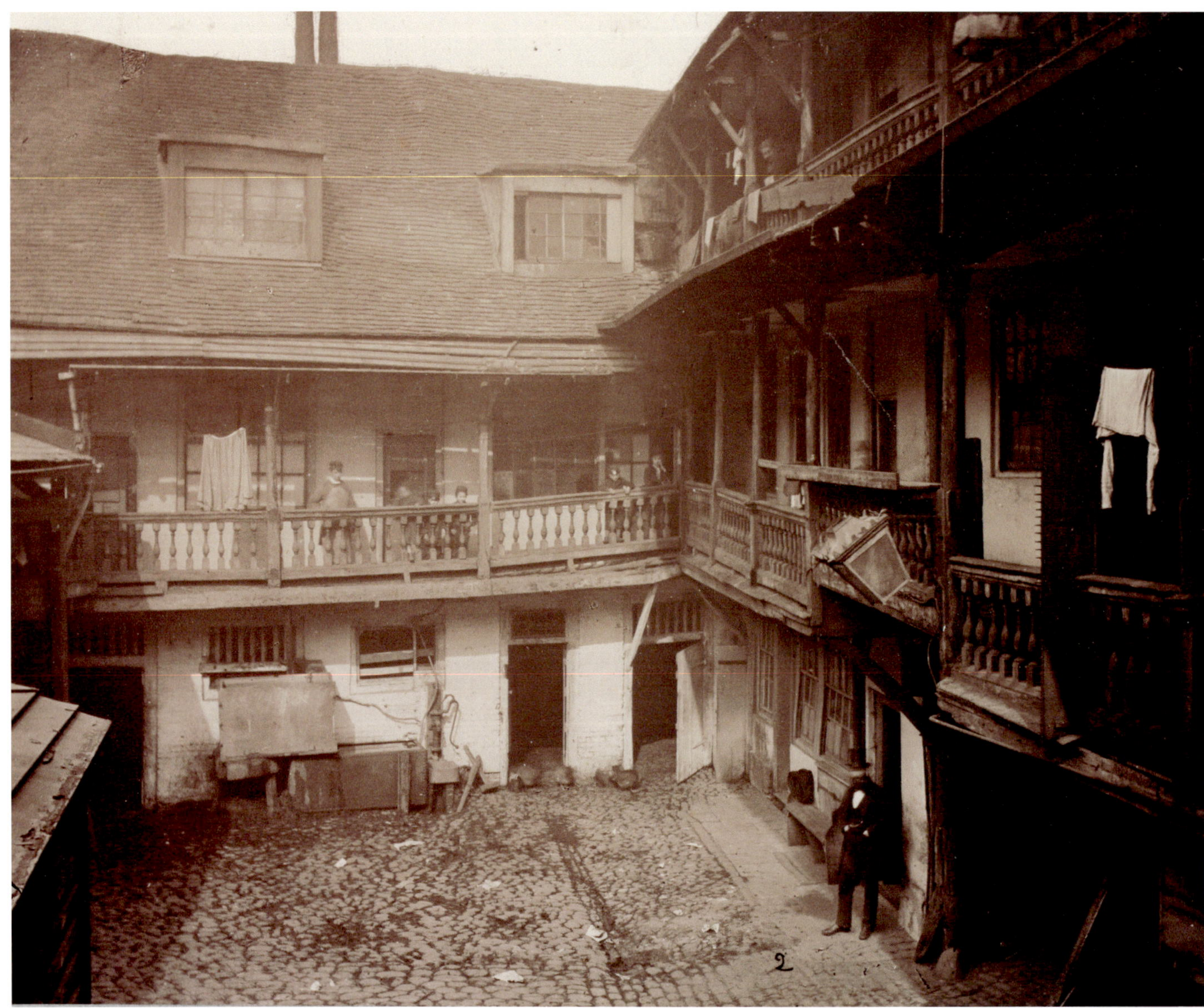

bringing in of plate to the Mint ruled that 'no person keeping an inn or tavern or selling wine should expose in his house any silver except spoons'. In fact, from the mid-1690s the Old Bailey Sessions records thefts from Middlesex taverns of several hundred silver tankards and punch bowls, showing how widely these touches of modest luxury were used.

Innkeepers attracted official hostility only when they kept disorderly alehouses. Worcester innkeepers were ordered in 1597 to pay for eight men to patrol the city streets from 9pm to 5am to maintain order. Some justices conducted petty sessions in inns: in 1706 the justices at Wingham in Kent alternated their custom between the 'Red Lion' and the

'Black Dog'. Eighteenth-century landowners, who were often justices too and issued licences, speculated on alehouses, renting them to tenants and allocating licences with partiality. A Kent gentleman protested that his new tenant could not obtain a licence as the gentleman who owned the neighbouring inn did not want his profitable monopoly disturbed. While the numbers of alehouses gradually declined in the eighteenth century, inns expanded. In Sussex their number trebled.

As a dietary staple when water was polluted, ale could not be regulated out of existence. A fresh brew of ale did not last long. Until the fifteenth century most was brewed and

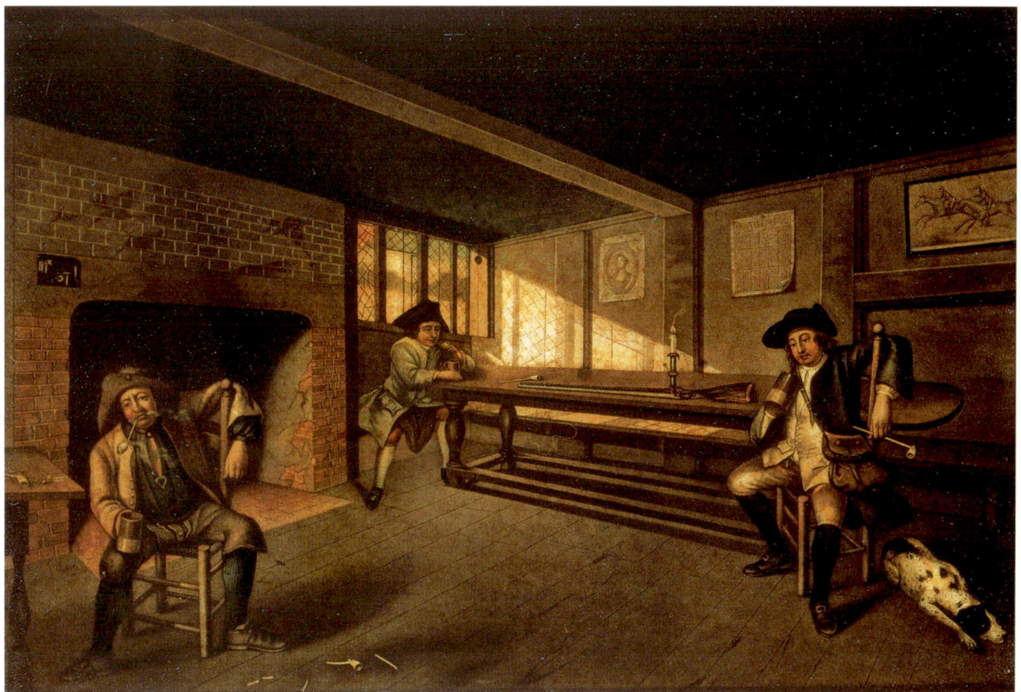

sold on the same premises. Brewing was largely a female occupation: 'ale-wives' gained a secondary income, selling ale surplus to family needs. Regulation was haphazard. Some brewers took short cuts and sold a sickly brew that was not fully fermented. Others in larger towns, keen to offload their perishable ale, sold to 'tipplers' who offered the ale for resale at inflated prices, thereby denying basic nutrition to poorer families.

An 'ale-stake' decorated with branches or garlands of leaves above a door was an invitation to the 'ale-conner' to test its home-brewed ale. Wearing his essential trade tool, his leather breeches, the ale-conner sat for half an hour in a sample of ale poured onto a bench or stool. If he could then stand without his breeches sticking, the barley sugar in the ale was fully fermented and ale was ready for tasting. The taste test dictated at what price the ale might be sold and allowed the brewer to display the ale-stake (a forerunner of today's inn-sign) until that brew was sold.

Flemish weavers coming to England in the 1400s are traditionally credited with bringing the controversial beer, brewed with hops. Beer took longer to brew and required more equipment, but it lasted longer. In the 1680s barrel measures for ale and beer were unified, a recognition of beer's commercial ascendancy. Frothy beer required the 'thurdendel', a pewter measure for selling from the barrel directly to table. Slightly taller than earlier ale measures, its official capacity marked below the rim allowed the froth to sit on the top. Sold by the quart, beer was a daily necessity, whose price was controlled for centuries. When the malt tax was increased in times of war (it went up four times between 1800 and 1810), the brewer had either to adulterate his product or reduce his profit.

Brewing in bulk was more efficient, encouraging the growth of retail-only outlets, the ancestors of the modern pub. London brewers already had a large share of the retail market by 1600, exporting beer to the Baltic. Around 1720 they developed a new dark, keeping beer, called London porter, while the Burton upon Trent brewers specialized in brighter ales, advertised in Calcutta by the 1780s. Large entrepreneurs such as Whitbread effectively drove out small brewers in southern England by the early nineteenth century. However, private brewing continued: in 1801 the naturalist Sir Joseph Banks was still brewing 25 times a year for his visitors and staff on his Middlesex estate, about 4,000 gallons at a cost of just over 1_d._ a pint, including the excise duty, at a time when a quart in a tavern cost 3_d._ or more.

Legislation doubling taxes on beer, wine and brandy and allowing distillers to operate without a licence brought gin to the fore. Introduced by William III in 1690 and 1701 and Anne in 1702, these laws were effectively trade sanctions against France, encouraging grain imports from Holland and increasing revenue from duties. In the 1730s the drinks trade supplied a quarter of English revenue from excise, customs and licences, as well as taxes on serving and storage vessels of glass and stoneware. The distilling lobby was strong in the City and lobbied Parliament effectively. Before 1750 half the wheat sold in London was converted into alcohol and almost a quarter of central London houses were dram shops selling spirits. Drunkenness reached epidemic proportions. From this well the government added water to its legislation and diluted its potency. But public and medical opinion gradually found a voice, aided by Hogarth's visual polemic (p.16). The Gin Act of 1736 ordered retailers to obtain an annual licence for £50 if they intended to sell less than 2 gallons of spirits at a time and pay a further £1 on every gallon sold in lesser quantity. From 1751 the government began to gain control of public drinking.

Today, microbreweries across Britain are reviving the concept of distinctive local beers, sold as a delicacy for the discerning drinker and deliberately priced above the widely advertised and distributed commercial beers and lagers that have dominated the market in the past century. A symptom of the growing demand for specific, often organic, food and drink, this is also a return to the spirit of earlier practices, when local brews were savoured as part of social activity.

The 'Rose and Crown', Saffron Walden, was designed in the 'Queen Anne style' by William Eden Nesfield in 1872.

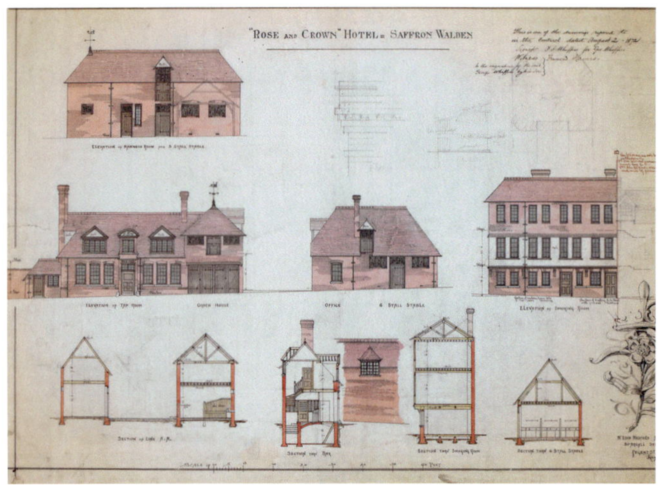

The Black Friar: an Arts and Crafts pub

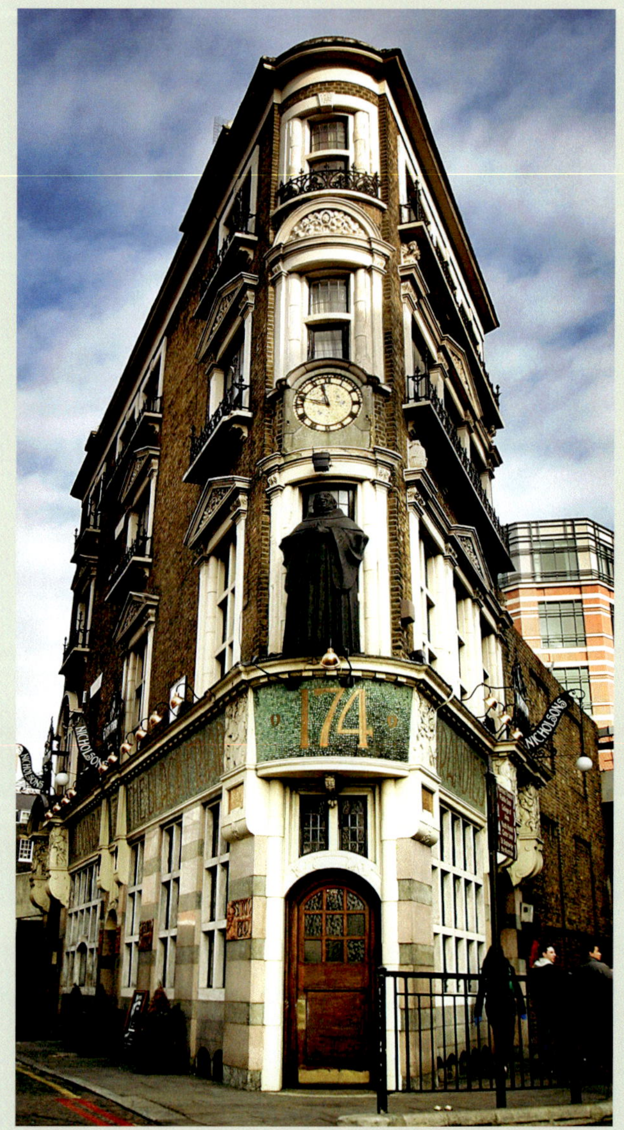

Exterior and interior of a distinctive Arts and Crafts pub, 'The Black Friar', opposite Blackfriars Station in the City of London. It is on the site of a medieval friary.

'The Black Friar' pub, close to Blackfriars Station in the City of London, is memorable for its exceptional design scheme, both inside and out. It was lucky to escape demolition in the mid-1960s, when neighbouring buildings made way for redevelopment. The building takes its name (and the inspiration for its interior decoration) from its proximity to the site of a medieval friary. Although the pub was built in 1875, its design scheme dates from 1904–5, with the intimate smaller saloon added between 1917 and 1924. The scheme

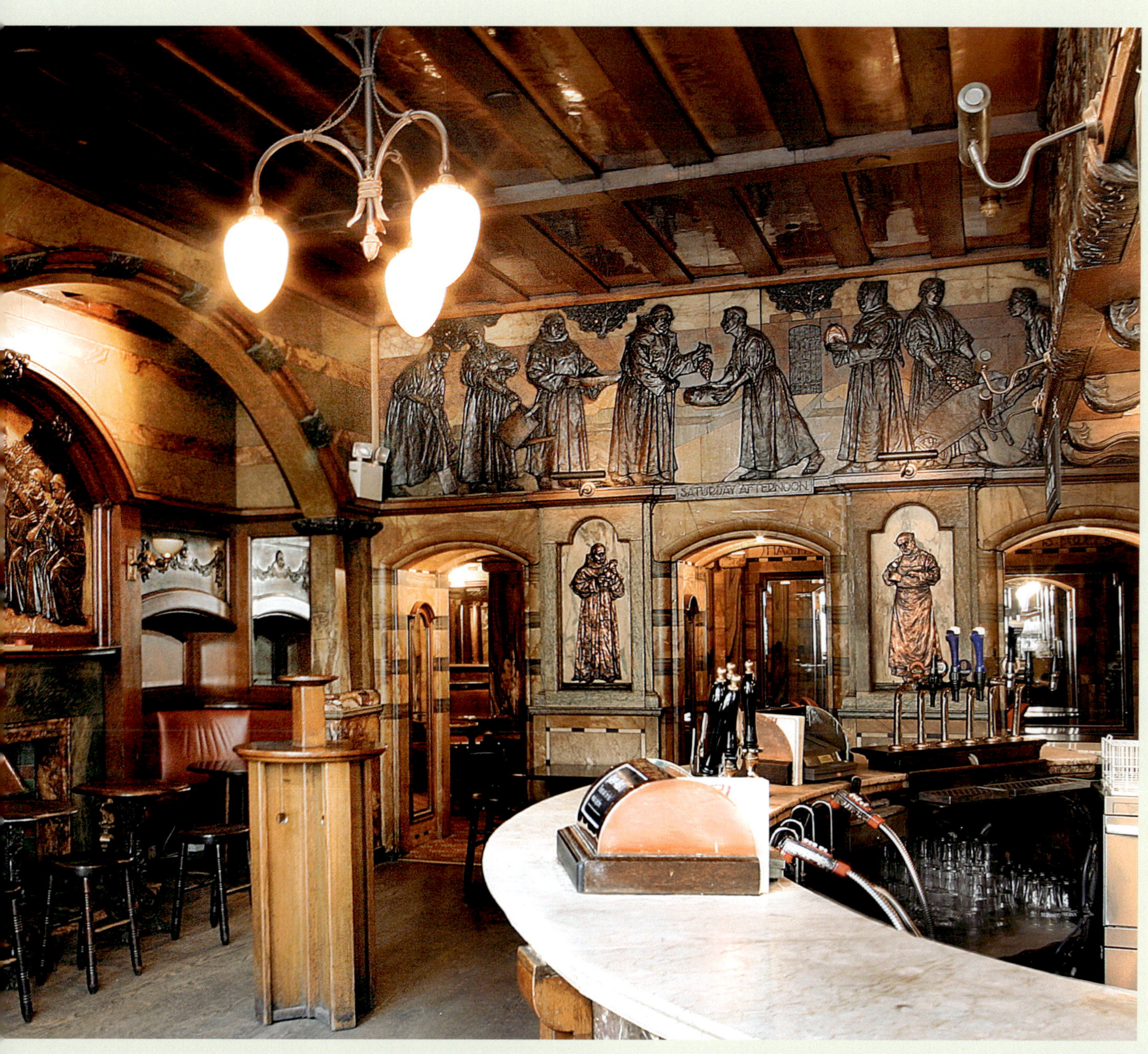

was overseen by Herbert Fuller-Clark, an architect not known for pub interiors but whose experience designing music halls may account for the wonderful theatricality of the interior. Mosaic ceilings, marble decoration and the artwork commissioned for the pub all work together to create a unique drinking experience.

The interior decoration draws on a nostalgic interpretation of England's medieval past, characteristic of Arts and Crafts design. Inside and out friars adorn the walls, pointing the way to the saloon bar or accompanied by lettering spelling out maxims for drinkers'

contemplation, such as 'DON'T ADVERTISE IT: TELL A GOSSIP' and 'FINERY IS FOOLERY' (where a rotund friar is shown wearing a fine collar and pendant). Leading sculptors of the time, Nathaniel Hitch, Henry Poole (a former apprentice of G.F Watts, an artist best known for his paintings) and Frederick Callcott supplied relief work for the scheme. Above the bar, a copper relief shows friars catching fish (the fish picked out in mother-of-pearl). Beneath, a reminder of restraint but a comfort for any drinker, are the words 'TOMORROW WILL BE FRIDAY'.

Assemblies

Formerly the country ladies were stewed up in their fathers'
old manor houses and seldom saw company, but at an
assize, a horse race or a fair.

J. MACKY, *A JOURNEY THROUGH ENGLAND AND SCOTLAND* (1722/3)
ON THE APPEAL OF ASSEMBLIES

Assemblies, a new English social phenomenon around
1700, offered women a social venue, a chance to dress
up and an antidote to the tedium of rural life. Restrictions on
hard drinking, an entry fee including wine and non-alcoholic
refreshments and decorous, non-drunken behaviour
characterized these novel public events. Occasions for
socializing, assemblies offered the opportunity to
promenade, dance, listen to music, chat, flirt and play cards.

Masquerades were essentially assemblies where those
attending were masked and cloaked or dressed in exotic
costume. Count Heidegger, an impresario who became
theatre manager at the King's Theatre, Haymarket, London in
1711, introduced masquerades or *ridotti,* on the Venetian
model, at a guinea a night when there were no
performances.

A Long Room masquerade in February 1724 was
exceedingly large beautifully adorned and illuminated
with 500 wax lights … on the sides are diverse beaufets, over
which are written the names of the several wines therein
contained, as Burgundy, Canary, Champain, Rheinish etc,
each most excellent of its kind, of which all are at liberty to
drink what they please.…
There were no spirits, beer or punch on offer.

Since assemblies started early in the evening, after the
normal dinner hour, neither hot food nor strong drink was
available. A major factor in instilling polite behaviour, they
contrasted sharply with male drinking, gambling or
cockfights, the 'gross sensual amusements to which the
hours of vacancy had before been devoted', as a Bath
historian summed up their impact in 1801.

In private houses large rooms were specifically designed
for the easy movement of guests who wanted to enjoy such
gatherings. At Wynnstay, Denbighshire, a Great Room was
added for the coming of age celebrations of Sir Watkin
Williams-Wynn, Bart., in 1770. Originally intended as a
temporary structure, it was so convenient that it was
retained, equipped with an organ, and subsequently
redecorated for a ball in 1828.

As assemblies caught on, assembly rooms were built in
both cities and small market towns, as well as at the spas. An
annual subscription allowed attendance at assemblies,
masquerades and a winter season of balls. Elaborate
structures, added to inns or purpose-built, offered the
essential imposing spaces lit by large chandeliers with
mirrors to reflect the light, combined with smaller rooms for
card playing, conversation and refreshments.

In York the magnificent assembly rooms were designed by
Lord Burlington in 1730 with two rooms either side of a very
large hall with marbled columns, a garland frieze and
clerestory lighting. Several hundred people bought tickets for
the York assemblies; in Race Week 1736 the drinks on offer
included Champagne, French red and white wine, 'mountain
wine in a negus' (mild punch) and arrack punch. In Newark,
Nottinghamshire, the town hall, built in 1774–6 by Carr of
York, has an imposing first floor assembly room with pairs of
giant Corinthian columns screening off niches, appropriate
for private conversations, and a ceiling and doors in the style
of Robert Adam.

Smaller towns had more modest assemblies, such as those
attended by Jane Austen. As a young woman of 17 living at
Steventon Rectory in Hampshire, she attended balls in the
Basingstoke Assembly Rooms, above the town hall, and at
the 'Dolphin Inn', Southampton. When visiting Lyme Regis in
1804 she enjoyed a few dances at the Lyme Assembly
Rooms. Two shillings for a subscriber or four shillings for a
non-subscriber gave access to a card room, billiard room and
chandeliered ballroom, with music for dancing every
Tuesday provided by two violins and a cello.

Refreshments were tea, coffee, chocolate and light wines,
rather than any stronger alcohol, to prevent unseemly
behaviour and to protect young women from unpleasant
advances. However this convention did not prevent those
attending from drinking beforehand; Miss Austen commented
on a quarrel between a drunken, married couple in the New
Assembly Rooms in Bath.

The Bath rooms epitomized the architectural, decorative
and functional qualities of assembly rooms. Created by John
Wood the Younger in 1772 with a ballroom, octagon card
room and tea room, the rooms were specifically designed so
large groups could enjoy dancing, card playing and
refreshments simultaneously.

In the ballroom the lower walls had tiers of seating for
chaperones and older subscribers. The tea room, for both
concerts and refreshments, had a bar behind the colonnade.

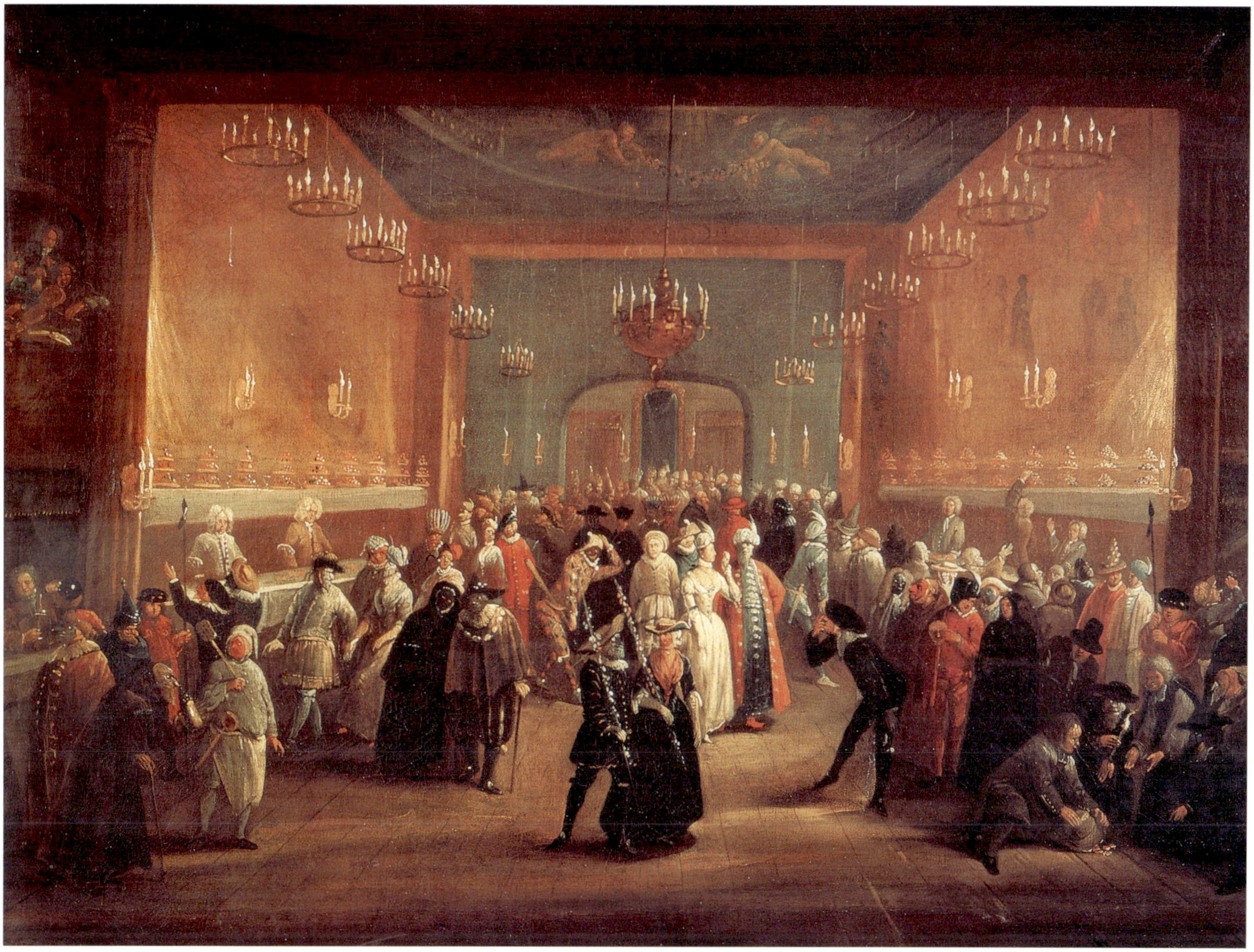

Above: Grisoni's *Masquerade at the King's Opera House* shows bars selling wine erected along the walls and the temporary dance floor, an enjoyable new venue for polite society.

Below: A genteel way to entertain a large gathering, assemblies in private houses were less costly and alcohol-fuelled than a seated dinner. The Wynnstay assembly room offered a setting for female grace and male elegance.

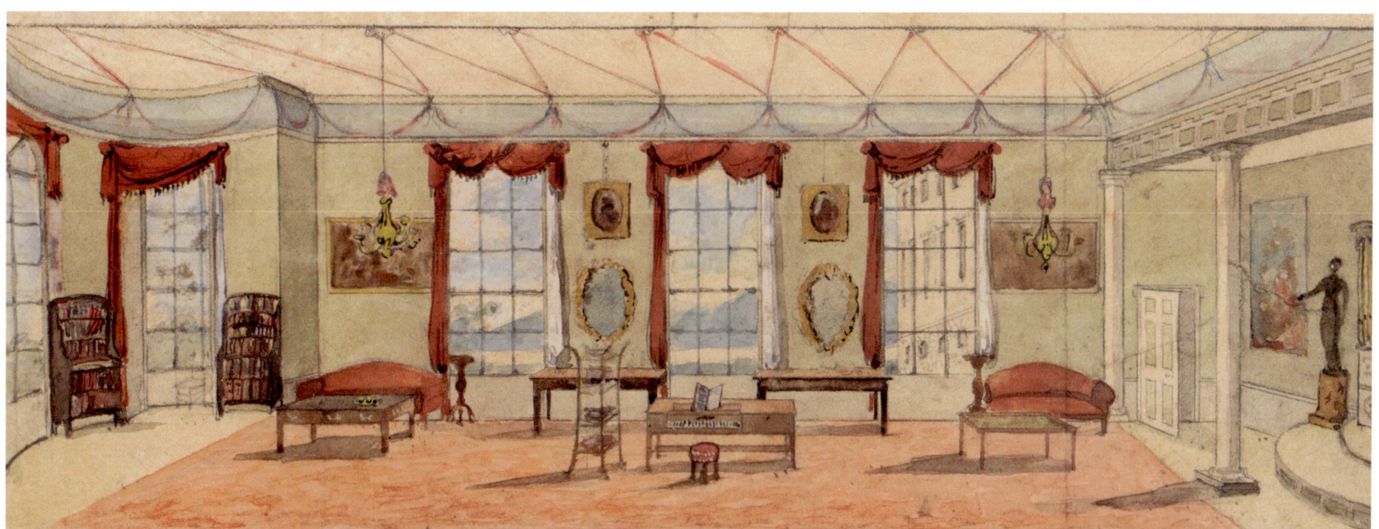

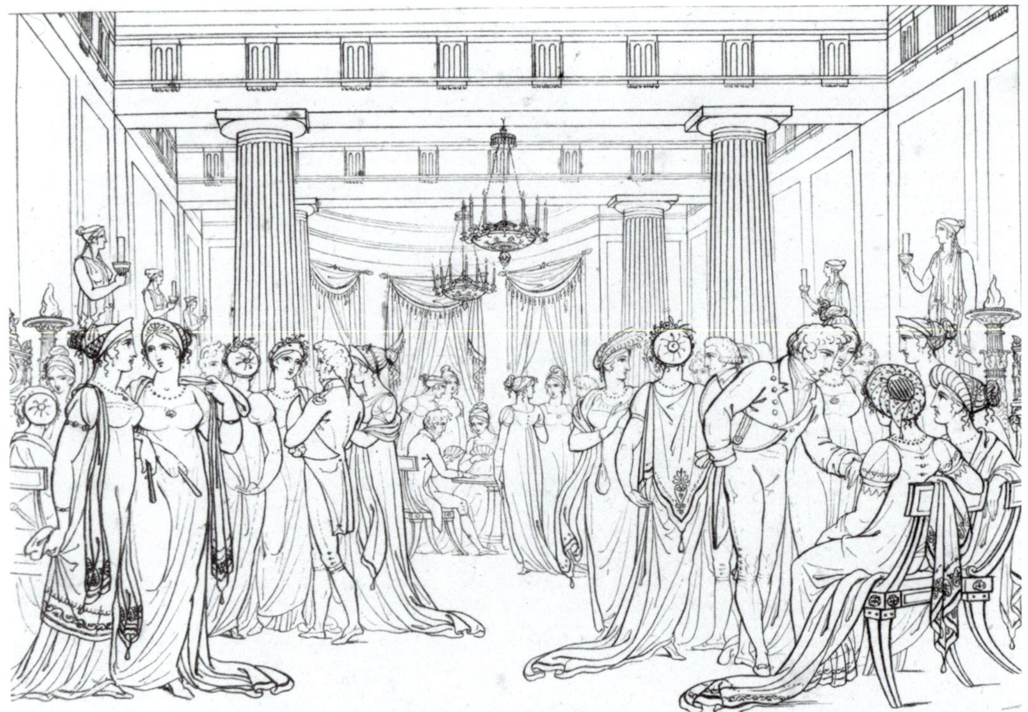

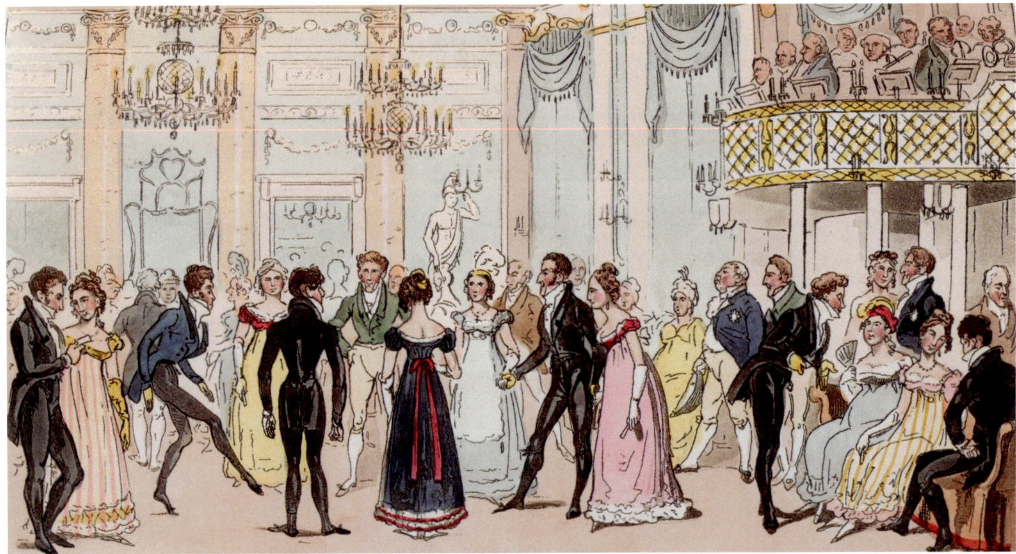

Both rooms were double height and newspapers were available in the coffee room. Careful preparations for the evening jollities included instructions to contractors for setting up platforms for the orchestra at one end of the ballroom, arranging the furniture and providing sufficient water for tea and coffee. While subscriptions for ball tickets included the cost of tea, visitors to the building on Sundays were charged for drinks. The steward managed the catering and the wine cellar.

As the New Assembly Rooms became more popular, so their social cachet fell, particularly in the early nineteenth century when balls became much more crowded. Changing social habits brought different events such as lectures, but concerts remained popular. Refreshments at a 1873 reception, from 9 pm to midnight, included tea, coffee, ices, Champagne cup, claret cup and sandwiches.

In London the most fashionable assemblies were held at Carlisle House, Soho Square, Wyatt's magnificent Pantheon in Oxford Street built in 1772, and Almack's Assembly Rooms. Carlisle House was rented in 1760 by Mrs Cornelys, a Viennese opera singer and courtesan famed for her celebrated assembly rooms. Those who could afford her subscriptions

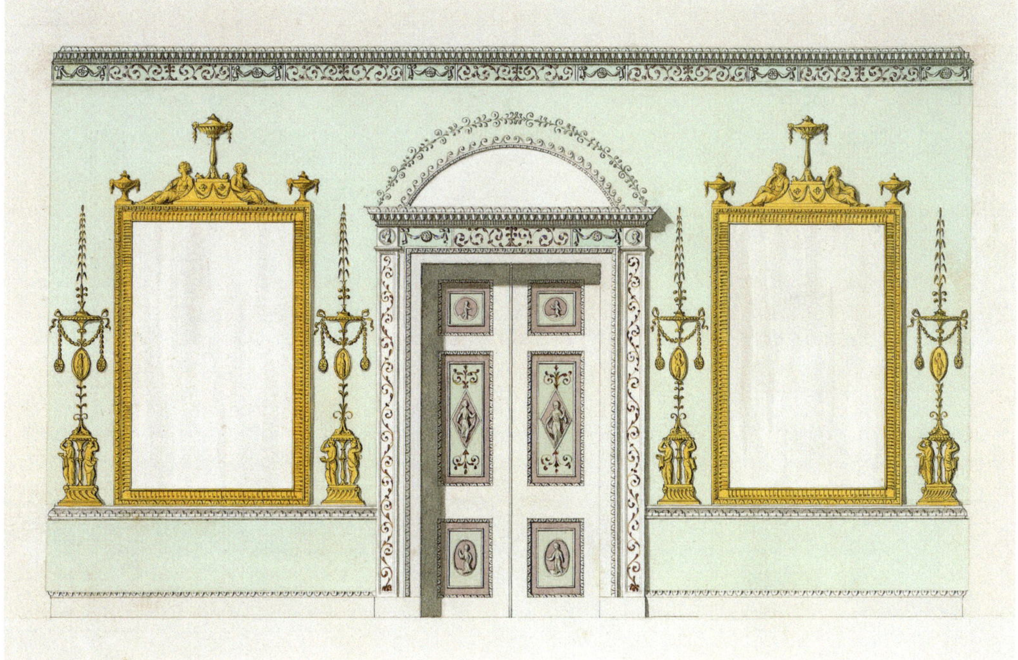

attended for dancing, cards, operatic concerts and masquerades. Her lavish furnishings included a large amount of furniture at a value of over £1,000 from the cabinetmaker Samuel Norman. Mirrors and light fittings were essential for interiors lit largely by candles and Norman supplied several in exotic styles including a girandole decorated with a sheep in chinoiserie style. Fanny Burney, visiting Carlisle House in 1770 aged 18, praised the magnificent interiors and splendid lighting but was disappointed by the crowded and overheated rooms. Refreshments at Carlisle House included tea, coffee, chocolate, lemonade, orgeat (a non-alcoholic drink of orange-flower water with barley or almonds), negus (port or sherry and hot water, sweetened and flavoured with lemon and spice) and milk.

Facing competition from the newly established Pantheon and Almack's, Mrs Cornelys extended Carlisle House, adding a suite of Chinese rooms in 1769, but her clientele became less exclusive. She was declared bankrupt in 1772, Carlisle House being assigned to her creditors, one of whom was the cabinetmaker Thomas Chippendale. Carlisle House continued as a venue for public entertainment with concerts and masquerades on weekdays. On Sunday evenings visitors could promenade through an indoor garden with shrubs, a mineral spring and a grotto.

The Pantheon was designed by James Wyatt as a winter alternative to Ranelagh, the fashionable pleasure gardens in Chelsea. The main space, a large and beautiful rotunda based on Hagia Sophia in Constantinople (Istanbul), with smaller vestibules, card rooms, tea and supper rooms, opened on 27 January 1772. Described by Horace Walpole in glowing terms, it became a fashionable but short-lived venue for assemblies, masquerades, fêtes and concerts. Already declining in popularity by 1780, the price of admission was reduced from 12 guineas during the season to 6 guineas, and the Pantheon was used for other purposes, including an exhibition of Lunardi's balloon in 1784 and 1785. In 1791 it was altered into a theatre and almost completely destroyed by fire in 1792.

Almack's Assembly Rooms, originally established in Pall Mall by William Macall, later Almack, moved to more magnificent premises in King Street, St James's, in 1765. Considered the most exclusive club in London, membership cost 10 guineas, with vouchers or tickets issued for particular evenings at the discretion of the despotic lady patronesses. They were aristocrats such as Lady Sefton and Lady Jersey, or wives of foreign diplomats such as Princess Lieven and Princess Esterhazy, whose husbands were the Austrian and Russian ambassadors respectively. The ballroom, about 30 m (100 ft) long and 12 m (40 ft) wide, with gilded columns, classical medallions and very large mirrors, was originally candle-lit and subsequently by lit gas in enormous cut-glass lustres. Some 200 of the beau monde attended a ball and supper every Wednesday during the Season (early March to early June). The main function of Almack's was to launch debutantes into society; only light wines or ladylike drinks such as orgeat or ratafia (a very light liqueur flavoured with fruit or almonds) were served. From about 1835 Almack's declined as assemblies fell out of fashion.

Clubs, Dens and Political Drinking

[T]here are regular clubs which are held in coffee houses and taverns at fixed days and hours. Wine, beer, tea, pipes and tobacco help to amuse them at these meetings … everything necessary for convivial jocundity. Each member drank what he liked, in quantity which he thought proper. The master of the house had nothing more to do than to produce a fresh supply of licquors as soon as they were emptied.…

JEAN PIERRE GROSSLEY ON LONDON CLUBS, 1765

In early-modern England inns were central to collective life. Postal services, internal trade and all travellers, from the highest to the lowest, depended upon them as places for stopping off, refreshment and changing horses. Royals and rebels used them as rendezvous; both Oliver Cromwell and the Duke of Monmouth stopped at 'The George' at Norton St Philip during their campaigns, while Charles I took his last draught of freedom at the 'Saracens Head' in Southwell where he was given up to the English by the Scots.

Beer and ale sustained existence at a time when water was dangerous to drink. As beer replaced ale as a necessity, the local alehouse or tavern became an everyday venue. According to one contemporary report, by the 1630s there was a drinking place to every 95 people in England.

Taverns and inns, often the largest public buildings in a market town apart from the church, offered warm and well-furnished venues, decorated and embellished far beyond the modest amenities of home. Special vessels to drink and eat from, such as silver, pewter or stoneware cups and tankards or 'fun' objects like painted 'merry man' plates and wager or puzzle cups, added to the visual and tactile pleasures. Many venues offered musical entertainments, ranging from a simple fiddler to a concert programme. Inns on main routes such as the Great North Road offered travellers local antiquities or curiosities. The 'Scole Inn' in Bedfordshire had a round bed large enough to sleep 20 couples at a time, it was claimed, twice as many as the famous Great Bed of Ware, now at the V&A. Ripon mayor and innkeeper Christopher Hunton exhibited the famous Wakeman's Horn and Belt embellished with miniature silver devices and badges of past officers, in his 'common room' in 1686/7 to show guests 'the honour and worship' of his place. The silver was stolen, to his chagrin.

Inns and taverns were the setting for many activities. John Bunyan, writer of *The Pilgrim's Progress,* preached his fiery sermons of moral reformation at 'The Swan' in Bedford. Famously the haunts of highwaymen, villains, whores and

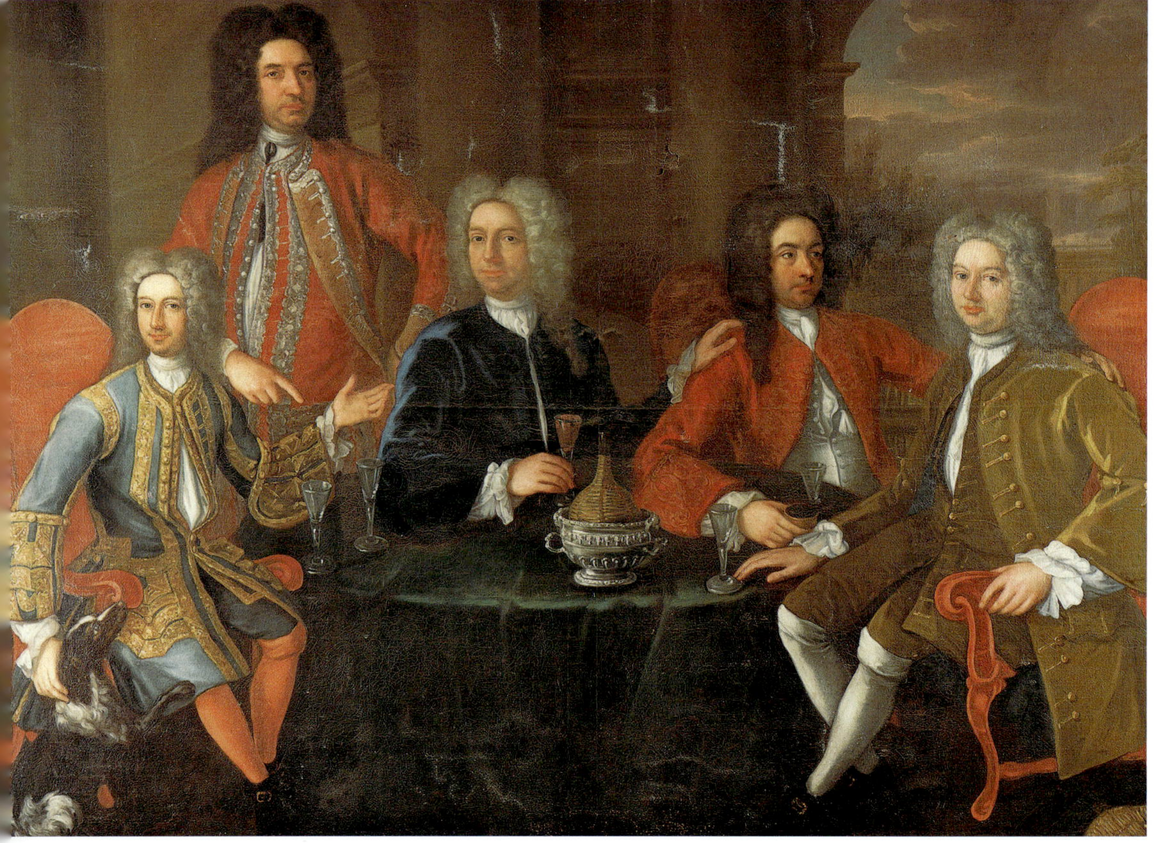

A meeting of the Hell Fire Club, Dublin, about 1735. Toasting at the Kit-Cat Club, which met at a Temple Bar tavern, focused on an annual Belle of the Year; 'her name is written with a diamond on a drinking glass … to remind her her value is imaginary and her condition is frail'.

scoundrels, taverns were also venues
for judicial hearings, while sharing a
convivial dinner was an essential part of
the assize. Socializing was mixed with
business. The great eighteenth-century
entrepreneur Matthew Boulton wrote
home to the Soho Manufactory,
despairing that he had succeeded only
in making himself unpopular when he
suggested to the worthies of Cornwall,
with whom he was trying to do
business, that they postpone their
drunken feasting until *after* their
meetings had taken place.

Innkeepers, usually from the ranks of
the middling sort and with responsible
roles in their communities, equipped their premises lavishly.
The drinking room of a Southwark brewer in 1724 had red
curtains, mirrors and framed prints, and he served his
customers from porcelain and silver punch bowls. A gilded
sign hung over the door and musical instruments were at
hand for parties. He offered porter, beer, spirits of various

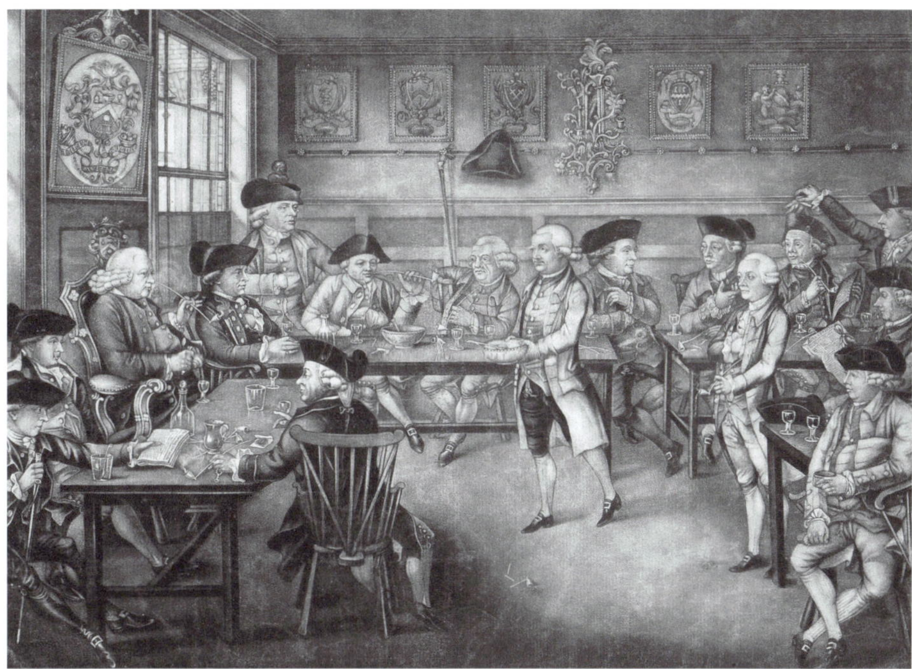

Above: The Court of Equity, 1779. Printers, booksellers and other trades
shown in their regular tavern meeting room, hung with the arms of their
respective guilds.

Below, left: Beer jug urging abolition of slavery, about 1800. Home-brewed
beer was always appreciated. Elihu Yale, governor of Madras in 1682, was
sent 'four Rundletts of Sandpatch Ale from Erddig in Denbighshire.' His
return gift was a Japan screen and local mango chutney.

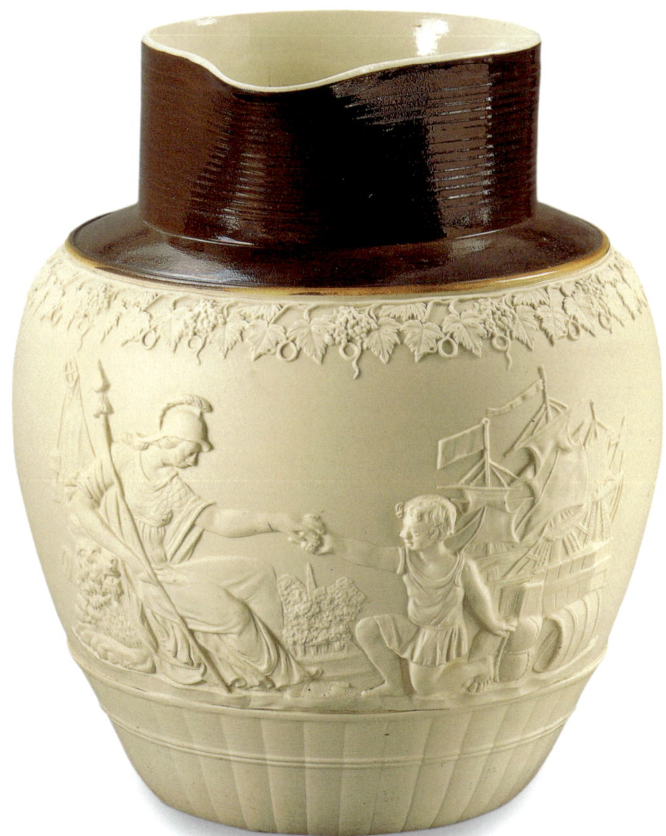

kinds and wine, as well as cordials. He also gave the all-
important credit facilities to his customers and suppliers,
from a few pence to several hundred pounds; the best of
society were not always the best at paying their bills.

These establishments were the natural setting for a new
form of civil conviviality. Thousands of clubs and societies,
set up for every conceivable purpose, were a feature of the
English-speaking cultural landscape from the late
seventeenth century. Foreshadowed by medieval religious
and trade fraternities, drinking clubs for educated men
emerged in the time of James I. Ben Jonson was a founding
father; he and the 'sons of Ben' were key to the tradition,
consciously basing their rules, practices and purposes on
classical examples such as Plato's *Symposium*. It was
common to accuse someone of drinking 'like a Greek' – that
is to excess.

During the Civil War and especially after the Restoration
drinking clubs and associations grew enormously in number.
By the 1720s, one contemporary remarked, London had 'an
infinity of clubs or societies for the improvement of learning
and keeping up good humour and mirth.' There were plenty
of locations for meetings – St Giles-in-the-Fields had

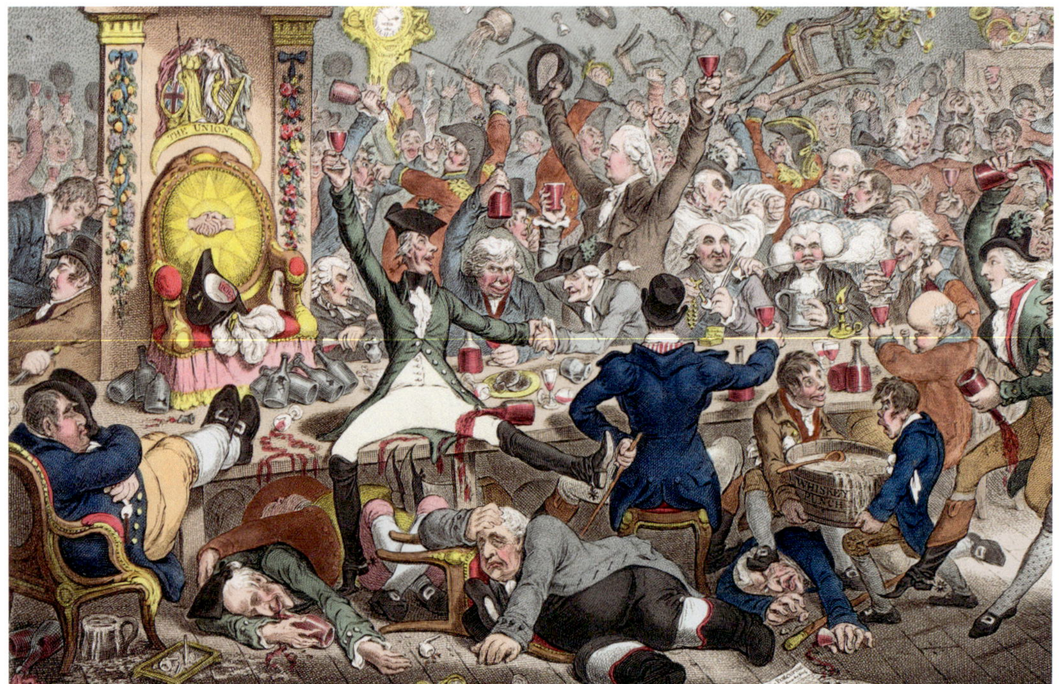

Loss of dignity, damage to furniture, vessels and clothes, corruption of servants and the degradation of identifiable public figures show the impact of too much drink on a London club meeting. Symbolically, the President's chair is empty and all restraint has vanished.

between 2,000 and 3,000 houses, of which 270 were licensed as inns or alehouses, and the metropolitan divisions of Middlesex, Finsbury, Holborn and the Tower had 2,470 licensed houses in 1723 – and providing a room and refreshments for club meetings was good business.

Peter Clark has calculated that about 25,000 clubs and societies were set up across the English-speaking world by the end of the eighteenth century; several thousand and over 130 different types operated in Georgian Britain. Although clubs sprang up in towns all over the country, the greatest numbers were found in London, Edinburgh and Dublin. Whatever the purpose of such clubs – they included Alumni, Artistic, Benefit, Debating, Gambling, Horticulture, Literary, Masonic, Medical, Musical, Neighbourhood, Philanthropic, Political, Professional, Religious, Sporting, Scientific and Tradesmen – they almost always involved alcohol. Drinking, feasting and singing were the primary purposes of the Hell Fire Club, which met at the 'George and Dragon' in West Wycombe, and the foxhunting and feasting club called the Shropshire Fraternity of the True Blue, which met at the 'Raven Inn' in Shrewsbury. Horace Walpole claimed that while the nominal qualification for membership of the Society of Dilettanti was having been to Italy, 'the real one [was] being drunk'. This burgeoning social activity led to a veritable industry of special vessels, objects and art that enhanced rituals of good fellowship and social cohesion.

Not all clubs were élite, drunken or based in London. In Sheffield a club of cutlers held weekly meetings and in

Birmingham sober debating societies included a famous political club; Freeth's Coffee House met in political balladeer John Freeth's 'Leicester Arms Tavern' in Bell Street. In 1774, at the time of the Wilkes and Liberty campaign, this club became involved in a political dispute between the pottery workers at Worcester and the Corporation. The pottery workers, incensed at the Corporation's illegal attempts to remove their voting rights, offered a gift of a large silver plate to the local Wilkesite candidate Sir Watkin Lewes. The images on it celebrated popular political causes of the previous 12 years including temperance, represented by a satyr tipping liquor out of a goblet and attacks on taxes recently imposed on Cider and Perry – which had been bad for the farmers and trades around Worcester. It is no surprise that commemorative 'Wilkes and Liberty' teapots and other politically-focused objects emerged from the Worcester potteries.

Clubs were normally male preserves, although at least one gambling club was licensed to a female aristocrat and lower class, provincial clubs, such as Freeth's, did admit women. In the eighteenth century most clubs met upstairs in a simply furnished 'Chamber for Company', apart from the general hub-bub of the public rooms, open to anyone who could pay for entrance (sometimes as little as 6*d.*) and who shared members' interests. By the nineteenth century élite clubs withdrew from this over-public forum and built or hired their own palatial premises, becoming 'private' institutions, whose histories are now being written. Practices such as 'black-balling' developed – based on medieval Venetian voting

The Toby Jug

Of all British drinking vessels produced in the eighteenth century, none have come to symbolize the stubborn island breed as much as the toby jug. Like Sir Toby himself, its pedigree is both long and entirely unambiguous: for the inventor, place and date of its birth can be pin-pointed. Even without documentation, its origin would inevitably be laid at the door of one of the talented Wood family of Burslem in Staffordshire, most probably of Enoch Wood whose triumphs included the well-known bust of John Wesley, which he modelled from life. Fortunately, however, the matter is proved beyond reasonable doubt by the surviving sales account books of the potter John Wood for the period 1783–7, which include the first mention of a single toby jug supplied in September 1785, an example with expensive over-glaze enamelling priced at 4s. The seed was sown, the imagination of the public captured, and two months later consignments of enamelled, colour-glazed and plain bluish-white

This popular stereotype of the jolly English yeoman, strong from drinking home brew, became anachronistic after 1800, as unemployment, poverty, enclosures and wartime taxation undermined traditional rural life.

'China Glazed' versions were being sent to the Oxford Street retailer George Phillips priced at 3s., 2s. 6d., and 2s. respectively. No doubt in order to satisfy a Christmas demand for this novelty, Phillips ordered more in December, and a further 36 in January 1786 – by which time its huge popularity, no doubt coupled with stiff competition from the many other contemporary Staffordshire potteries, had forced the price of the coloured version down by 3d., or 10 per cent.

The jug itself, brilliantly linking liberty, hunting, beer and tobacco with a form that was both functional and decorative, was not however an entirely original creation of the Wood family; rather, it represents a translation into three-dimensions of the engraving by Robert Dighton of Sir Toby Philpott, which accompanied verses entitled *The Brown Jug* by the Rev. Francis Fawkes, published in 1761. Here, the uncompromising, hard-drinking hunting squire was depicted cradling an archaic globular brown stoneware mug filled with foaming ale, together with the inevitable clay pipe, his blunt politically incorrect attitude specifically designed to appeal to lovers of a Merrye Englande threatened not only by the Age of Enlightenment, but also by the approaching French Revolution. Even his end had a certain rugged nobility, when 'His breath-doors of life on a sudden were shut, And he died full as big as a Dorchester butt'. In popular culture his antecedents may perhaps be traced back a further 150 years to Sir Toby Belch in Shakespeare's *Twelfth Night*, which in turn must surely have informed the choice of Uncle Toby as the hero of Lawrence Sterne's *Tristram Shandy* of 1768.

The aura of flawed but dogged patriotism that these jugs exuded was later to be exploited whenever Britain was under threat, substituting the figures of later heroes such as the sailors of the Napoleonic Wars, or the political, military and naval leaders of the First World War. It is safe to say that production of this old favourite is here to stay.

practices – leading to would-be members being rejected. This represented a considerable social pressure for the aspirant upper classes, since to be 'black-balled' could spoil a young man's career or marriage prospects

Few clubs have left many traces of their existence. Clues emerge as they were satirized in the press, mentioned in passing in letters and diaries, or because tickets or sermons record that they had feasts. Some of the strangest clubs – such as the Knights of the Golden Fleece, the Purple Society, the Lumber Society, the Cathcembytes – survive only in name. Others, such as the Bachalmotegottingeld club, which met in various Holborn taverns from 1683 to1762, have left merely an account of their rules.

Early political clubs such as the Whig Kit Cat Club or Mug-House clubs are better understood. Members of the Kit Cat Club were immortalized in portrait and print by one of their members, Sir Godfrey Kneller. Clues to the Whig 'Mug House' clubs, which participated in the violence that erupted all over London when the Hanoverian George I became king, appear in the court records of the Old Bailey. Stoneware mugs were hung as identifiers in the windows of Mug House establishments. But perhaps one of the most curious, drunken and violent clubs was the Mohawks. Young bloods who drank to excess and then went out into the streets to attack innocent passers-by with the utmost cruelty, they sound like extreme versions of today's 'happy slappers'.

Gardens and Grottoes

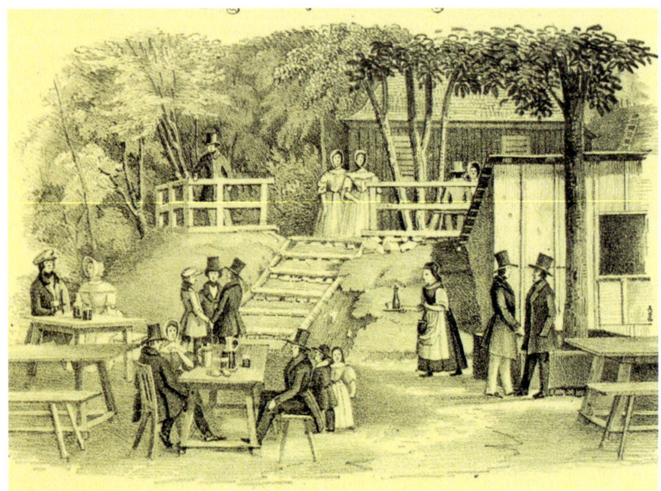

We went to refresh ourselves in a cave or grotto cut into the rock, where they keep at all times several pipes of a very good beer, which for that reason is deliciously cool, and which they serve in the English fashion in wooden tankards … an ancient curved copper horn … many visitors drink from this horn as a curiosity.…

A Dutchman, William Schellinck, visiting Dover Castle, July 1661

*D*rinking beer or wine chilled, in a time before refrigeration arrived in the mid-nineteenth century, was a pleasure requiring some effort. Cellars where the beer was maturing in barrels, grottoes with wells attached for cooling wine or a garden pavilion with a fountain and marble cooler were all settings described by appreciative and thirsty travellers.

William Schellinck admired the Hampton Court banqueting house, 'a pleasant octagonal summerhouse which stands upon a higher level, from there one has a view over the whole garden…. Below this place a deeply vaulted wine cellar'. Because barrels of beer and wine kept well in cool conditions, cellars often doubled up as informal drinking locations, as another traveller, the gently-born and inquisitive Celia Fiennes, commented, when she visited Nottingham in 1697: 'for Cellars they are all dugg out of the rocks and so are very coole, at the Crown Inn is a Cellar of 60 Stepps down all in the Rock like archworke over your head, in the Cellar I dranke good ale'. At Newby Hall, she noted 'very good Cellars all arched and there I dranke small beer four years old – not too stale very clear good beer well brewed'. Daniel Defoe confirmed her high opinion of Nottingham's

beer, and commented on the benefits of its geological oddity, deep tunnels and caves where beer kept well.

Hospitality was customarily offered to visitors in cellars. When at Audley End, Suffolk, Schellinck was 'led by the keeper into the cellars which have very high vaults resting in pillars…. The cellarmaster entertained us with very good old beer and showed us in another room a model of the whole house, very prettily made of paper as a hobby by a nobleman'.

Banqueting houses for summer parties were often located on a river bank as at Hampton Court or somewhere high up to catch a refreshing breeze and purer air, away from the smells of a large household, such as the small rooftop chambers at Longleat or Lacock, or Wilton's 'Pavilion in Italian style, its fount with marble statues … the stone railing on top like shells to let the water flow'. Grottoes ornamented with shells became popular as cool places for relaxation in the early seventeenth century; they might be attached to the house and facing north, as at Woburn, or

Above left: Scene of outdoor drinking, Austria, about 1825, from a small poster for a wine garden.

Below: Pleasures of conviviality. The glass of beer, banded to show each drinker's share, is followed by a small glass of wine and accompanied by oysters, typical of the salty piquant snacks that stimulated more drinking.

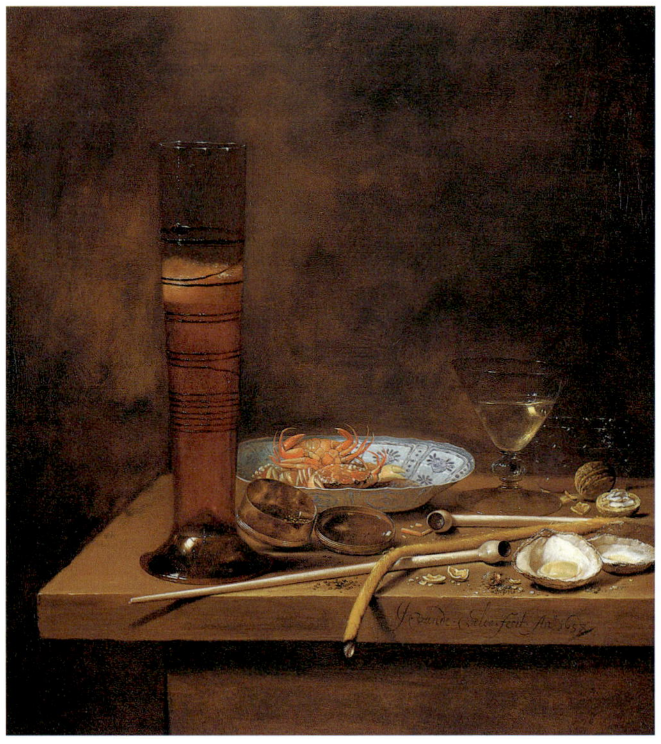

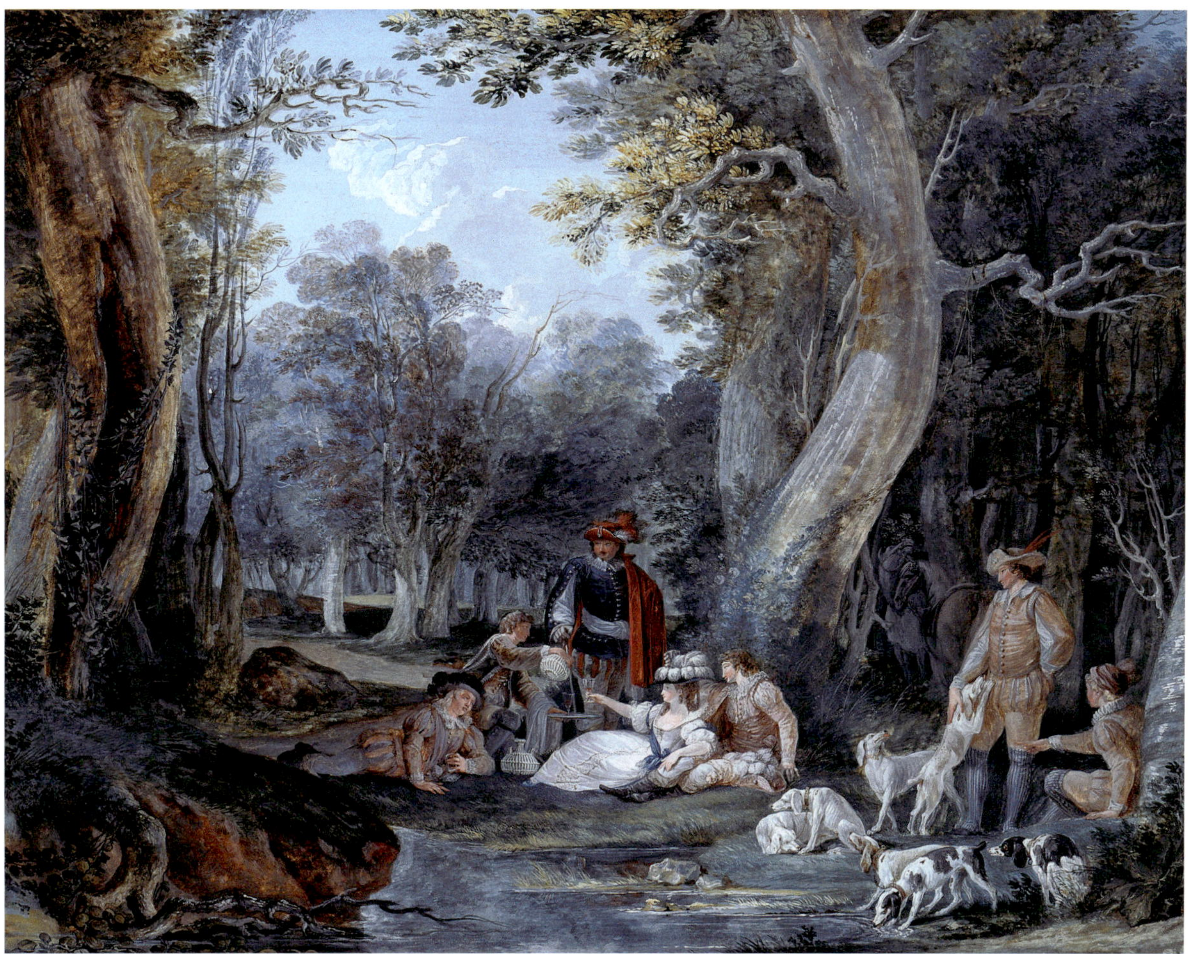

Party by a pool. At Wanstead House, Surrey, in 1735, eight pleasure houses – including a toy castle by a lake – offered places to take refreshments.

created in the garden, as Mrs Delaney did for the Duchess of Portland at Bulstrode.

Drinking out of doors is always a pleasure and although descriptions of specific occasions are rare, garden pavilions at Versailles, Marly and Mme de Pompadour's châteaux were the settings for light refreshments after dinner, which was taken at mid-afternoon. The theme of a garden, or its temporary evocation in treillage, was typical of French fêtes and public events. For example in December 1745 for the engagement of the Dauphin, the Place Vendôme was filled with halls of treillages with buffets offering goose, bread and wine, lit by chandeliers, where the citizens of Paris danced in celebration.

Strolling out to a fantasy structure, a porcelain Trianon, a dairy, or a tree house, to drink some wine and talk was a refined delight, and one that stimulated its own artefacts. A miniature version in silver-gilt of a three-storey hunting tower built for the Dresden court around 1610 formed a rather unwieldy drinking vessel for the elector to share with his male intimates. For the Viennese court in the 1720s, the porcelain maker Du Paquier created a novelty drinking toy of an elephant surrounded by dancing peasants holding cups. Inside the elephant is a reservoir for liqueurs with a spigot in his trunk, and each figure could be swivelled around to fill the cups in turn. Now in the Hermitage, it was probably a present to the Tsarina around 1739, at the time of the treaty of Belgrade. This delicious toy, descendant of Augsburg automata of a century earlier, may well have been devised for drinking Tokay in an 'Indian' garden pavilion.

Tudor citizens enjoyed cider and syllabubs in commercially-run gardens around the City, and in the eighteenth century Londoners took boats down to Greenwich or up towards Chelsea. In 1762 Vauxhall Gardens offered drinks at all prices and for all tastes, from 'Table Beer a Great mug' at 4*d.* a quart to Champagne at 8*s.* Burgundy was the only other French wine; Old Hock, with or without sugar, was 5*s.* a bottle. A quart of arrack, that is punch, cost 8*s.*; it was this potent mixture that drowned the hopes of Becky Sharp in *Vanity Fair*.

Even today the leafy bowers for drinking young German wines are a delightful feature, escaping the noise and bustle in the outskirts of towns in the wine-growing regions.

On the Move: hunting and travelling

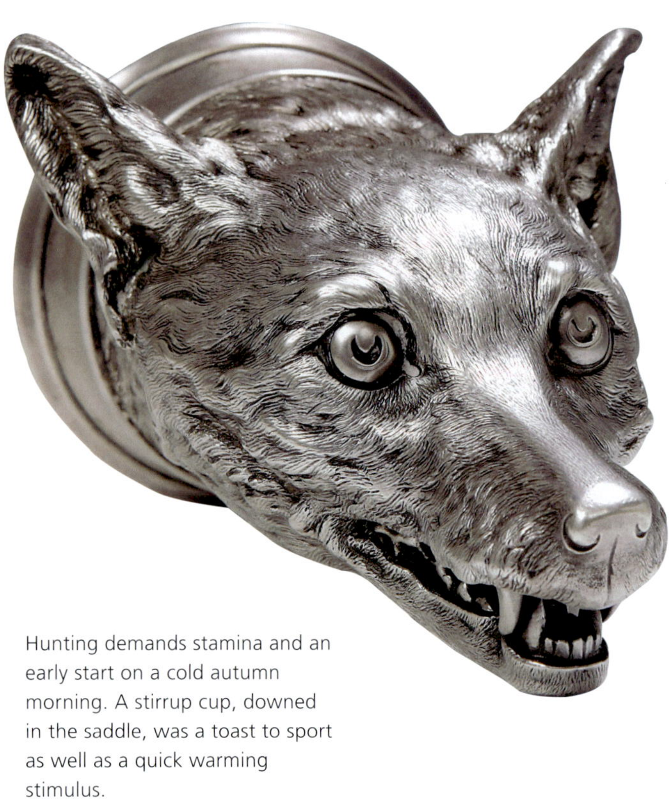

Hunting demands stamina and an early start on a cold autumn morning. A stirrup cup, downed in the saddle, was a toast to sport as well as a quick warming stimulus.

I am heartily sorry that the Burgundy I have sent your Ldshp has proved so very bad. When it went from hence it was thought exceeding good by everybody that tasted it … but what has occasioned this wine to take this unlucky turn I can't imagine….

JAMES UNDERHILL, LORD FAIRFAX'S WINE MERCHANT, TO HIS CLIENT, 1763

From antique depictions of Bacchic processions with revellers carrying barrels on poles wreathed with vines to the New York stretch limousine complete with its well-stocked drinks cabinet and refrigerator, alcohol has been consumed on the move. Some drinks, particularly more sought-after kinds, might also travel a considerable distance before they reached the consumer. This applied particularly to wine and spirits, whether based on grain or grape. In Europe, beer was generally made locally; 'small beer' was usually brewed by the household for its own use, and drunk with meals very like water. Some beers, though, were regarded as luxury products, and exported accordingly. Good quality English ale was being shipped to France and beer from Brunswick and Bremen was in demand in the East

Indies in the late seventeenth century. In the 1760s the enterprising Samuel Whitbread sent his beer overseas not only to Dublin, Amsterdam and the Baltic, but also as far as the British settlements in India and Canada.

Because vines in Europe flourished only south of the mouth of the river Loire, wine from Spain, France and the Mediterranean had to be transported to northerly markets. Although the Romans had used pottery amphorae and Spain continued the tradition of storing wine in massive jars, wine could be transported most easily in the barrel. So it was invariably drunk young, since there was no reliable means of protecting it from detrimental contact with air. The older the wine, the more likely it was to be oxidized. In 1500 a cask of old Bordeaux cost only 6 livres tournois compared to 50 for new. Barrels, made of wooden staves bound with iron, were not perfectly airtight, making it imperative that the contents reached their destination swiftly. In 1539 the Duke of Mondejar advised Emperor Charles V that if too large a quantity of wine were bought for the navy there was a risk that it would 'turn into vinegar', making it preferable that the barrels 'remain with their owners rather than your Majesty'.

Fortified wines owed their origins and popularity to the complications of transporting wine. Port, sherry and Madeira, for example, could cope with longer periods in the barrel and the vagaries of sea passages. This also explains why the English fleet in the West Indies adopted rum as a daily drink in the 1680s. Beer spoiled after a few weeks at sea and wine soured. In the seventeenth century, wine was transferred from the barrel to a stone or (from the 1640s) glass bulbous bottle with a loose cork stopper by the household, and then decanted again into something more elegant for the table. However, in the eighteenth century, the concept of vintage wines was becoming established. Following the introduction of tight corks, bottles gradually changed from onion-shaped to the familiar modern profile, both to allow them to be stored in horizontal racks and so that they could be safely shipped and stored for the wine to mature in the bottle.

Once drink in the barrel or bottle had reached its final destination, there were plenty of occasions that entailed refreshment on the move, and specialized equipment evolved to facilitate this. Travelling could be a risky, chilly and tedious business, often with no guarantee of either comfort or cleanliness en route. It was safer to take one's own dining, drinking and personal equipment, particularly for those of higher rank for whom the correct observance of ritual and etiquette defined their public status. Specific travelling services were one response to this, furnished with everything needed for personal comfort, including drink. By the late seventeenth century their popularity in aristocratic circles was such that in Augsburg in south Germany specialized workshops were producing such sets, usually combining silver-gilt, Meissen porcelain and glass, all contained within a purpose-built case. As the eighteenth century progressed, the services increased in size and complexity. One made in Paris for Queen Marie-Antoinette in 1787–9 had 117 objects, ranging from a bedpan and pestle and mortar to tableware. By the nineteenth century, the apogee was reached by manufacturing retailers such as Biennais in Paris, whose cases combined the functions of

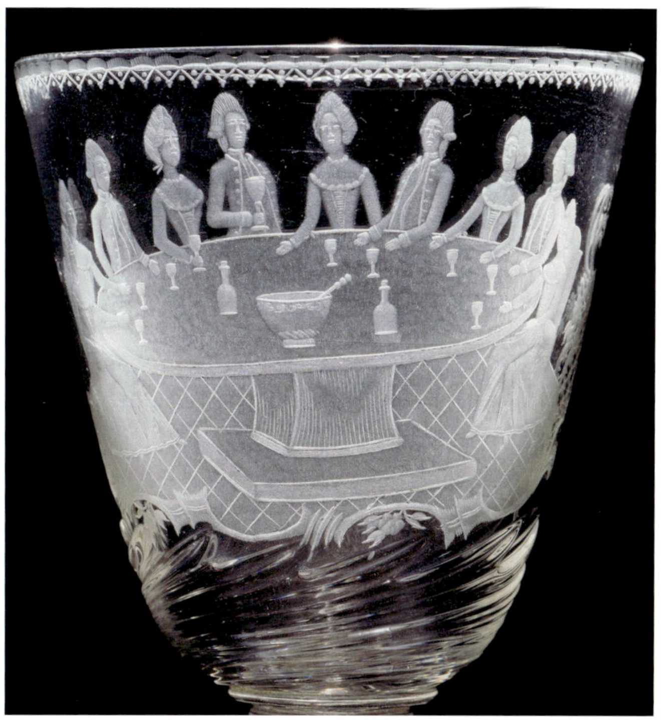

Plebeian drinking habits were transformed by the legal right, granted in the 1830s, to buy small quantities of drink to take away. These stoneware bottles for gin, formed as popular reform-minded politicians, could be carried in a pocket and refilled.

Right: This Norwegian glass shows a party drinking out of doors without servants, a rare occasion of informality.

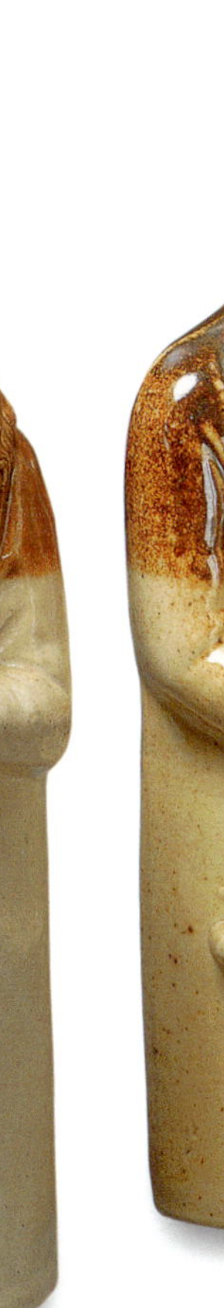

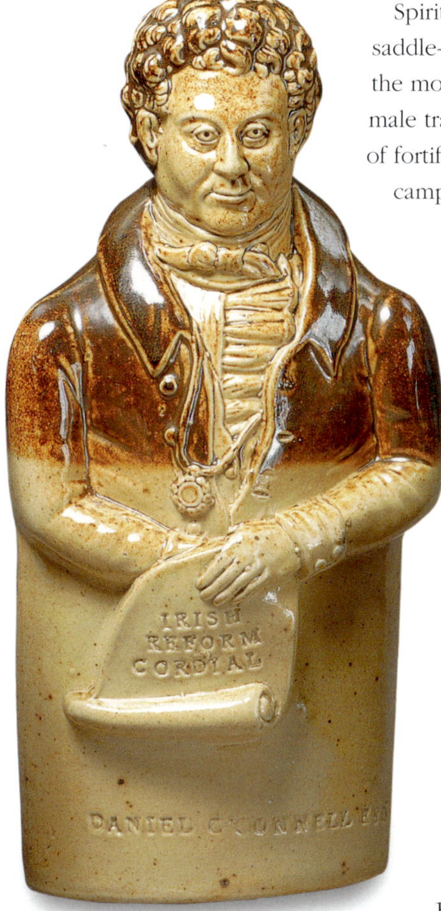

dining room, bedroom, bathroom, office and bank in miniature.

Spirit flasks, small enough to be carried in the pocket or saddle-bag, were another response to the need to drink on the move. The grandest were made in silver, to enable the male traveller, huntsman or soldier to take sustaining swigs of fortified wine or spirits, whether journeying or on campaign, without the need for a servant, glass or corkscrew. An early English example, from around 1690, belonged to Charles Beauclerk, 1st Duke of St Albans and the illegitimate son of Charles II and Nell Gwyn, now in the V&A. This has a removable gilt-lined sleeve, to act as a beaker. By the nineteenth century, these flasks, now known as hip or pocket flasks, had become an indispensable accessory, produced in silver, pewter or even tinned brass. At a lower social level, novelty stoneware flasks for gin appeared in the 1830s, as part of an expanding market for drinking away from home or the tavern. Costing a few pence, they could be taken to retailers to be re-filled and slipped into a trouser pocket.

Alcohol also played an essential role in hunting parties and picnics. Paintings give some sense of the elaborate nature of hunt banquets although the illusion of informality was created through spreading a white damask tablecloth on the ground. Both red and white wine and beer would be drunk

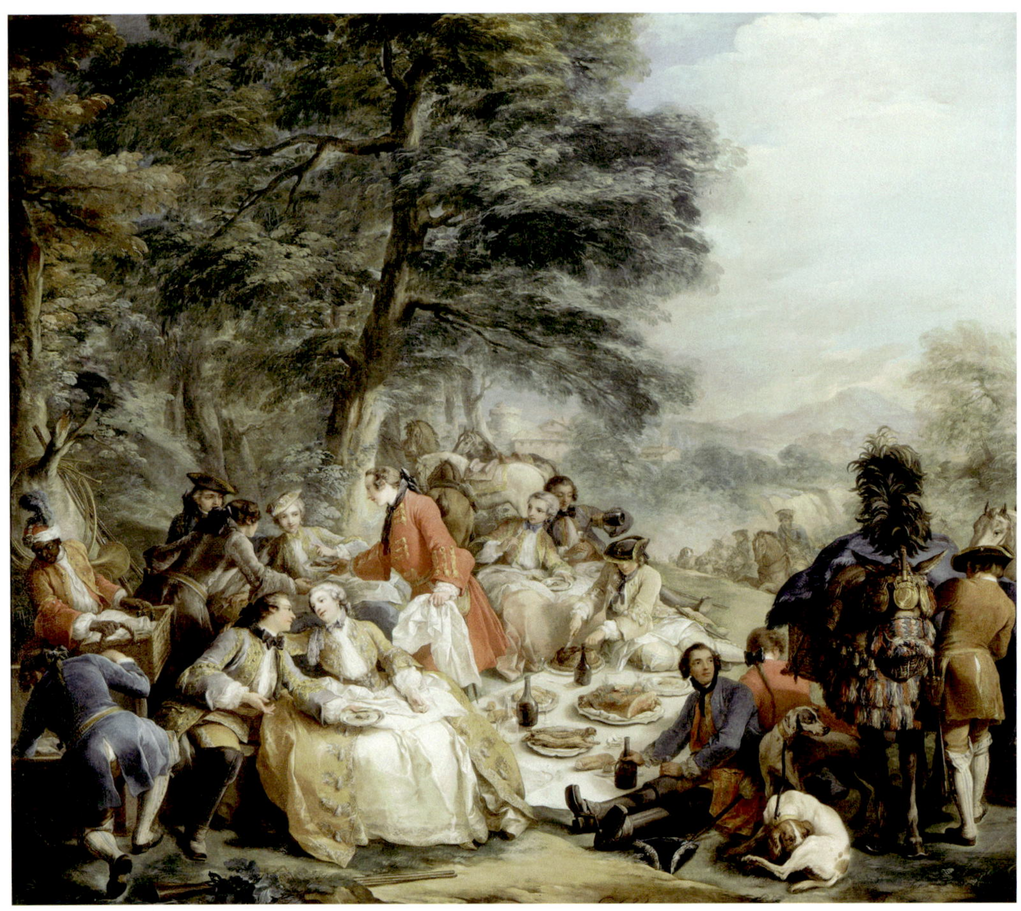

Left: A French hunting party stops in a glade to eat and drink. Attendants bring wine and eatables, to be enjoyed without the formal constraints of the dining room.

Below: Peasants under a vine trellis are preparing a picnic, a joyous subject for a fan leaf, taken from a painting by Teniers.

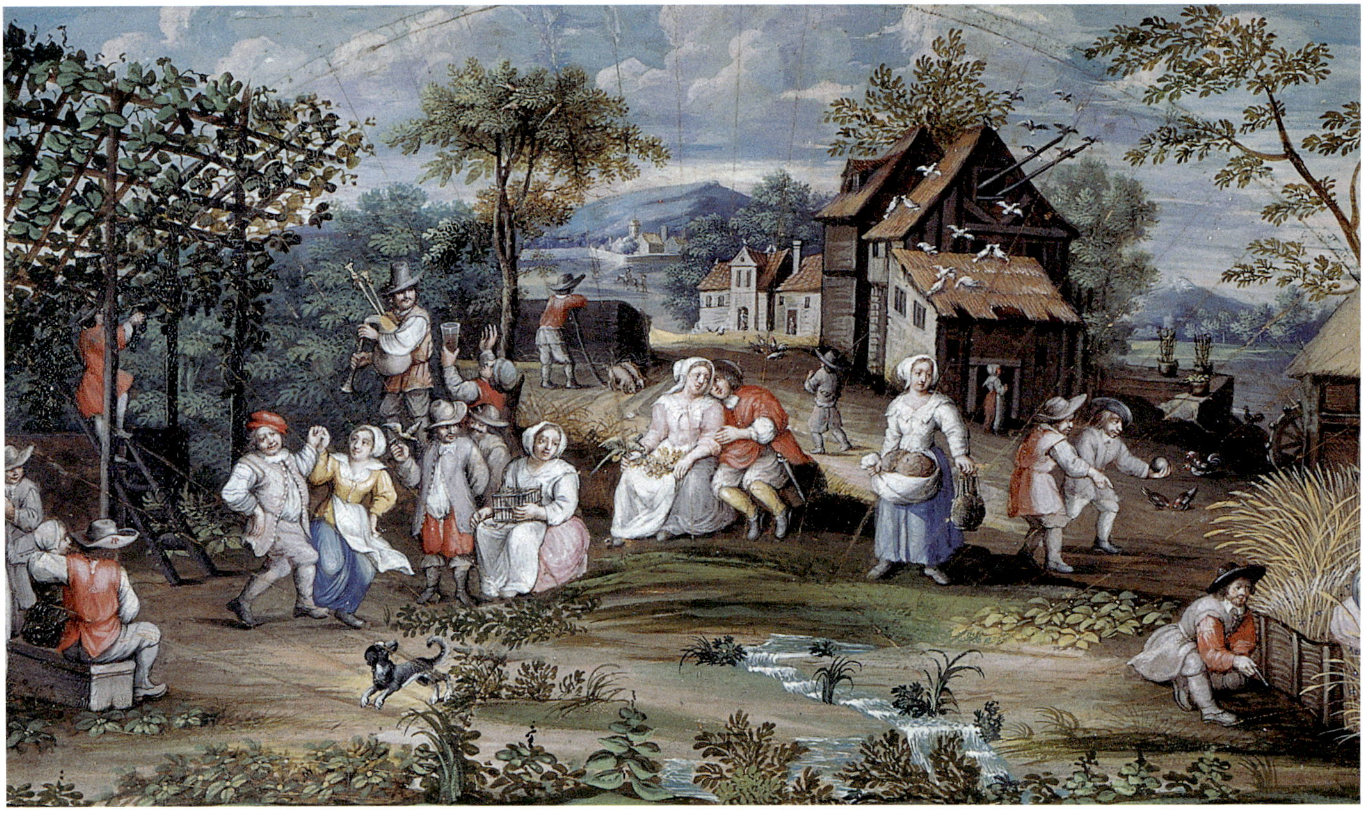

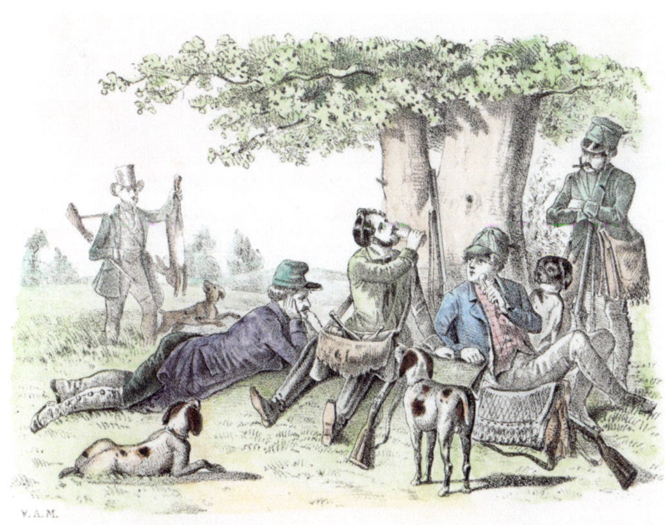

in the 1890s echoed the concept of the hotel on wheels. One writer likened a train on the Pennsylvania Limited service to 'a snug retreat … [which] serves all the purposes of a club', complete with refreshment buffet serving a range of drinks, smoking car and library. Such sophistication seems a far cry from the modern equivalent – the packed bar of a twenty-first-century intercity express train.

Left: A German shooting party. In 1828 an Irish hunt breakfast for 'six or seven sturdy squires, they do not think much but their life is all the more gay and careless', included whisky, wine, eggs, beef steaks, mutton, kidneys – good fuel for an October day.

Below: Travelling spirit flask and cup made for the Duke of St Albans, around 1690. Brandy, distilled from the cheap, poor white wines of France, was more portable, more profitable and kept better than wine.

on such occasions. William III had a supply of Champagne carried out on hunting parties, imported for him by the Duke of Celle. The tradition of a sustaining meal half way through, or a celebration at the end of a day's sport, continued into the twentieth century, as can be seen in photographs of hunting parties in late nineteenth century France where the hunstmen mingle with followers over hampers containing bottles of wine. The shoot lunch is the modern equivalent. In Britain, fox hunting, which became increasingly popular in the late eighteenth century, created its own traditions, such as the stirrup cup, offered to riders at the meet before the hunt started and, as its name suggests, taken on horseback. Usually consisting of sherry, Madeira or port, the intention was to fortify the participants for the rigours ahead. This led to a market for personal cups, often in appropriate novelty forms. The pleasures of the chase also spawned a trend for commemorative cups in which to toast exceptional achievements on the field. One rare porcelain survival, made in Meissen in 1741, depicts the culmination of a notable stag hunt of 1739 on its cover.

By the late nineteenth and twentieth centuries, more and more people were travelling, creating more opportunities for peripatetic drinking. In the case of the railways, for example, it was possible to travel in great luxury with every conceivable comfort. Private carriages could be attached to public trains, and in 1853 the American Harriet Beecher Stowe observed that it was possible for the traveller to have 'all his domestic manners and peculiarities unbroken. In fact, it is a little compact home travelling about.' Travellers in Pullman carriages

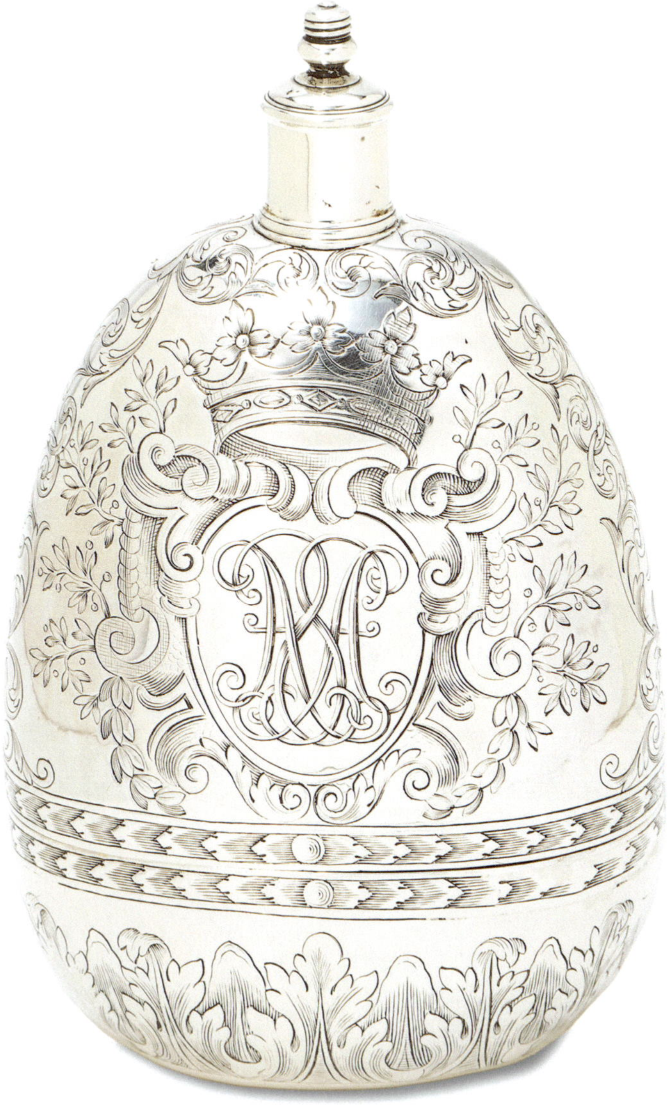

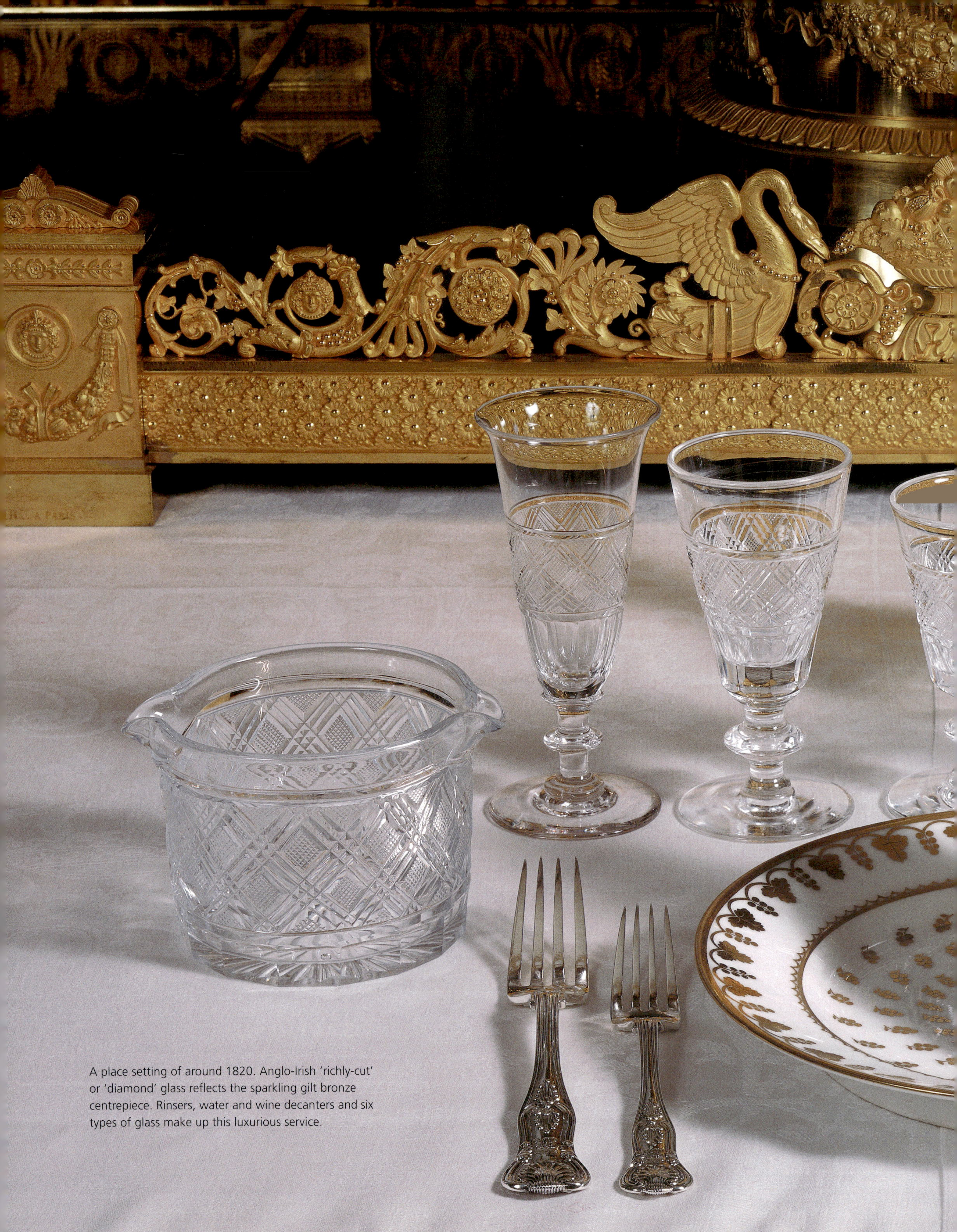

A place setting of around 1820. Anglo-Irish 'richly-cut' or 'diamond' glass reflects the sparkling gilt bronze centrepiece. Rinsers, water and wine decanters and six types of glass make up this luxurious service.

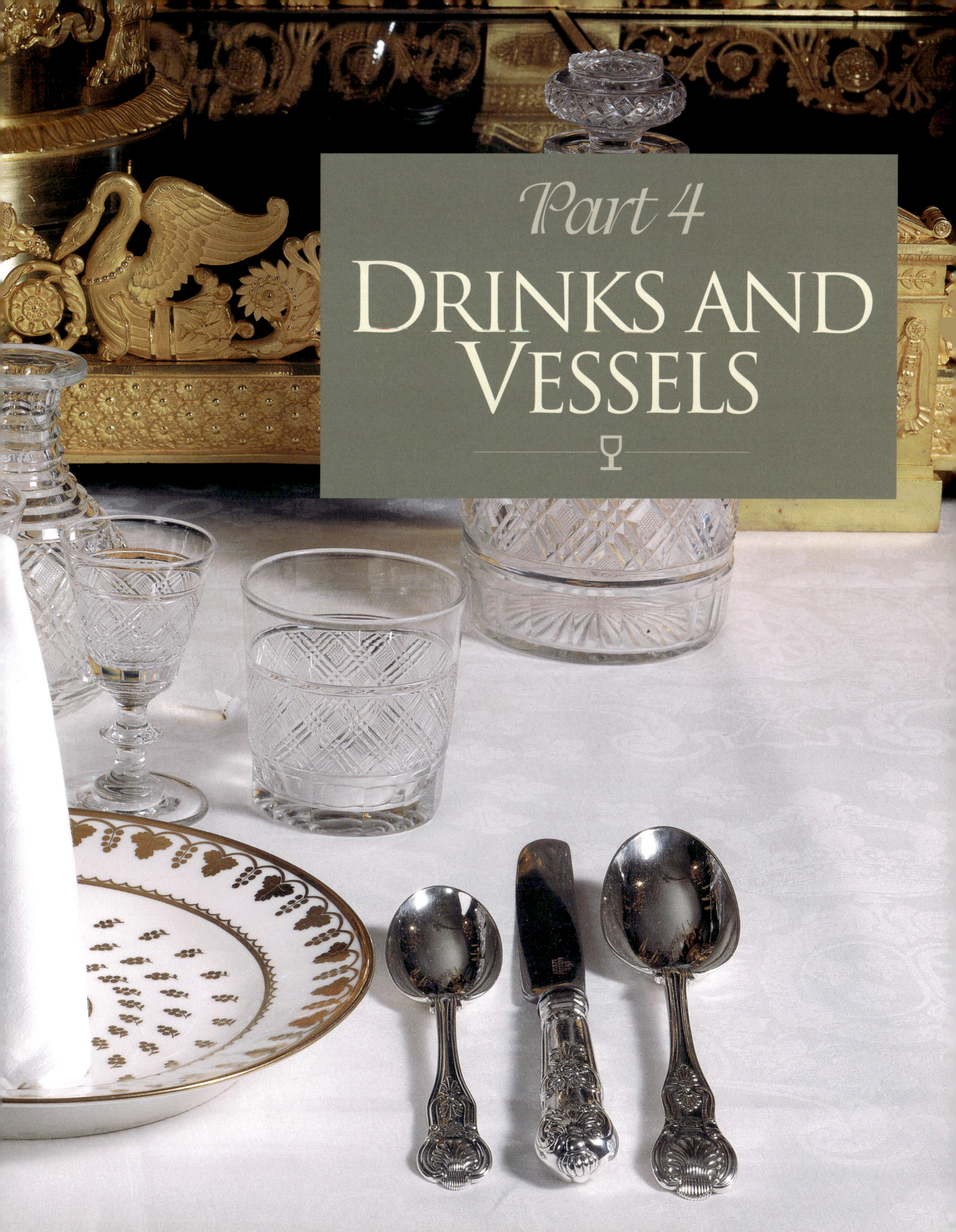

Part 4
DRINKS AND
VESSELS

Which Glass?

Certain drinks, whether alcoholic or not, demand vessels that match their characteristics. In time the association becomes so strong that receptacle actually seems to affect taste. Such refinement, however, was largely unknown to Tudor England where crude greenish *waldglas* (forest glass) was a comparative luxury, the Venetian *cristallo* was reserved for the wealthy, and brown stoneware beer mugs from the Rhineland with local earthenware copies and turned ashwood cups sufficed for the rest. The English imported most luxury goods, including wine from France, Spain, Portugal and Italy; so that when Cromwell's sober austerity was succeeded by the conspicuous consumption of Charles II, a repressed desire for 'designer' glass immediately expressed itself. No fewer than 400 different types were ordered by the London dealers Michael Measey and John Greene from their Venetian glass supplier Allesio Morelli in the years 1667–72.

After George Ravenscroft developed expensive but brilliant and durable lead-glass in the 1670s, the range narrowed rather than expanded; he supplied only glasses for beer, claret (the generic name for red) and sack (white) to the Glass-Sellers' Company in 1677. From this severe rationalization, simply the result of difficulties with working the new slow-cooling molten glass, evolved the classic English baluster glass.

The origin of wine consumed in England changed according to the politics of the day. Hostilities with France encouraged alternative trade with Spain and Portugal, culminating in Methuen's Port Wine Treaty in 1703, which reduced the tariff to £7 per tun compared with £55 for French wine. Apart from Champagne and of course brandy, French wine remained comparatively scarce. After Waterloo, while port and sherry retained their popularity, enthusiasm for the vintages and the cuisine of France kept pace with the meteoric rise in Britain's wealth. Fittingly, in the late nineteenth century wine connoisseurs made the first historical studies of English glass.

Naturally the drinks available at different periods had a profound effect upon glass-makers. In the eighteenth century, to compensate for the odd and limited selection of wines imported in barrels and subsequently bottled, blended or adulterated by wine merchants, glasshouses developed shapes and stem-forms, both moulded and cut, exploiting lustrous lead-glass and unique to Britain. But this miscellany was replaced towards the end of the century by a new bourgeois lust for conformity, for smart matching table settings of bright-cut silver flatware, English blue-and-white

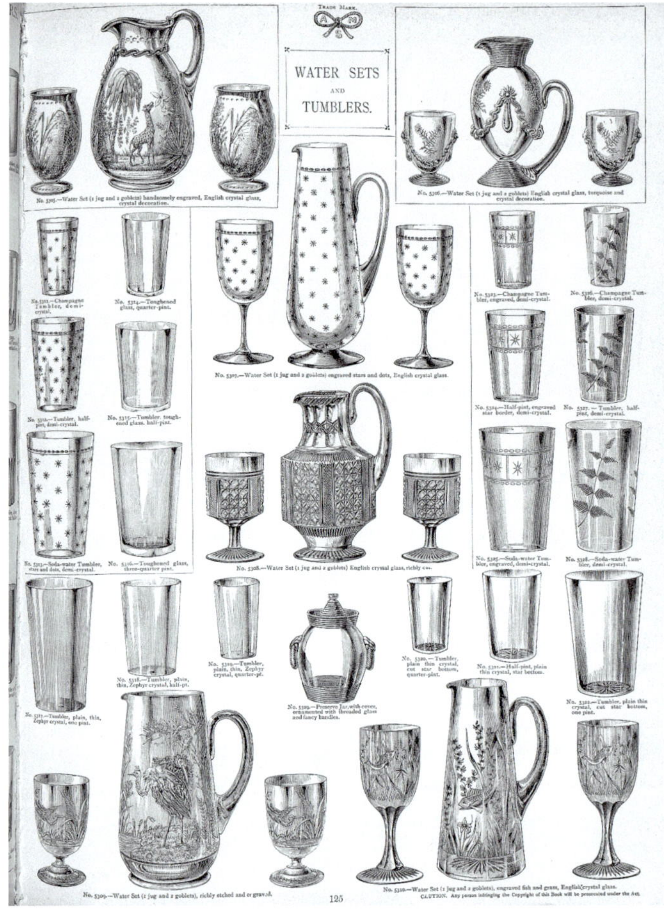

Page from a Silber & Fleming retail catalogue, c.1882, showing a set of matching glasses.

porcelain and Wedgwood's creamware. Glass-makers responded from about 1780 with elegant suites of drinking glasses and decanters, with identical cut patterns, and standardized shapes and capacities. With the application of steam power to glass-cutting, and triumphs such as the magnificent table service made for the Prince of Wales in 1806–8 (p.33), these sets tended to grow ever heavier until the massive *broad flute* style of the 1840s perfectly complemented the heavy dining tables of the period. In the 1870s French tastes were reflected in thin glass and delicate engraving, with comforting Regency solidity reviving later.

Measey & Greene noted the exact purpose – claret, white wine, beer, cider, brandy – of every type they ordered by 1670. From about 1685 London glassmakers applied the starkly simple baluster stem equally to small spirit glasses or the huge lidded ceremonial goblets for communal drinking at the Stuart and Hanoverian courts – the 'State Glasses &

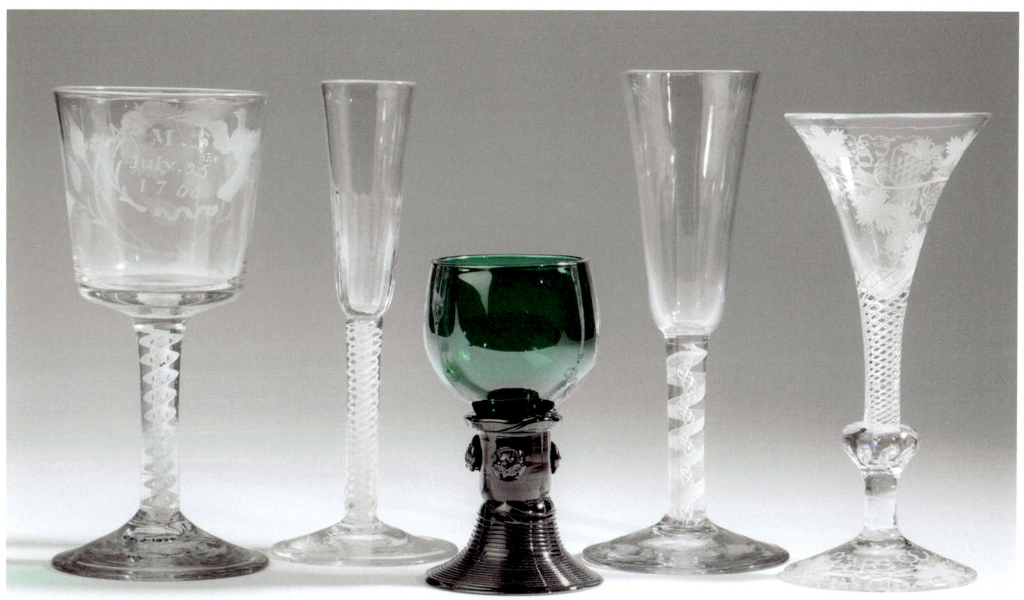

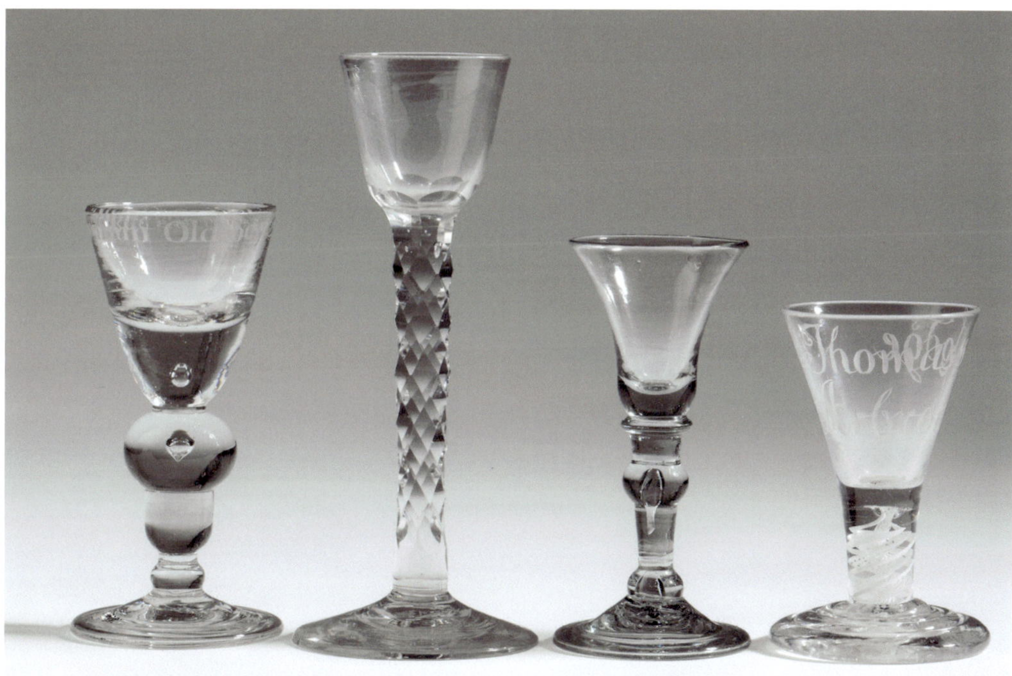

Covers' listed at Hampton Court in 1736. But from about 1715 the hollow mould-blown Silesian stem ushered in different stem types, such as the air-twist of about 1730, either as a two-piece glass with drawn trumpet bowl or a three-piece with separate bowl, stem and foot. Many thousands of cheaper plain glasses at under a shilling must have been sacrificed in the mass breakage that accompanied parties.

By the 1750s bell or trumpet shaped bowls were being replaced by egg or bucket, with air-twist, white enamel-twist and facet-cut stems. By about 1800, the simple neo-classical ogee profile was adopted for practical and neat glasses with plain cut or short knopped stems. As today, the smallest were for port and sherry.

Tiny eighteenth-century wine glass bowls, hardly deep enough to accommodate an engraved vine-leaf border, sufficed when servants with salvers constantly re-filled glasses from the sideboard. Later in the century, decanters or neat cylindrical bottles were placed on the table for guests to circulate. For wine diluted with mineral water imported from Belgium and Germany, special large glasses were used.

Red or white would generally share the same glass, to be rinsed at the table either in a communal monteith or towards the end of the eighteenth century in a small double-notched rinser. One notable exception, however, was Hock, traditionally drunk from special goblets based on the greenish German *Roemer*. A minority taste for Rhenish wine is clear from a number of apparently unused Ravenscroft *Roemers* while every one of his mass-produced wine or beer glasses has perished. Champagne, too, was a special case; at 5*s.* a bottle in the 1750s the most expensive wine of all, it demanded a flute glass to conserve its *petillant* sparkle. Bottle-fermented *méthode champenoise*, developed in the nineteenth century, increased the fizz; this may have triggered the invention of the wide-bowled *coupe* glass, an instant success after it was introduced in the 1820s.

Spirits had specific glasses, for example the brandy tumblers illustrated by Measey & Greene, the small but sturdy baluster dram glasses, and the weedy gin glasses that so eloquently evoke the desperate poverty of Hogarth's *Gin Lane*. Special glasses with heavy feet, firing glasses, were made for reinforcing toasts by banging on the table, echoing the roll of drums and trumpet blast of royal toasts. Rum, imported in large quantities from the West Indies, particularly Jamaica, from the late seventeenth century, formed the essential anti-freeze for the Royal Navy as well as the basis of the ever-popular punch. Watered-down spirits may also have been drunk out of rummers, capacious short-stemmed glasses enduringly popular from the 1770s for beer and cider: miniature versions, with square moulded feet, may

be the 'punch glasses' listed by London dealers in the early nineteenth century. The cult of whisky drinking came in the nineteenth century, along with the soda-syphon, the brandy balloon, the Tantalus, the Havana cigar, the gentleman's club and the Indian army colonel. As early as 1800 Bristol glassmakers mounted together sets of coloured glass decanters labelled in gold as 'Rum', 'Hollands' (i.e. gin), and 'Brandy' (p.102). These replaced cordials drunk in the eighteenth century from tiny glasses by fashionable women at assemblies and tea drinkings.

The history of ale is as old as that of wine. When beer was created by adding hops to preserve and enhance the flavour, the resulting nutritious drink was consumed at all social levels. The globular Rhineland stoneware mug sufficed until John Dwight at Fulham unravelled the secret of salt-glazing in 1672. Then heavy pint and quart tavern mugs were made in vast numbers in London, and their finely-turned counterparts in Nottingham. Pint mugs, costing no more than 2*d.* in the mid-eighteenth century, were often incised with the publican's name and an applied inn-sign to ensure customers returned them when empty. Giant 6- or 8-pint mugs with applied decoration, often including a bust of Queen Anne, were apparently made for the Jacobite clubs that thrived in the 1720s; their spirit was revived around 1800 by the Kisheres at Mortlake with their invention of the hunting jug.

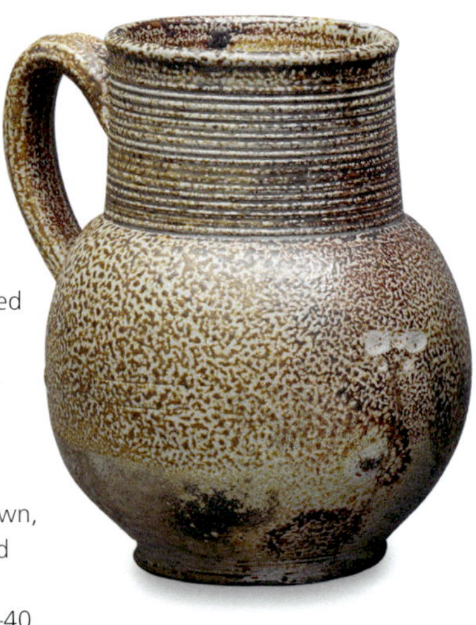
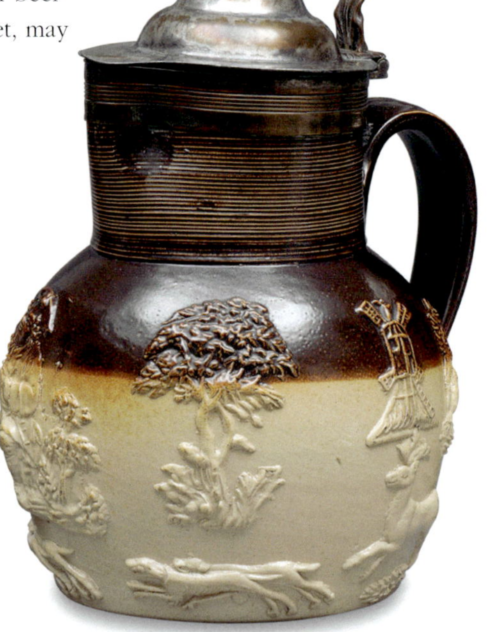
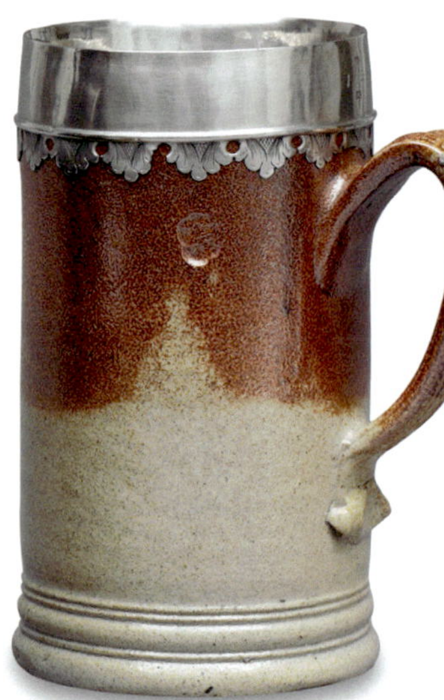

Stoneware for tough tavern use: tavern mug, *gorge*, salt-glazed stoneware, London, about 1700; hunting jug for beer, salt-glazed stoneware, Mortlake (probably Kishere factory), about 1795–1800; tavern mug, quart capacity, marked 'WR' and crown, the silver rim inscribed with its Troy weight, London, about 1720–40.

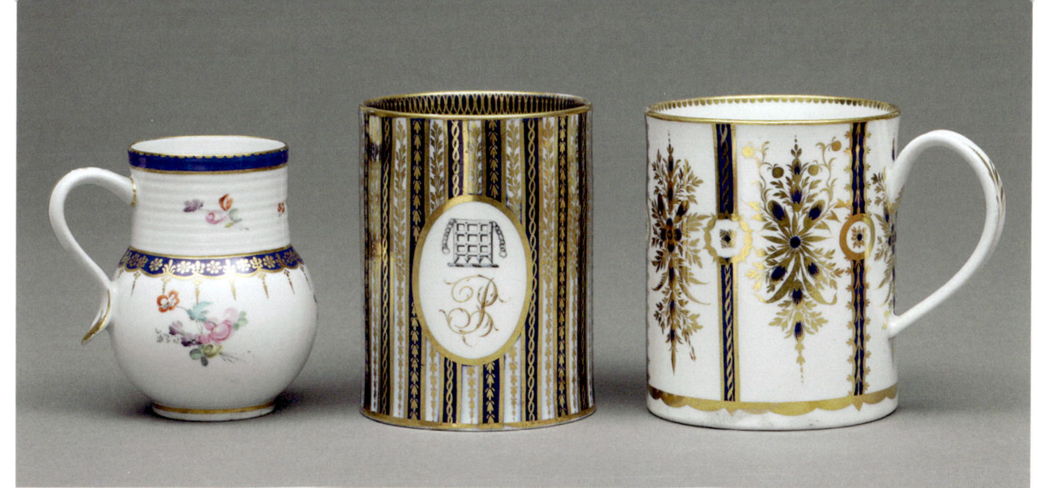

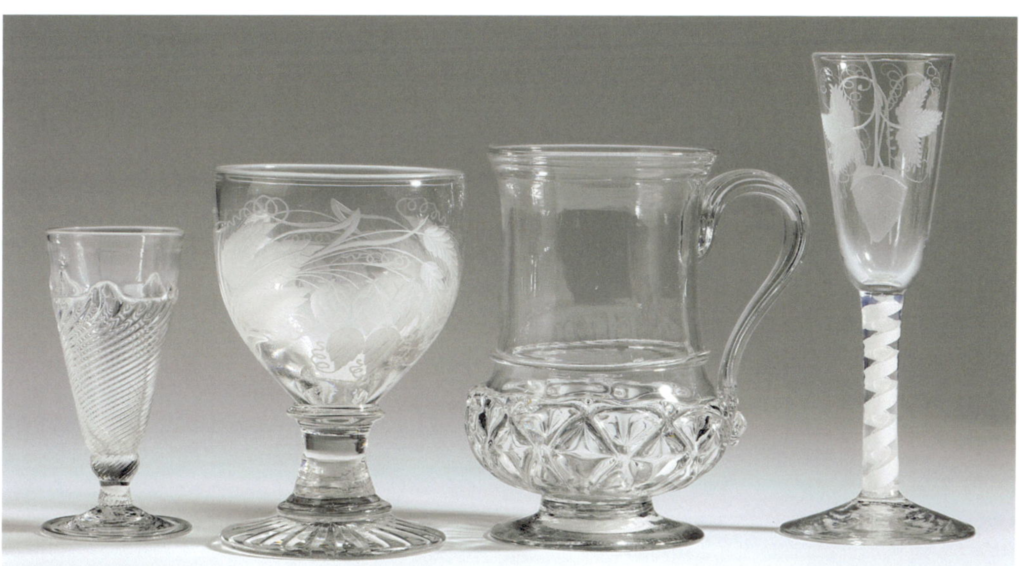

An Act of William III in 1700 required that the capacity of tavern mugs was to be approved by stamping with a crowned 'WR' – a simple matter for pewter, but potters had to stamp the marks into the wet clay before firing. Further standardization came in 1824 when the Weights & Measures Act replaced the Elizabethan ale gallon (still in use at American petrol stations) with the larger Imperial Measure, rendering all existing mugs illegal. Traditional stoneware mugs were giving way to cheap colour-banded Mocha Wares made in the beer-producing area around Burton upon Trent, at first bearing applied pads impressed 'Imperial Pint' but later marked with stencil and sand-blaster.

Beer drinkers in the late eighteenth century, famously nostalgic, could drink strong beer at home from smart porcelain jugs and updated versions of the old cann and gorge made by John Dwight fully a century before. By 1800 half or quarter pint white stoneware cans with handles were being made for drinking the newly fashionable dark porter at the dining table. Bottled at first in corked stoneware bottles, then in the fully-moulded glass bottle of Henry Rickett's 1821 patent, beer was served without a corkscrew after Henry Barrett's hard rubber screw stopper of 1872, the swing-stopper of 1875, and finally the crimped metal crown cork patented by an American, William Painter, in 1892. The classic beer bottle of the twentieth century had arrived.

Glass was preferred at home for drinking strong bottled beer, for which ladies probably selected the tiny dwarf ale. After about 1730 these become elegant flutes barely distinguishable from Champagne glasses, continuing in shortened form until the mid-nineteenth century, after which they grew to become tall heavy 'pub' glasses with cut dimples. Glass tankards were made, sometimes enclosing silver coins in their knopped stems, but again their archaic moulded nip't diamond wayes ribbing, which harked back to the 1680s, suggests use mainly by wealthy antiquarians. For public houses where durability was paramount, matters were rudely brought up to date with the introduction of cheap pressed-glass, including cans with handles, in the early twentieth century.

Labelling the Bottle

SAMUEL PEPYS DESCRIBING THOMAS POVEY'S HANDSOME CELLAR,
19 JANUARY 1663/4

Today the metal identifiers of bottles for drinks, sauces or condiments that graced British dinner or dessert tables are called wine labels. When they first appeared, before 1730, bottle tickets was the term used. Anonymous dark green bottles brought to the table as decanters concealed the identity and colour of, for example, Mountain, Caracavella, sherry or portwine, the most popular grape-based drinks of the time. Identifiers of prosperity and polite behaviour, bottle tickets hinted at the ownership of a well-stocked cellar, a butler and a lavish standard of entertaining. A single bottle ticket cost about half an artisan's weekly wage

Labels attract collectors, both as social and as silver history. The Vintners' Company and the V&A hold large and diverse collections, and the Wine Label Circle publishes research in its journal. J.H. Fitzhenry gave the V&A in 1903–5 some 56 wine labels dating mainly from the late eighteenth century. But Percy James Cropper's bequest of 1,625 labels in 1943 makes the V&A's collection probably the most comprehensive.

Cropper had no connection with the wine or restaurant trade that might explain his unusual hobby. A Nottingham lace manufacturer with an interest in local history, at the turn of the twentieth century he was active in amateur sport and an energetic supporter of the Mechanics Reference Library. In 1908, Cropper went to Philadelphia where he successfully established himself in the lace trade, returning to London a decade later. Here he formed his collection.

Ranging from some of the earliest known labels, from about 1730, to the virtual demise of the industry in the early 1900s, it includes labels of silver, Sheffield Plate, electroplate, Battersea enamel, mother-of-pearl, enamel and porcelain, shell, ivory and bone. Bottle tickets or wine labels were predominantly a British phenomenon although Cropper

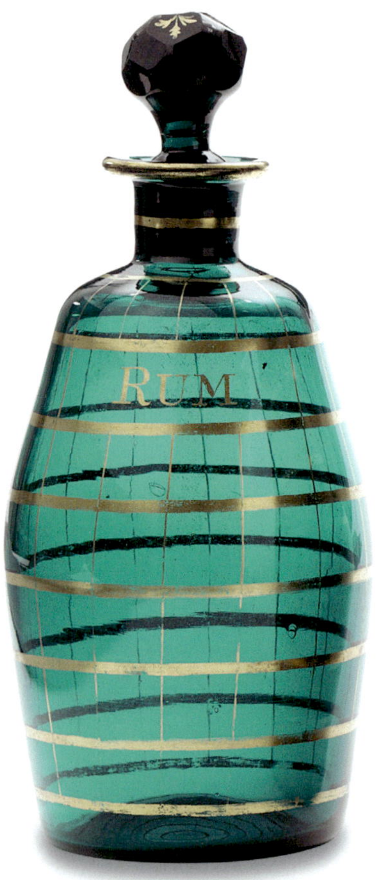
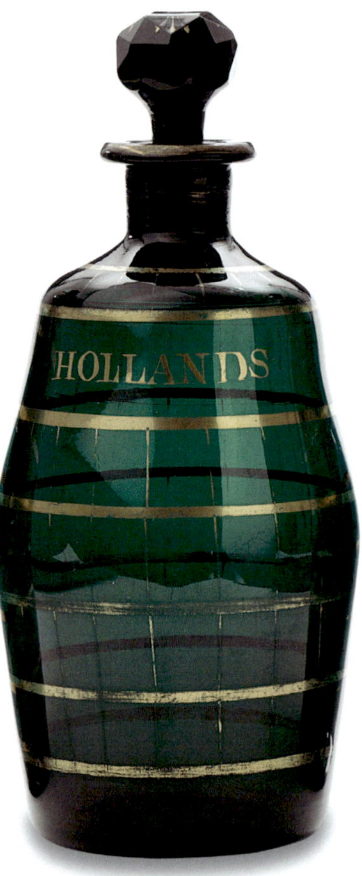
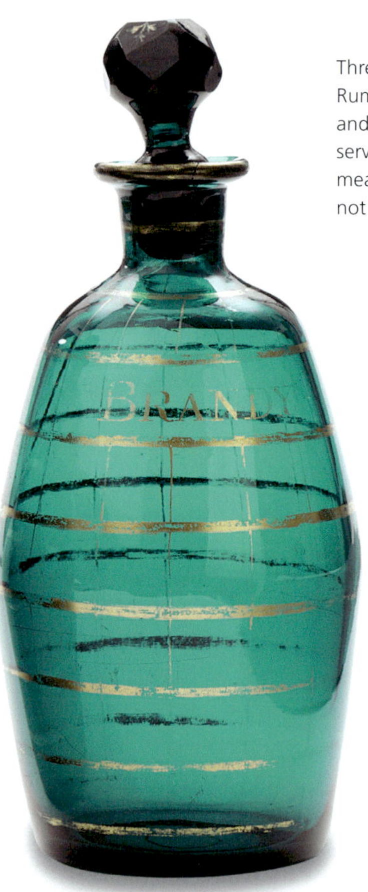

Three green glass decanters for Rum, Hollands (corn-based gin) and Brandy, labelled for self-service at the dessert stage of the meal, or for men drinking toasts, not for use during meal service.

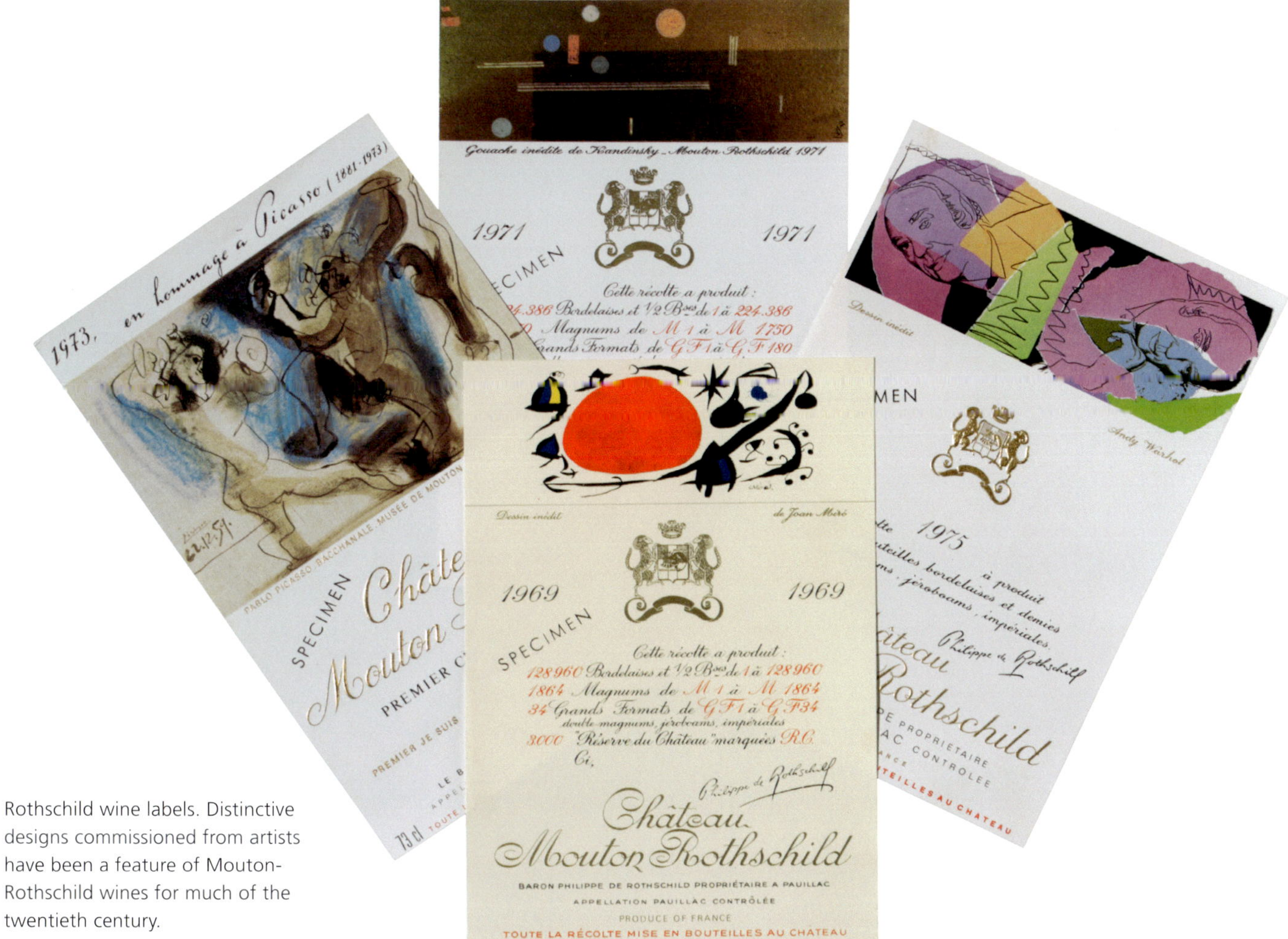

Rothschild wine labels. Distinctive designs commissioned from artists have been a feature of Mouton-Rothschild wines for much of the twentieth century.

found a few French examples and one Danish label from Copenhagen. Where the British colonized, principally India and Hong Kong, local craftsmen mounted boars' tusks and tigers' claws in silver to form labels.

Bottle tickets show patterns of consumption between the 1720s and 1900. Because purchasers chose what was fashionable, as drinking patterns changed so did the names on the labels. Whisky, for example, became popular in the south only from the 1820s, and the first labels appear at that time.

The Parker and Wakelin ledgers in the Museum's Archive of Art and Design show the economics of these modest luxuries. Between 1747 and 1790 this West End retailer sold 1,171 labels for a total of £335. 17*s*. 6*d*., an average of 64 per year. The average price remained steady through 40 years. Ledgers show that for an individual label the charge for workmanship was 1*s*. 3*d*., the silver cost the same, while the retail price was 5*s*. 9*d*., leaving a profit margin of 3*s*. 3*d*. When the silver substitutes of Sheffield plate (*c*.1743) and electroplate (1840) became available, both significantly cheaper than sterling silver, bottle tickets were amongst the first products in these new materials.

Customers expected fresh designs. In the late 1720s, the earliest labels, made entirely by hand, were simple rectangles of silver with the only decorative concession being a scalloped or ogee border. The name of the drink (or sauce) was engraved or pierced. From about 1760, elements of neoclassical design, urns, vases, cartouches, swags and festoons, appear. Decoration was essentially two-dimensional with borders occasionally enhanced with bright-cut engraving to give a glittery effect. But innovation in the Sheffield plating industry soon permeated the mainstream silversmithing trade. Discovered in 1743 by a Sheffield cutler, Thomas Boulsover, this method produced a laminate of a very thin silver surface fused to a much thicker copper backing, giving the weight and appearance of sterling silver but at a third the cost.

An innovation in the newly established Sheffield plating trade was die-stamping, made possible by Benjamin Huntsman's development of crucible or cast steel. With steel dies, cut to form a moulded pattern to a sheet of metal under great pressure, a semi-skilled workman could stamp out objects quickly, cheaply and in quantity. An ideal

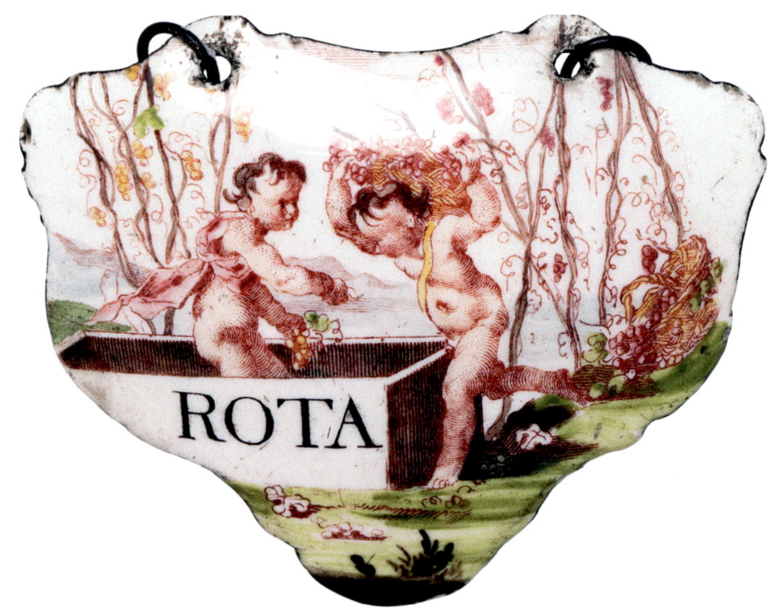

process for the manufacture of wine labels in large numbers, it quickly became popular amongst silversmiths.

Die-stamping, emulating the appearance of traditional hand-chasing, encouraged a three-dimensional aspect to wine label design. Neoclassical delicacy, now viewed as flimsy and insubstantial, gave way to Regency appreciation for rich surface effects and gilding, both exploited in many heavy, decorative wine labels made between 1800 and 1820. The royal goldsmiths Rundell, Bridge & Rundell sold to the Prince Regent some of the most extravagant wine labels ever produced.

Victorian attitudes to drink, the matching of specific wines to particular stages of the meal, and the advent of the commercial labelled bottle drove wine label production in another direction. Increasingly, wine labels became regarded as novelty items, spawning a fashion for a single letter such as W for whisky or C for claret. From 1840 electroplating, the electro deposition of silver onto a base metal, usually nickel silver, a brass alloy, became commercially viable when the Elkington brothers of Birmingham registered patents for the process. Old dies preserved from the eighteenth century

Above: Spanish sherry, such as Rota, was drunk in the eighteenth century not as an aperitif but as a celebratory, stimulating drink, dark and strong to taste (a descendant of the popular Spanish sack used in possets). Sherry was easily available in Britain due to favourable trade links.

Right: Indian claret label, comprising a tiger claw mounted in silver, late 19th century. The host advised guests that drinks were available on the sideboard. In 1743 Lady Grizell Baillie instructed her butler at the dessert to 'set down what wine is called for, with the silver marks upon them, in bottle boards and a decanter of water and glasses to everyone round'.

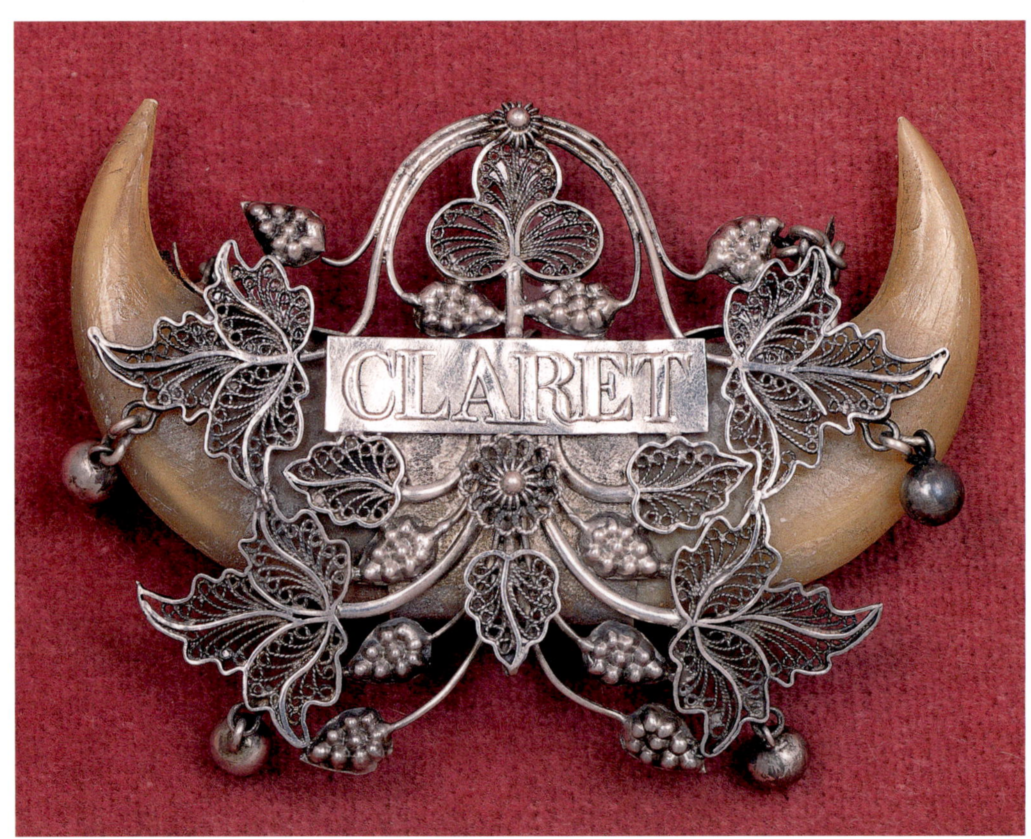

were brought back into use, making the dating of unhallmarked silver labels hazardous today.

In the later nineteenth century, as more spirits were drunk at home and distinctive decanters for sherry, claret and other wines appeared, labels simply identified spirits. Sales of metal bottle tickets went into decline. In 1861 William Gladstone, Chancellor of the Exchequer, reduced import duties on a whole range of wines and the Single Bottle Act enabled grocers to obtain a licence to sell wines by the bottle for consumption off the premises. Until then, the retailing of wine had been restricted to wine merchants who sold only in wholesale quantities or to publicans who seldom sold wine acceptable to the genteel consumer. The Single Bottle Act both gave rise to the term 'Off Licence' and required that the contents of each bottle sold be identified by a printed paper label. Bottle tickets were no longer needed.

Changes in the wine and spirits industry hastened the bottle ticket's decline. Just as the Methuen Treaty in 1703 had reduced duties on port wine from Portugal to Britain, the Excise Act of 1823 reduced the duty and standardized the licence fee for distilling Scotch whisky, making illicit distillation immediately unprofitable. Then the large Scotch whisky corporations, William Haig, Johnnie Walker, John Dewar amongst others, exploited the collapse of the French wine industry, ravaged by phylloxera. By the 1890s they were exporting their blended products worldwide. The image of the kilted Highland piper on the Dewar's label reinforced its authentic Scots origins. A paper label could convey more information, both subliminally and directly, to reinforce a brand image. Wine and spirit producers and importers, forever looking for ways of expanding their markets, were fully aware of that. And today, one of the most respected Burgundian wine merchants, Robert Drouhin, defends the deliberately antiquated wine labels of the region, stressing its medieval heritage, 'We don't sell wine, we sell luxury.'

To emphasize their sophisticated cultural credentials, Château Mouton Rothschild has taken the unusual step of commissioning contemporary artists to design its wine labels. Until 1924, as in every vineyard in the Médoc, wine was sold in casks to a Bordeaux merchant who was responsible for maturing, bottling, labelling and distributing the wine. With no rights over the finished product, the owner had little interest in the bottle's appearance. Baron Philippe de Rothschild took the then revolutionary decision to bottle the entire harvest before it left his property. The label now had an important function. It was not only the trademark but also the proof of origin, the guarantee of quality and the signature of the vineyard. Rothschild commissioned the famous poster designer Jean Carlu to design the label for the 1924 vintage; it remains one of the greatest examples of Cubist influence in commercial art.

In 1945, Rothschild dedicated that year's vintage as the 'Année de Victoire' (victory year) and commissioned the young painter Philippe Jullian to produce a label design based on the 'V' sign made famous by Winston Churchill. Thus a continuing tradition was established. Each year an artist is commissioned to design a label and instead of a fee, is given cases of wine from two different years, one being the year they provided the label. The list of artists commissioned over the years reads as a roll call of contemporary art and includes Marie Laurencin (1948), Georges Braque (1955), Salvador Dalí (1958), Henry Moore (1964), Joan Miró (1969), Andy Warhol (1975) and Francis Bacon (1990). A sample of these labels is held in the Prints and Drawings collections of the V&A.

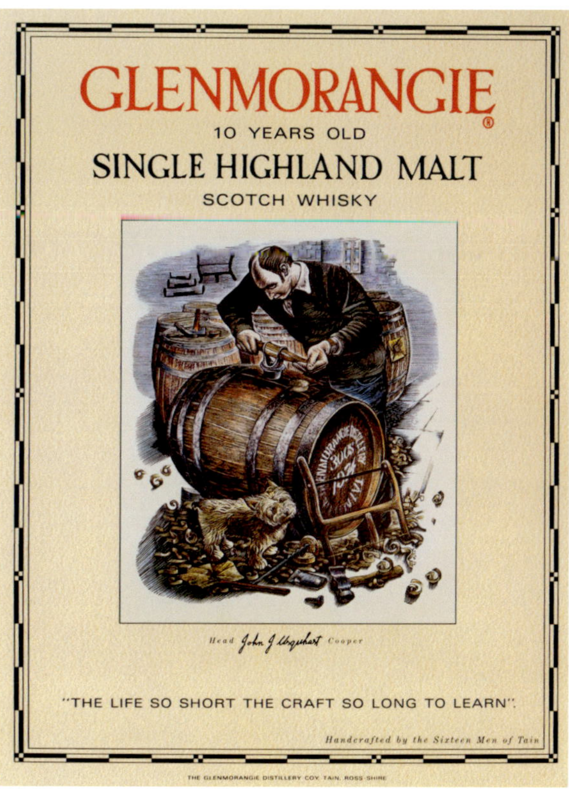

Above: This image of a craftsman coopering a barrel emphasized the tradition of malt over Scottish Lowlands whisky, which was already being produced on an industrial scale by the late 19th century.

Corkscrews

*E*veryone loves the festive pop of a cork, but some might not realize that the corkscrew, 'this useful engine', has a close link with guns and pistols.

The earliest devices for removing corks from bottles were simply shaped hooks. By the seventeenth century, however, a special tool was developed called the worm. Devised to extract the charge from the barrel of a gun and also to clean it, the worm consisted of a wooden rod with a short helical screw attached to the end. Worms and bottle screws are mentioned in the 1600s but Nehemiah Grew makes the earliest specific reference to a corkscrew in 1681 in the *Museum Regalis Societatis.* He cites the use of a steel worm for drawing corks from bottles. The link between corkscrews and gun tools seems clear; indeed, some of the earliest dateable corkscrews are to be found on combination gun tools.

The problem of effectively sealing a container for wine but allowing it to breathe and mature occupied vintners from classical times. Pliny mentions the use of cork to seal casks in his *Historia Naturalis* of AD 77. Cork (*Quercus suber*) is the inner bark of the cork oak, which grows principally in Spain, Portugal, Italy and the Mediterranean coast. Cork as an effective temporary seal for stoneware or glass bottles was already preferred in the sixteenth century to the alternative of a wooden plug wrapped in cloth or parchment and soaked in oil or wax. Early corks were cone-shaped and projected, so that they could be pulled from the bottle by hand. In 1609 Sir Kenelm Digby recommended corks to seal home-brewed beer in bottles, but these were made of salt-glazed stoneware, not glass. At Claydon House in 1651, the Verney family, re-equipping their home after exile in France, ordered bottles of wine from a London vintner

In a Browne Hamper 2 dozen of stone[ware] Bottles with White Wine. They are all seald with Black Wax & by one Seale, I pray observe if the Seales are whole, & set them into sand in the Wine cellar by themselves & sometimes cast Water (that's well salted) upon the sand.…

Vintners gradually discovered that wine kept better in an airtight corked bottle, rather than in a barrel. Once the barrel was tapped, the air could spoil the remaining wine. From the 1630s, advances in glass-making ensured a steady supply of strong bottles in dark green glass. The earliest English dated glass bottle, of 1654, is in Northampton Museum. The consumer bought his wine by the cask as required, paid by the dozen quart bottles, and had the wine bottled by the vintner. He was charged both for the empty bottles, at 4*d.* a dozen, or was supplied with wine in returnable bottles and for the corks.

It became standard practice to allow wine to mature in bottles, with corks rammed well down, flush with the bottle rim, rather than in casks. The corks were coated with a protective layer of wax against damp and insects. Wine was decanted into glass bottles and jugs for serving so that the colour and aroma of wine could be enjoyed. For this higher social level, opening the many bottles drunk at a typical late

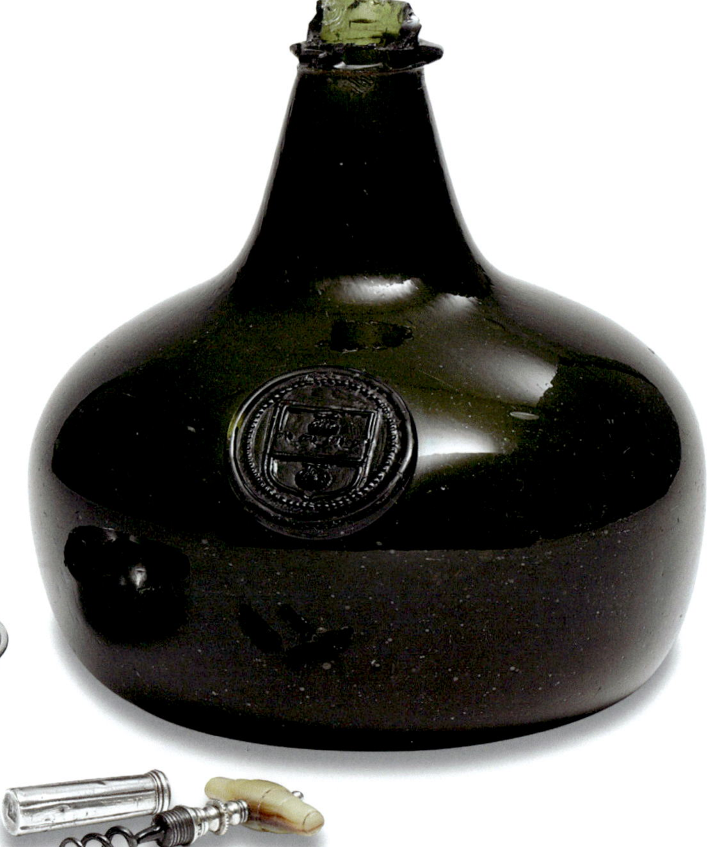

This late Stuart wine bottle, stamped with the owner's arms, would be filled by the vintner to order. The small corkscrew and shallow wine taster are accessories for a gentleman who expected to assess and sample a newly opened bottle before drinking.

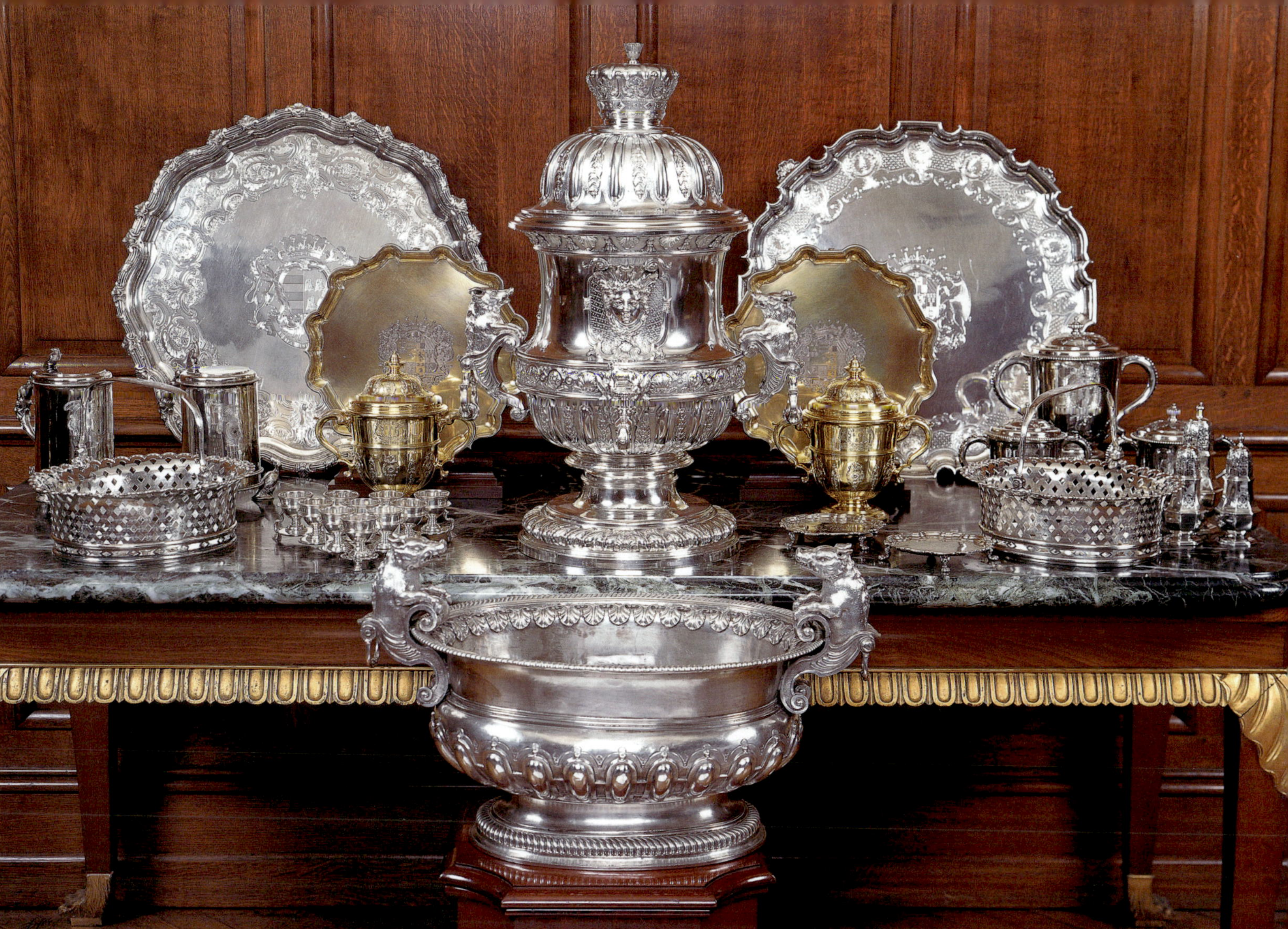

Serving wine required specialized kit, from the modest corkscrew to the massive cooler for bottles in ice and fountain for washing glasses. The large salvers on this marble topped serving table at Dunham Massey are for offering wine.

Stuart or early Georgian party was a task for a waiter, carried out away from the table with a large, practical and undecorated bottle screw.

Waste glass was a valuable and heavily taxed item. Cullet, or scrap glass, was recyclable and it is striking how little bottle glass is found on eighteenth-century archaeological sites. A Swedish horticulturalist visiting London in 1746 noted that asparagus cultivated in the vegetable gardens between Westminster and Chelsea was forced in the necks of broken dark green glass bottles. These protected the stems from the wind, concentrated the suns heat and kept them blanched. The quantities of bottles for wine, sherry, brandy and port were prodigious; at Sir Thomas Clavering's 1784 election dinner at Durham for 109 men, for example, 390 bottles were consumed.

A form of corkscrew that anticipated later patents by nearly one hundred years was developed in France, probably in the late seventeenth century. The screw was pushed into the cork, the cage fitted over the neck of the bottle and by turning the tap the cork was pulled from the bottle with a minimum of effort. These were much easier to use than the simple open-ring corkscrew and were the first examples of the truly mechanical corkscrew.

From the first, wine drinking was attended with ceremony and corkscrews were essential items of luxury for the wealthy. It was natural therefore that they should be made in silver and gold. The earliest dateable English silver mounted corkscrews date from the early eighteenth century. One is part of a silver canteen dated 1701, another has a ring handle mounted with a seal and has the mark of the silversmith John Bodington and the date 1702.

The small pocket corkscrew, popular throughout the eighteenth century, came in many different forms: the roundlet, in which the worm is folded into a barrel-shaped

Corkscrews

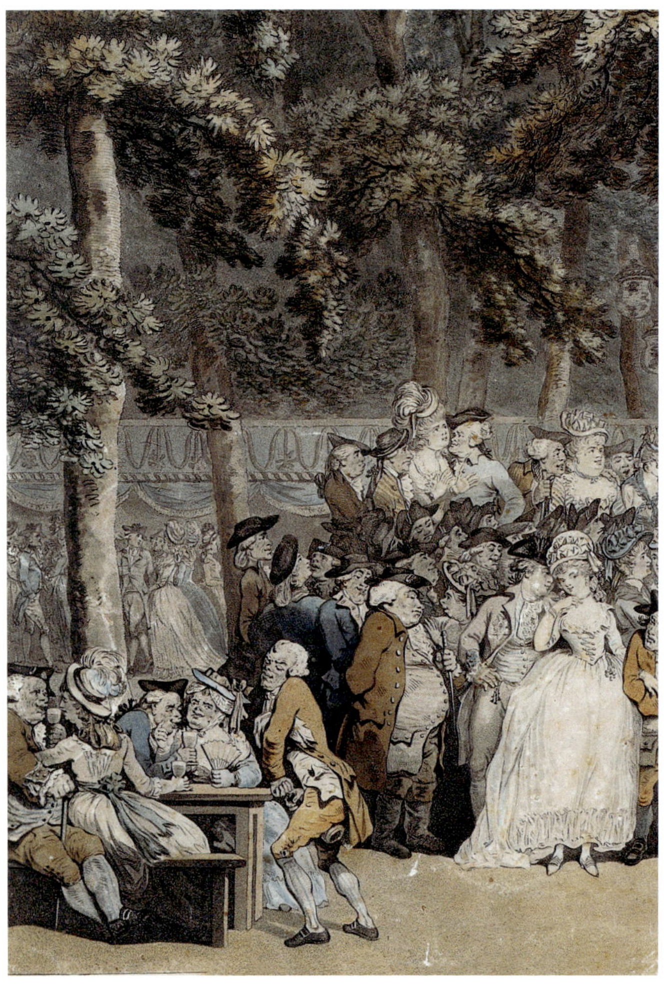

corkscrew, it was a great success and sold over 130,000 at prices ranging from one guinea to four shillings. Modern versions of the Thomason are still being produced.

A series of novelty corkscrews were produced in the twentieth century, one of the best known being a combined corkscrew and bottle opener made in the United States of America in the form of Senator Volstead, the prohibitionist, which would certainly take the joy out of that end-of-the-day glass of wine.

There is a battle to keep corks. Substitutes and reconstituted cork are already widely used; many wine experts say screw caps are just as good as corks, and Portugal is fighting to keep its cork industry afloat. If the corks go, so will the corkscrew, which will be a great pity as so much energy and ingenuity has gone into its manufacture for 350 years.

Left: Rowlandson's watercolour shows a waiter straining to pull the cork from a bottle at a London pleasure garden, without breaking the neck and cutting his hand. Machine made bottles of standard dimensions were produced by Ricketts from the 1820s.

Below: Two French cage corkscrews or tire bouchons à cage. Early examples are forged from iron and fretted out with a saw and file or cast in brass. The design is so successful that versions in wood are still being produced.

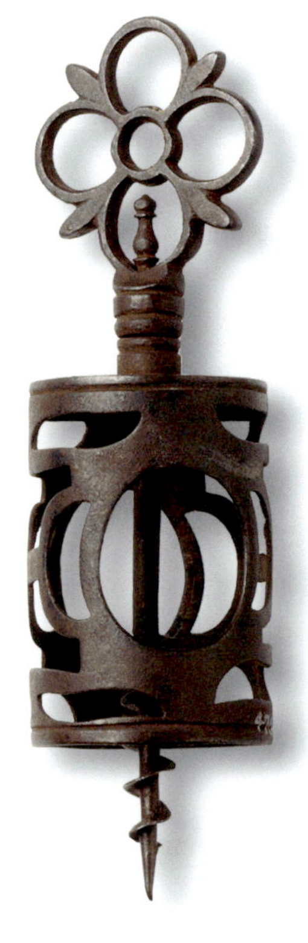

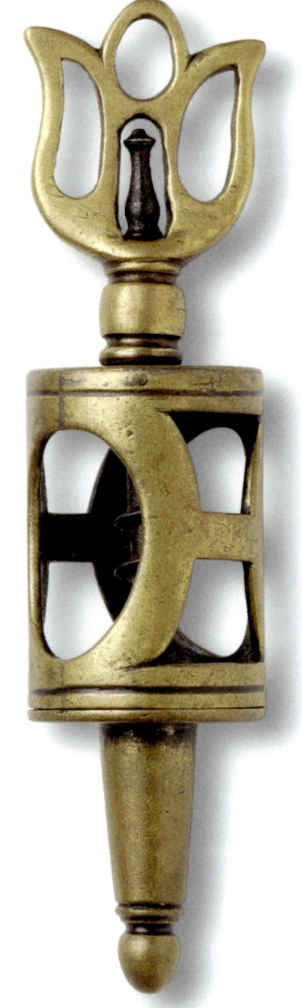

handle; the folding bow in which the worm folds back into a bow-shaped frame; and the peg and worm where the peg handle fits in the centre of the helical screw.

In eighteenth- and nineteenth-century Britain the design of corkscrews attracted attention from leading engineers and inventors. The Spitalfields clergyman Samuel Henshall took out the first English patent for a corkscrew in 1795, and over the next 200 years more than 300 more patents for corkscrews were registered. Henshall sought the assistance of the great entrepreneur Matthew Boulton in the manufacture of his corkscrew, which was known as 'Boulton and Co.'s Spiralytic Corkscrew'. Another inventor associated with corkscrews is Sir Edward Thomason (1769–1849). He had worked for Boulton and in 1802 took out a patent for a corkscrew that operated by a combination of three screws working together and had a brush set in the handle to clean the bottle. Patented as the Patent-NE-PLUS-ULTRA

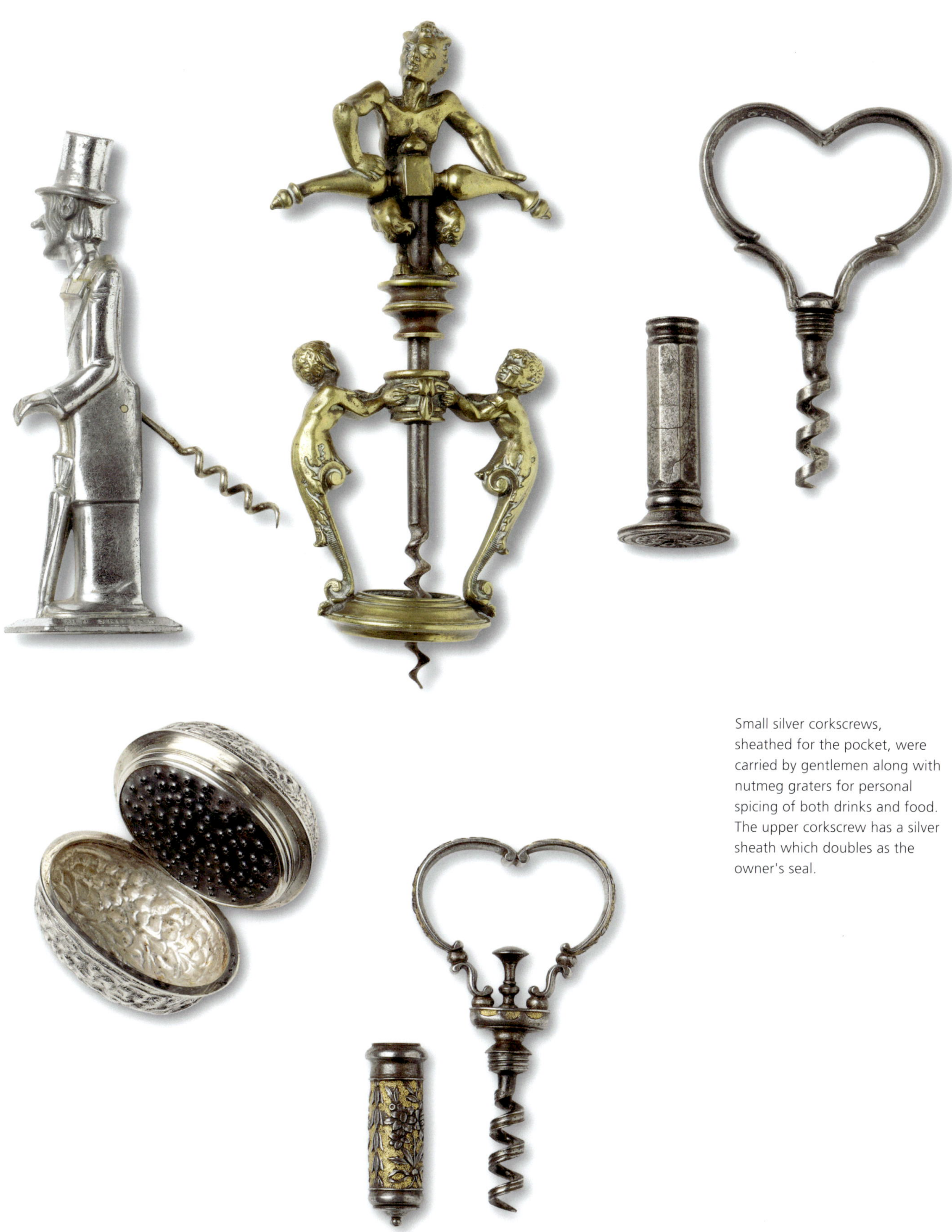

Small silver corkscrews, sheathed for the pocket, were carried by gentlemen along with nutmeg graters for personal spicing of both drinks and food. The upper corkscrew has a silver sheath which doubles as the owner's seal.

Drinking Games

Directly under my room was the tap-room, from which I could plainly hear too much of the conversation of some low people, who were drinking and singing songs, in which, as far as I could understand them, there were many passages at least as vulgar and nonsensical as ours.

KARL PHILIPP MORITZ, *TRAVELS IN ENGLAND* (1782)

Moritz's recollection of drunken revelry in a Windsor tavern rings true of any period. Rounds of song, toasts and teasing games expressed sociability and tested the dexterity and strong heads of participants. Games, associated with seasonal and family celebrations, were staple entertainment for men drinking together, whether at court, in a club or at the tavern.

Kottabus, one of the earliest known drinking games, enlivened Greek drinking parties in the 4th and 5th centuries BC. A *kottabus* player threw dregs of wine from a wide shallow wine cup (*kylix*) at a target, while reclining in the usual Greek fashion. The skill lay in throwing the wine with such force and accuracy that it would make a clear noise on impact.

Most games, however, have required the player to drink, the ever-likely possibility of inebriation being the central

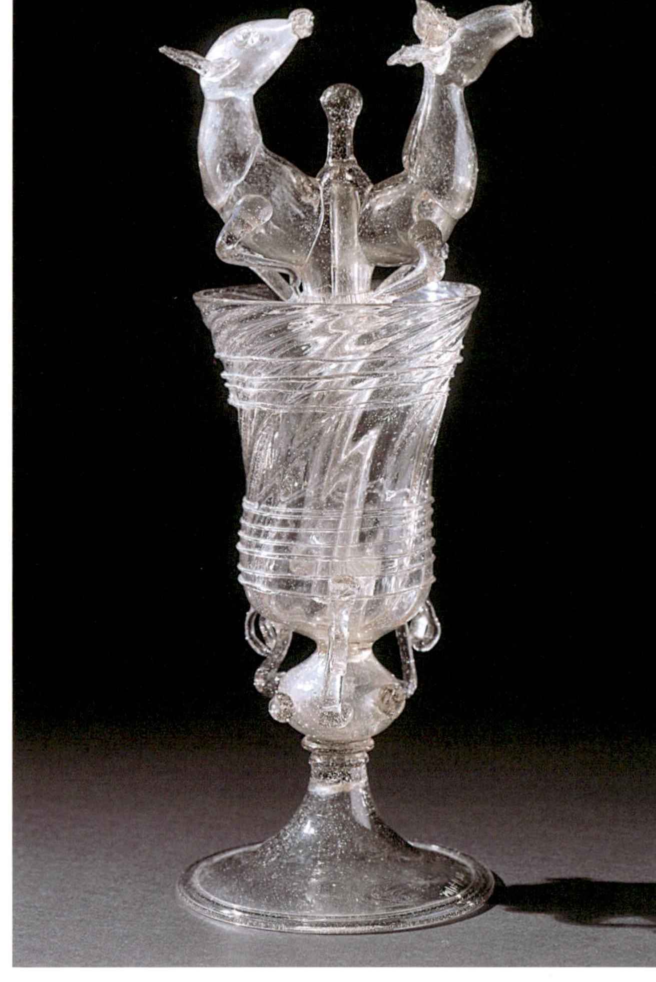

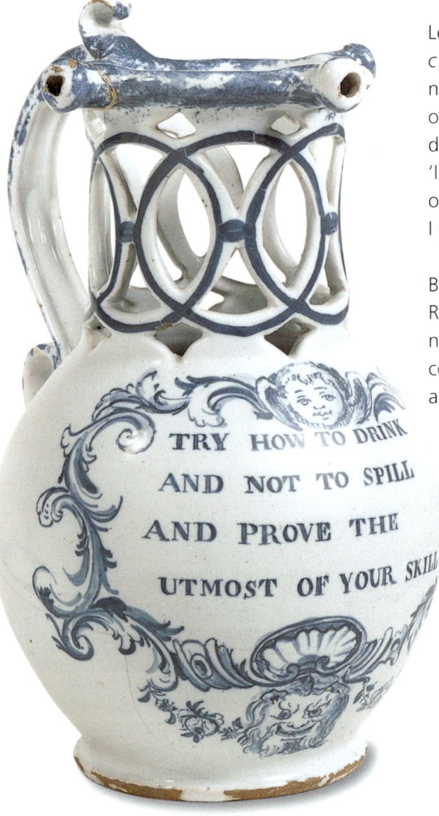

Left: Puzzle jug, Bristol c.1770–75, 'Try how to drink/And not to spill/And prove the/Utmost of your skill'. William Hickey described competitive drinking: 'I was always ambitious of sitting out every man at the table when I presided'.

Below: Italian drinkers from this Renaissance puzzle cup would need to know which spouts to cover with their fingers to prevent a humiliating soaking.

challenge. The *Ambras Trinkbuch,* a rare document of a sixteenth-century prince's drinking game, can still be seen in Schloss Ambras, a Renaissance castle just outside Innsbruck built by Archduke Ferdinand II (1529–95). This early guestbook records comments and toasts made by the archduke's guests after they had passed a bizarre drinking test in the Bacchus grotto close by the castle. Pinioned in a trick chair (*Fangstuhl*), each was released only after drinking a large amount of wine in one draft and then signing the book. If the attempt failed, the guest had to start again. Both men and women took part, listing their names respectively beneath Ferdinand's signature and that of his wife, Philippine Welser.

A popular Netherlandish drinking game demanded a tall glass beaker or *pasglas,* full of beer, decorated with rows of rings. The challenger had to drink down to the next ring and pass the glass on. If he failed to take his

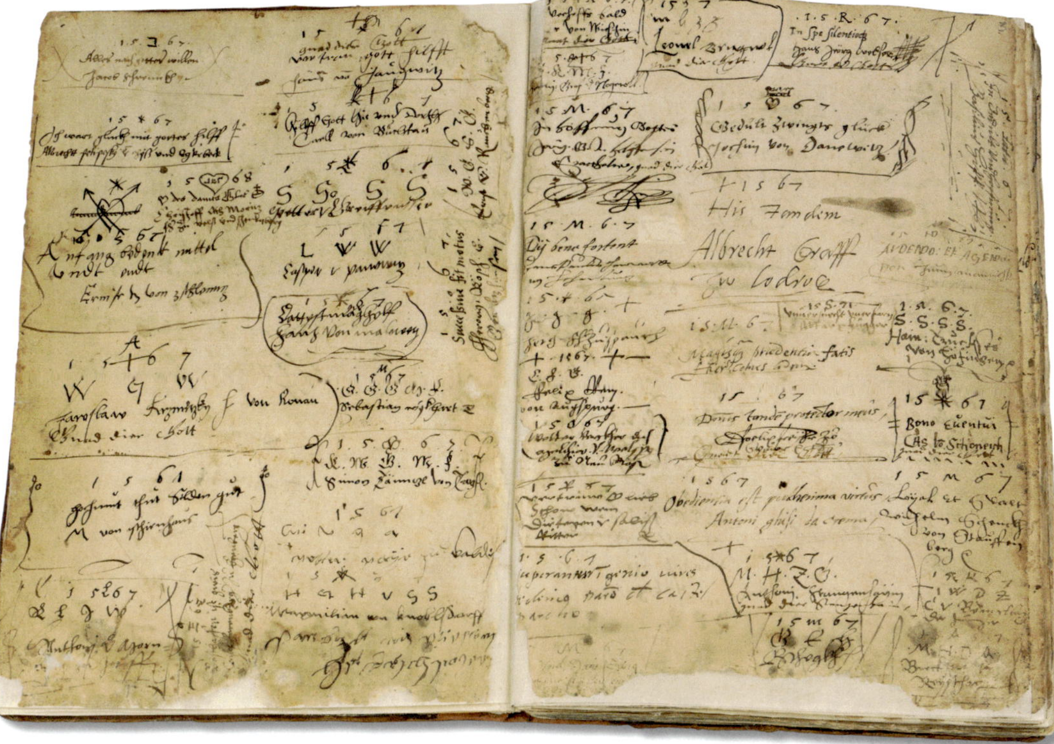

precise share, then he had to drink down to the next. In Adriaen van Ostade's oil painting *The Dancing Couple* (1680–85) a man at the back drinks from a *pasglas* while a seated companion looks on hopefully. In a *vanitas* still life of 1647 by Jan van de Velde, a very tall *pasglas* alludes to the transitory nature of earthly pleasures. Alongside a pipe and oysters the message is clear: indulge your senses but remember they are fleeting (p. 90).

Ingenious drinking vessels were devised to test the skill (and sobriety) of the drinker. An early sixteenth-century maiolica puzzle cup challenges the drinker to select the correct spout from the false ones. An early seventeenth-century Spanish trick-glass from Catalonia in the Corning Museum of Glass has interlinking bulbous chambers out of which wine would rush and douse the clumsy drinker. A Dutch windmill cup of *c.*1570 demonstrates how inventive novelty drinking vessels could be. The challenge was to drink the contents before the sails stopped turning.

Silver milkmaid or wager cups, popular in the Netherlands and Germany (and still made as wedding cups in the latter), took the form of a full-skirted female holding a milk pail above her head, which pivoted on her upraised arms (p.4). When turned upside down the skirt became a large cup, the pail a smaller cup and both were filled with wine. The Worshipful Company of Vintners in the City of London owns two rare London wager cups from the 1680s. By tradition every liveryman is expected to toast the prosperity of the Company from the main cup and the health of the Master from the small, without spilling a drop.

Puzzle jugs and fuddling cups provided merriment in alehouses and taverns throughout northern Europe. Earthenware puzzle jugs, made in the Netherlands, France, Germany and England from as early as the thirteenth century, had pierced necks, several spouts and tunnels within the handle and rim. If the drinker tried to drink the beer in the usual way, it would gush out through the holes in the neck. The only way to drink without mishap was to discover which spout connected to the tunnels leading to the beer, and which ones should be covered by fingers and thumbs to prevent a soaking. English fuddling cups – so-called to imply they would befuddle the drinker – consisted of a number of mugs (usually between three and six) connected by interlacing handles. Small internal holes between the mugs resulted in the participant spilling the contents of all when drinking from one. The challenge required him to adopt awkward positions in order to avoid a soaking.

Rum and Grog

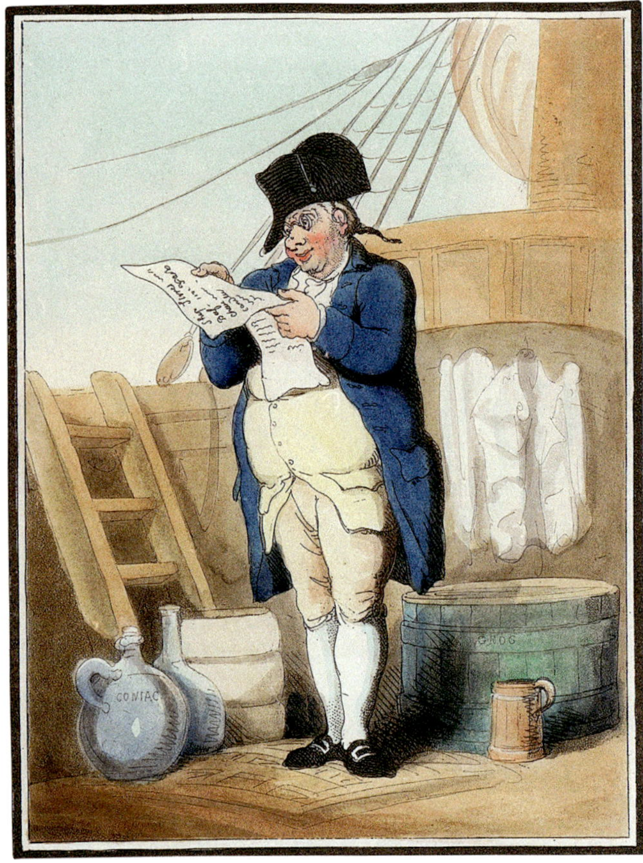

Smiling grog is the sailor's best hope, his sheet anchor,
His compass, his cable, his log,
That gives him a heart which life's care cannot canker,
Though dangers around him
Unite to confound him,
He braves them, and tips off his grog.
'Tis grog, only grog,
Is his rudder, his compass, his cable, his log,
The sailor's sheet anchor is grog.

Charles Dibdin (1745–1814)

For more than 300 years rum has been associated with the Royal Navy. As recently as 1970 a daily issue of the spirit in the form of 'grog', a carefully regulated mix of rum and water, was abolished. In the early days of sailing ships strong drink provided a vital morale-booster in the harsh conditions, a warming drink in cold conditions and a necessary compensation for the privations endured by seamen. By the twentieth century the technology of a modern warship required different skills and levels of

A naval purser standing by the grog tub. In 1687 a Jamaican planter offered to supply the West Indies fleet with cheap local rum. James II insisted on an experiment, to test the sailors' 'Health and Satisfaction' with this brandy-substitute, but lost his throne and rum became the naval spirit.

concentration. Over the centuries the daily rum ration was gradually reduced and issued in weaker measures, but 31 July 1970, the last day of the rum issue, has nevertheless gone down in naval memory as 'Black Tot Day'.

Rum was distilled in the West Indies sugar cane plantations as a fermented by-product of sugar production, and although the origins of the name are obscure, it may derive from saccarum, Latin for sugar. A cheaper spirit than French brandy, rum was first drunk aboard naval ships on the West Indies station after the British capture of Jamaica in 1655, as an alternative to the sailors' standard daily ration of a gallon of beer, which, like water, could not be kept fresh on long sea voyages. The beer was generally weak, around two to three per cent alcohol, and the measure was a 'wine measure' gallon (equivalent to a modern American gallon), roughly five-sixths of a British gallon.

Beer was issued to sailors because the water on board ship was often undrinkable. As one midshipman put it, 'water so putrid, thick and stinking, that often I have held my nose with my hand while I drank it strained through my pocket handkerchief'. Since the water was usually drawn from rivers and was not filtered, treated or purified in any way, but poured straight into wooden casks, often previously used for other substances, even the best of fresh water rapidly became tainted.

Beer took up as much storage space as water, but could not be replenished as easily. Even if it did not go sour or cloudy because the storage or brewing temperature was too high, the beer on board usually only lasted a month into a voyage, and from then on the sailors were issued a pint of wine a day or a half-pint of spirits. The wine was sometimes fortified with brandy to increase the length of time before it spoiled, and the spirits might be anything obtainable locally such as brandy, arrack or rum. In 1688, after lobbying by Jamaican merchants, Samuel Pepys authorized supplying rum instead of brandy to the fleet in the West Indies, although in the early days there was no official naval regulation governing the amount allowed to each man, which was decided by individual commanders. The 'proof' or strength of the rum could not be accurately measured until the invention of Sikes' hydrometer in 1816, but it was

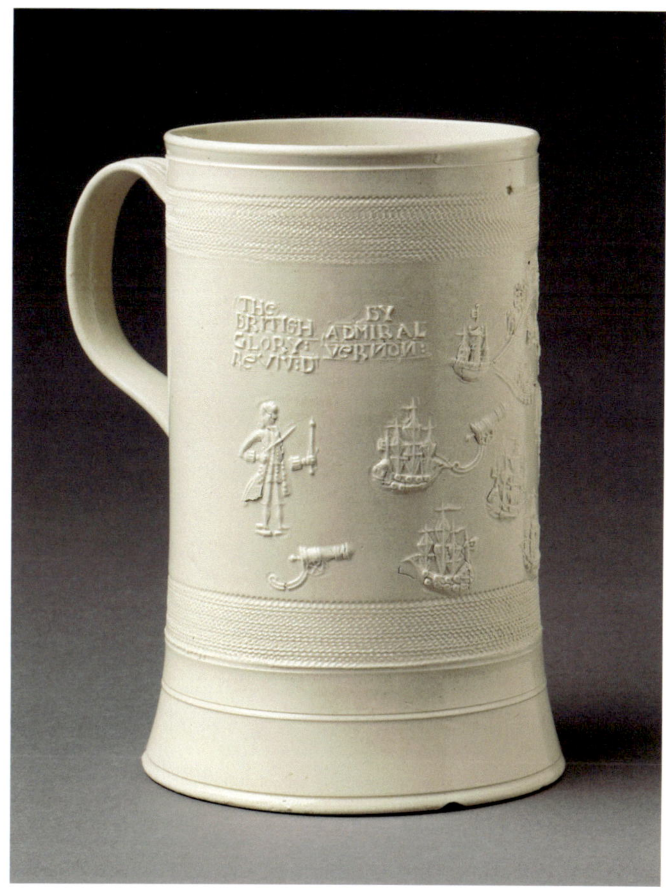

considerably stronger than rum today. The first *Regulations and Instructions Relating to his Majesty's Service at Sea*, published 1731, stipulated that on foreign voyages 'a pint of wine, or a half pint of brandy, rum or arrack, hold proportion to a gallon of beer'. The rum was originally drunk neat 'in drams', and the consequences of the half pint ration were inevitable.

'Grog' owes its introduction in the Royal Navy to Admiral Edward Vernon, Commander-in-Chief in the West Indies at the outbreak of war with Spain in 1739. His successful capture of Porto Bello was publicly commemorated by the production of innumerable medals and pottery tankards, bowls and teapots. Having seen the intoxicating effects of neat rum on his sailors, after consultation with other officers and naval surgeons, Vice-Admiral Vernon issued a general order to his captains from his ship Burford at Port Royal on 21 August 1740:

that the respective daily allowance of half a pint a man for all your officers and ship's company, be every day mixed with the proportion of a quart of water to every half pint of rum, to be mixed in a scuttled butt kept for that purpose, and to be done upon deck, and in the presence of the Lieutenant of the Watch.

This allowance was to be distributed half in the morning and half in the afternoon, and the sailors could add sugar and limes to make the grog more palatable. The name 'grog' was derived from the men's nickname for Vernon of 'Old Grog', after his habit of wearing a boat-cloak made from a coarse fabric called grogram (grosgrain).

A naval tradition had started that would continue with variations for the next 230 years. In 1756 the *Regulations and Instructions* first officially required the rum ration to be mixed with water. The daily rum ration was gradually adopted throughout the Navy and, after 1831, beer was no longer issued to the fleets. In 1824 the ration was halved to one gill (quarter pint) issued at noon, and in 1850 halved again to one eighth of a pint. Popular prints and later photographs record the drinking of grog and its related ceremonies that became to many sailors the highlight of their day. By the mid-nineteenth century a complex ritual had evolved in an attempt to ensure that everyone got a fair

share. Regulations had long prohibited all sale, loan, gift or barter of spirit, but inevitably those determined to obtain more could always find the means. From the harmless custom of friendly 'sippers' and 'gulpers' offered by shipmates, to crafty dealings in dispensing the precious supply, there were always opportunities to take advantage of the system.

'Up spirits' piped on a boatswain's call signalled the cherished daily ritual. The officers of the day collected the keys to the spirit room, where the rum was stored for safety and security reasons, and drew the spirit from its cask using a large copper pump. Chief Petty Officers and Petty Officers received their ration neat, after which the rum for the men's grog was transferred, using official copper measures, into a small padlocked barrico, or 'breaker' labelled 'Rum'. The grog tub, a lidded oak cask with the toast 'The King God Bless Him' or 'The Queen God Bless Her' in brass letters on the front, was set up on deck, and the mess hands were piped to receive the ration. Rum from the barrico was then mixed in the tub with twice the amount of water, and, after

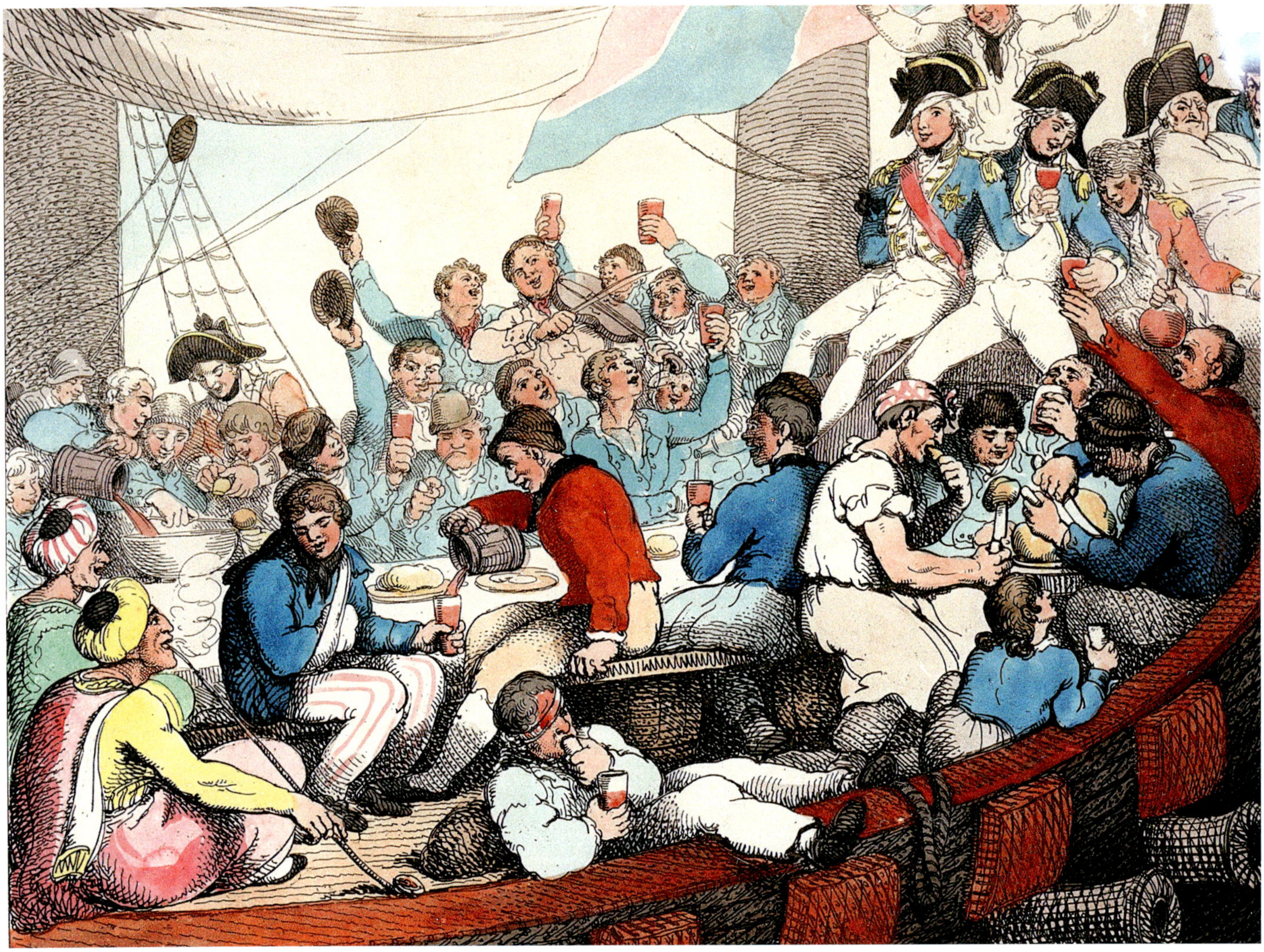

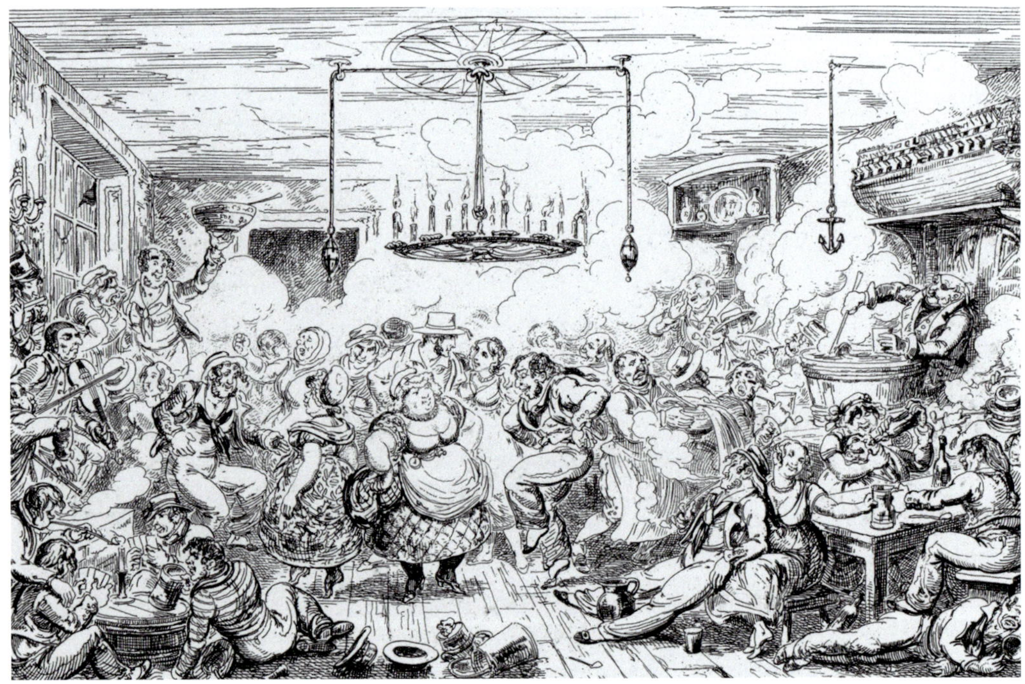

Above: Nelson recreating with his tars. Rowlandson shows how class distinctions were maintained, with wine for the naval officers and grog from a wooden barrel for the 'lower decks'.

Right: Sailors carousing in the Long Room, probably a Portsmouth tavern acting as a sailors' clubhouse and bedecked with ship models, anchors and other naval paraphernalia.

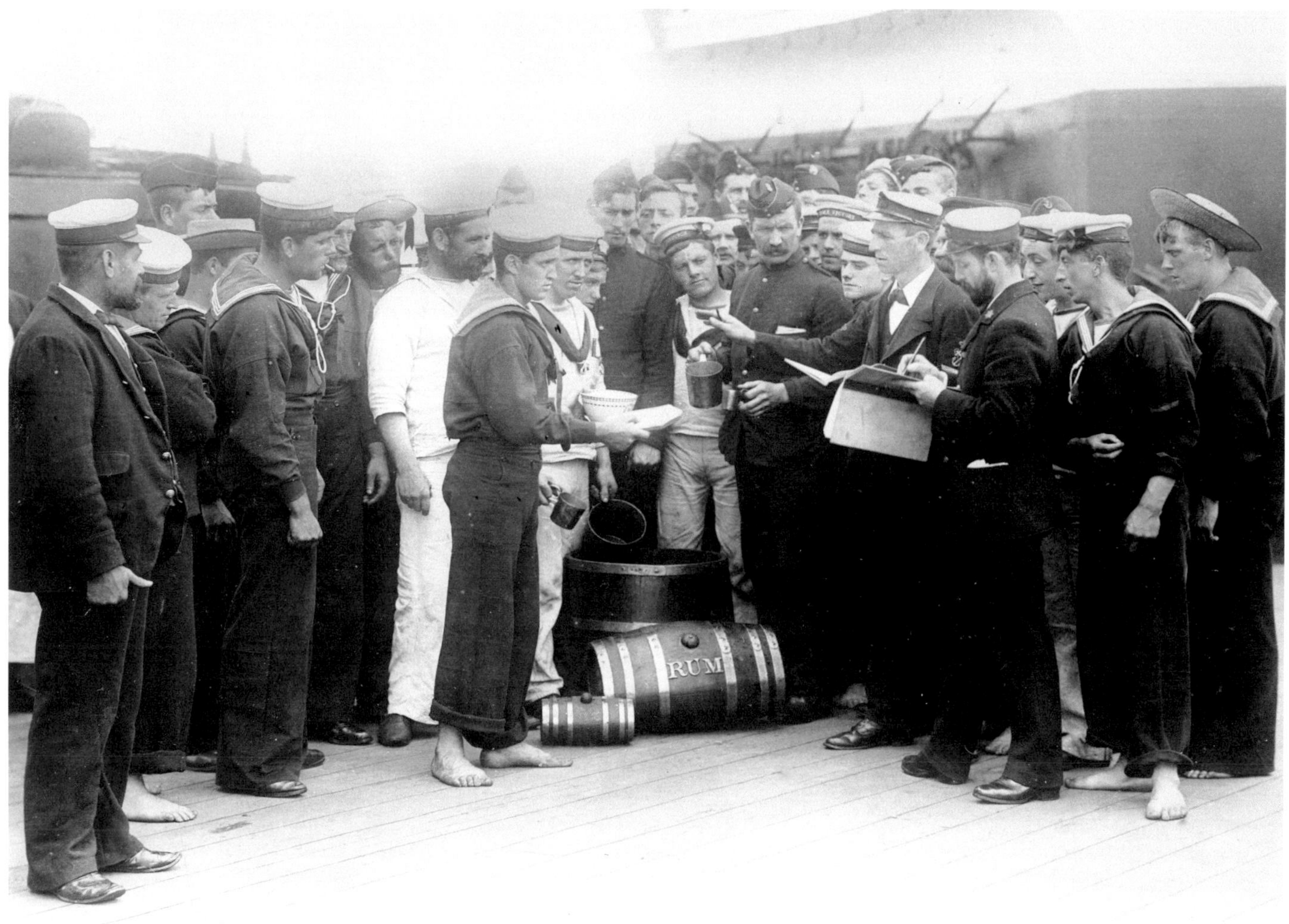

Grog issue. A half pint of rum was a principal component of a mid-20th-century Lincolnshire remedy for sore throat and quinsies, mixed with a quarter ounce of stone camphor and powdered alloes, 20 drops on a sugar lump.

careful accounting in the spirits book, the grog was measured into containers brought by mess representatives to carry back to their waiting shipmates. Any grog remaining in the tub was supposed to be thrown away, but devious ways were found to avoid wasting it.

In times of particularly arduous exertion or special celebration, the rare reward of an additional tot of rum might be announced by the order 'splice the mainbrace', an allusion to a virtually impossible task in a large sailing ship. The tradition continued into the twentieth century, being a much-appreciated comfort and reward in wartime for outstanding action, or to commemorate events such as a Royal Fleet Review. 'Splice the mainbrace' has even survived the abolition of the daily tot, although the chosen alcohol is not necessarily rum.

By 1970 it had become clear that the rum ration was an anachronism. Tinned beer was now carried to sea, and many seamen over the eligible age of 20 were opting for extra pay rather than rum. On 18 December 1969 the Admiralty sent a signal 'The Admiralty Board … has concluded that a daily issue of rum is no longer compatible with the high standards of efficiency required now that the tasks in ships are concerned with complex and often delicate machinery and systems, on the correct functioning of which people's lives may depend.' The last tot was issued on 31 July 1970 amid much ceremony and mock mourning. Aware of the importance sailors placed on this longstanding privilege, the Admiralty offered compensation by creating the Royal Navy Sailors' Fund (the 'Tot' Fund) to benefit the recreational facilities of sailors. Ten years after abolition of the rum ration, the secret blending formula of the dark naval rum once called 'Nelson's Blood' was released by the Admiralty rum broker E.D. & F. Man, to a new company, Pussers Ltd, and made available for public sale for the first time.

Liquid Refreshment

ATTRIBUTED TO SIR FLEETWOOD FLETCHER AND PRINTED IN THE *NEW YORK GAZETTE*, 13 FEBRUARY 1744

*P*osset, a drink widely enjoyed until two centuries ago, is almost extinct, surviving only in its simplest form of bread and milk. Its name is absent from contemporary cookery and cocktail books. This delicious warm, sweet and frothy egg-based alcoholic concoction lives on through our passion for Italian food. Zabaglione, a dessert now considered a treat when dining out, is a sibling of the good old English posset.

Simple ingredients, eggs, sugar, milk or cream, whisked over heat, made posset a staple of the daily diet. With the addition of 'bisket' or breadcrumbs it was an easily prepared and relaxing recipe for a private supper. It did not require more than a chafing dish, or an open fire, to prepare. Since many dwellings lacked an oven, or even a closed hearth, this was a useful recipe. If spices and sack (a sweet wine similar to sherry) or ale were added, the posset was transformed into a substantial celebratory drink. Normally served warm, its smooth glowing heat made it an ideal food for the infirm and its medicinal value was famed.

From the sixteenth century, housekeeping manuals recommended the addition of herbs and flowers such as dandelion and flaxseed to posset to help fight ailments and disease. The London herbalist Nicholas Culpeper advised that the juice from boiled daffodil roots be added to a posset

Below: *The Accomplished Ladies rich closet of rarities or the ingenious gentlewoman and servant maids delightful companion* (London: W. & J. Wide, 1691)

Opposite: Three earthenware English posset pots: maker's name Joshua Heath, dated 1707; painted white, c.1650–55; painted blue and white, dated 1697 (cover missing). Although the V&A's latest posset pot is dated 1728, they were made well into the 19th century.

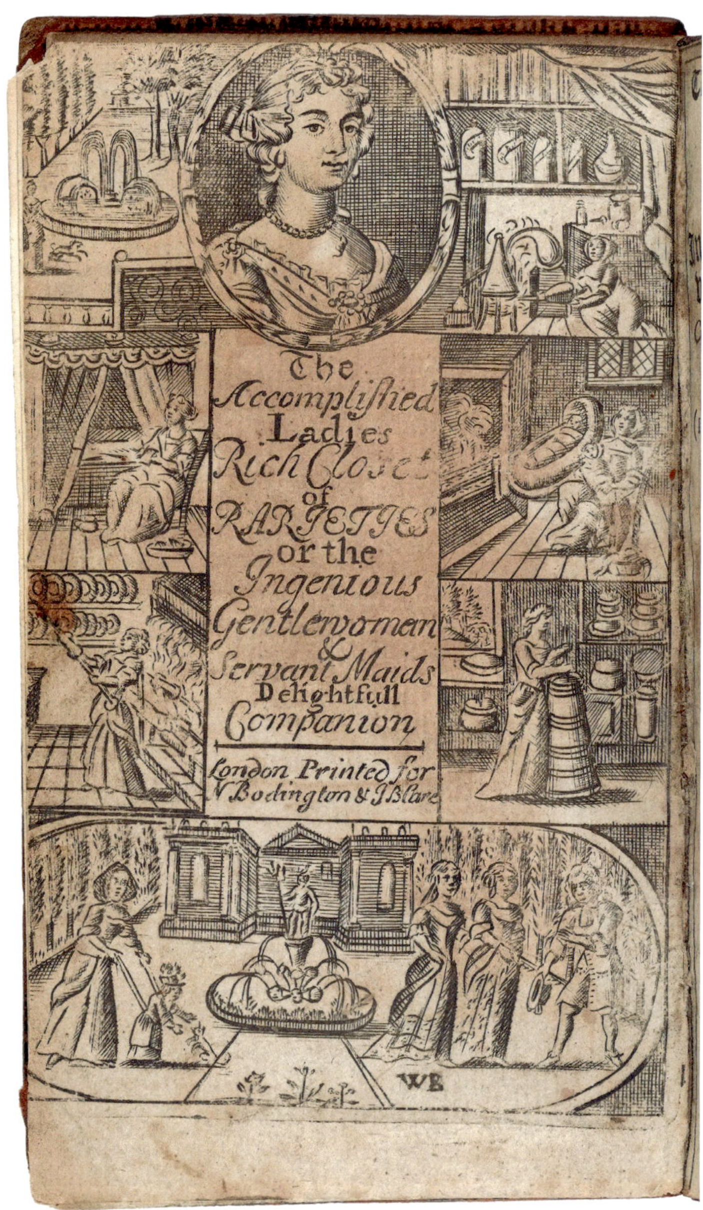

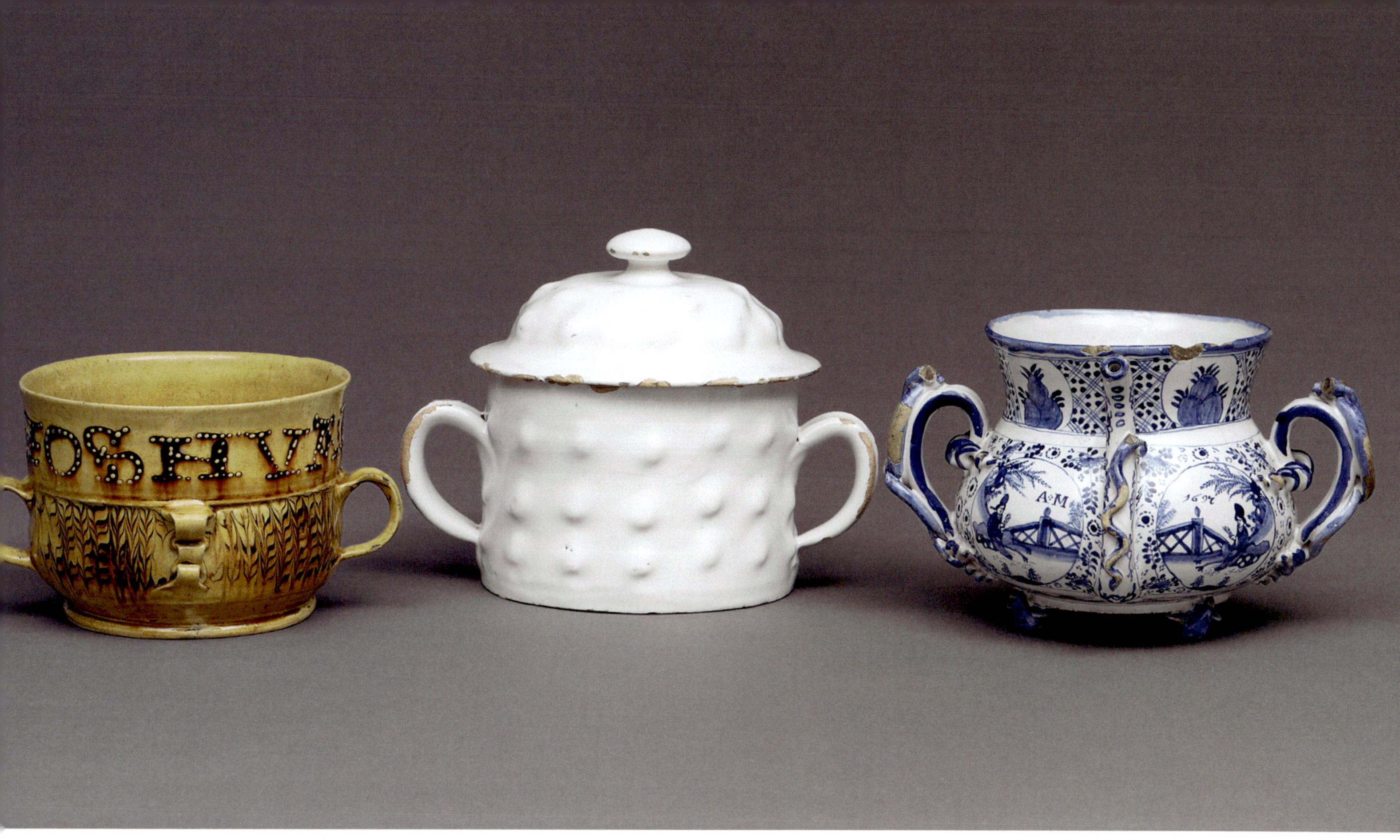

to cure the spring ague. Samuel Pepys when suffering from kidney stones was advised to take 'Castile soap in a posset'. There was a posset to suit every ailment, occasion and season. To make its ancient, equally sustaining but more modest cousin, the caudle or caudel, a thin gruel of oatmeal, broth or pounded almonds was mixed with warm ale or wine, to give comfort on cold days. Coddling, or cherishing behaviour, is a term derived from the effects of caudel. Today we are hard put to distinguish between vessels intended for posset and those for caudle, although the latter are likely to have been plainer, lacked spouts and did not necessarily have covers. Probate inventories refer to both objects, made of silver, delft, earthenware or porcelain, although caudle could easily be drunk from a porringer.

Possets had a wide social appeal. In Yorkshire, sheepdipping even in June was a chilly business; one farmer, Henry Best, encouraged his workers with a hot posset of ale, white bread, pepper and nutmeg, served in a wooden pail. 'This kinde of drinke doe wee give to our washers as they stand in the washe dyke, and when they are in the midst of their labour'. In polite society they served as dishes for supper, or even to dress a dinner table. In January, Elizabeth Moxon suggested in a 1764 cookery book, the middle of the table 'be dressed with a Lemon Posset', prepared in a

stoneware bowl and served in glasses. In March, she advised that the top of the supper table should contain a Sack Posset. A posset flavoured with gooseberry wine and almond paste was to be served in a china dish, probably to preserve the delicate flavour of the almond. A 1722 treatise on hot drinks recommended that pots for serving possets should be 'either of stone, or silver, being much better than Tin or Copper, which takes from it much of its flavour or goodness.' This comment is borne out by surviving posset pots. 'Stone' or rather tin-glazed earthenware posset pots exist in great numbers and several sizes. A Halifax example of the less costly local brown-glazed earthenware, dated 1737, holds almost two gallons. They range from the simple to the ornate, painted with flowers or inscribed. Glass, also inert and hygienic, was less popular, not only because it cost more but also because it was vulnerable to the heat of a posset. A mid-seventeeth-century edition of Kenelm Digby's cookery book tells us that 'My Lady Middlesex makes syllabubs for little glasses with spouts', illustrating the difficulty of reliably defining vessels once commonplace but now redundant. From the seventeenth century these small spouted cups might contain either posset or cool wet sweetmeats such as syllabub.

Silver posset pots with lids are rare today. An early

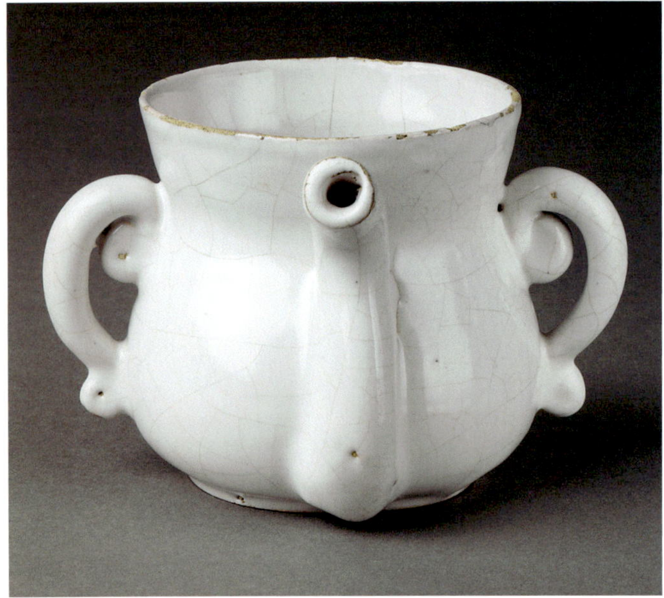

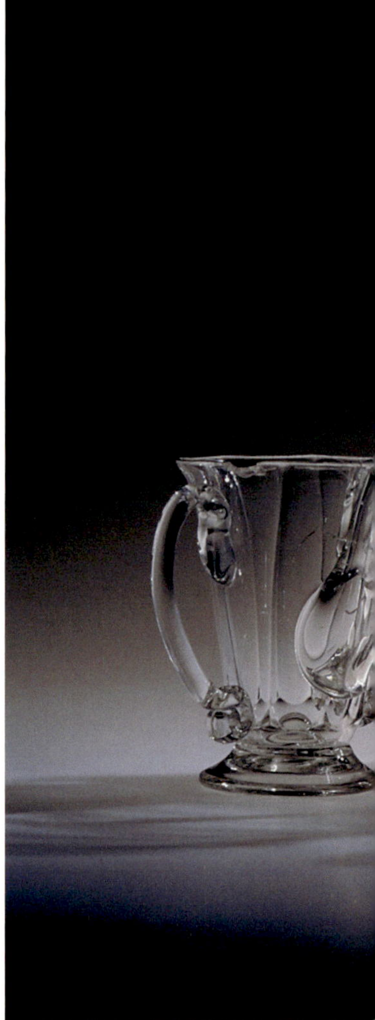

Left: This early 18th-century pot has two handles so that it is easier to hold and pass on. A stimulating version of the posset is the eggflip, prepared for an Oxford undergraduate's supper in 1835 from egg, hot sherry and sugar.

Right: English glass posset or syllabub pots, c.1680 to c.1760. In 1789, Archenholz described the unfamiliar English syllabub as 'a composition of red wine, milk and sugar'.

Below: When glass replaced silver, many silver spout cups were converted to small tankards, with the spouts removed.

description of a silver posset pot dates from 1606, 'Posset Cuppes car'ud with libberds faces and Lyons heads with spouts in their mouths, to let out the posset Ale'. But as the structure of dinner became formalized and ingredients for dessert expanded to include hothouse fruit, ice creams and brandied fruits, the homely posset slipped down the social scale. Silver posset pots were considered obsolete, however fantastic their design. They were 'turned in' to the silversmith to be melted down, perhaps to be made into a fashionable tureen or sauce boat. Silver spout cups, for example the small Norwich made tankard, may have continued in use for a medicinal posset. Free from the dictates of food fashion, this ensured its survival.

Quite apart from its decorative nature, the posset pot is a perfect example of function dictating form. A good posset had three parts: 'if it be right made it will be snow on ye top, thike in ye midle and clear liquor at ye bottome'. Some are straight-sided cylinders, others taper but most posset pots are of baluster shape with the central spout flanked by scroll handles. Stout and sturdy, the bulbous body contains the liquid and the narrow neck draws the froth away. The spout, which usually follows the silhouette of the pot, serves to suck liquid up from the base, which when drained leaves the residue to be eaten with a spoon. To help retain the heat of the posset, which could become stiff as it cooled, and to contain the froth, pots often have a simple cover. These do not always survive as they were easily broken or mislaid.

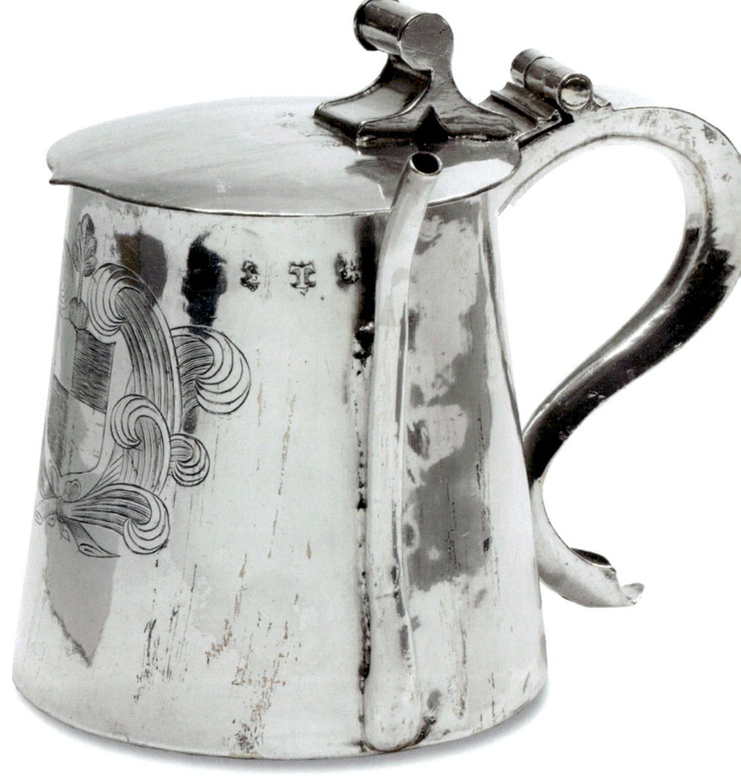

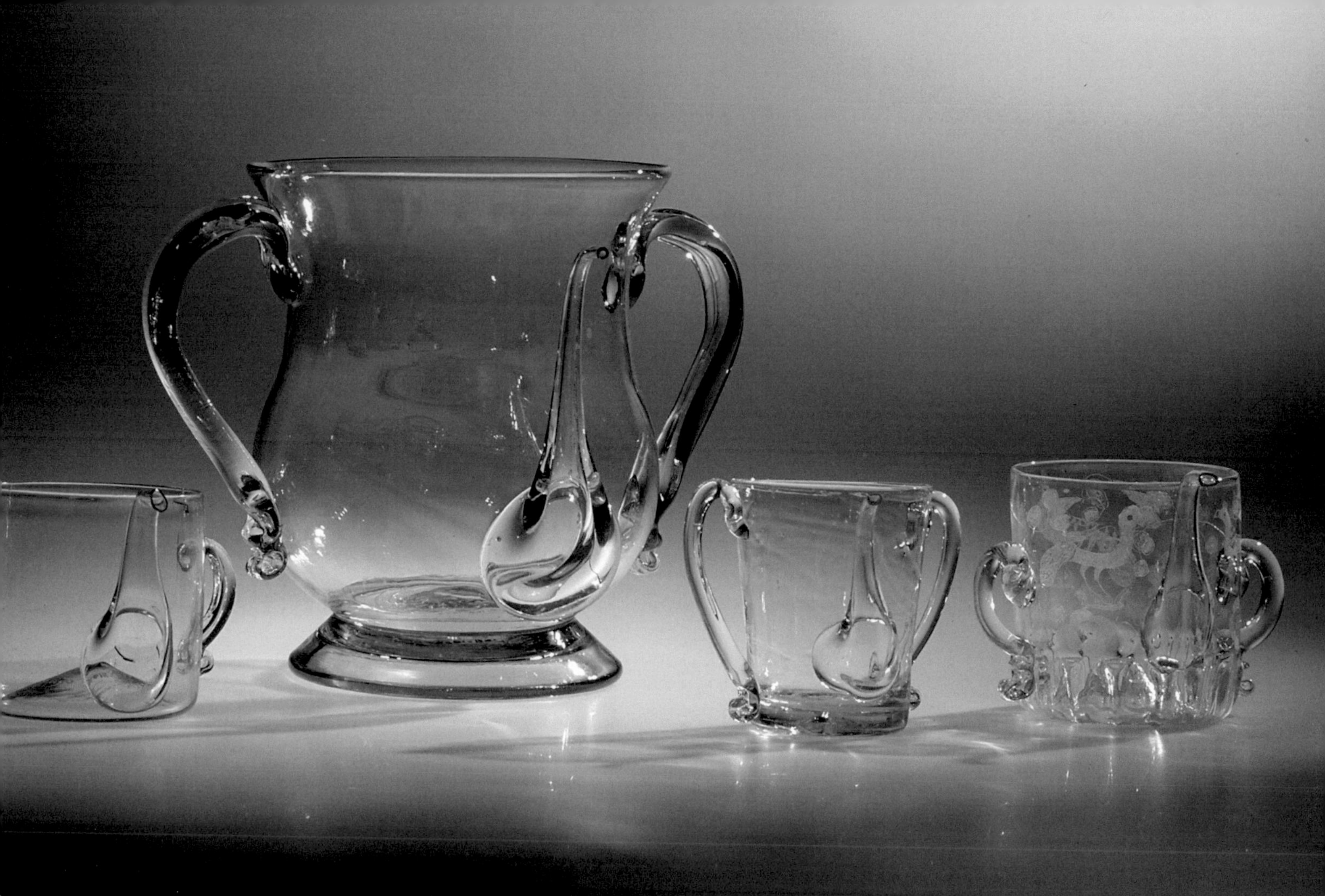

Some grander examples have (or once had) their own salvers for presentation, an essential aspect of seventeenth-century etiquette.

Large, highly decorative posset pots were brought out for celebrations, such as a wedding, Christmas or harvest feast. English delftware and slipware examples that are dated and inscribed with initials were almost certainly made to contain a benediction posset. In a tradition with medieval origins, a newly married couple would be presented with a consecrated posset, laced with wine or ale to drink before they left the wedding party. The commemorative posset pot would then remain with the couple, perhaps to be handed down as an heirloom. Multi-spouted lead earthenware tygs made in Kent and Staffordshire, and in Wales in the nineteenth century, were probably also intended for collective toasts at country weddings. The many handles made it easy to pass from guest to guest after the newly married couple had taken the first drink.

Sharing a posset at a great family wedding in the 1720s, the domestic pot could prove inadequate;

After supper, which was exceeding magnificent, the whole company went in procession to the great hall; the bride and bridegroom first, and all the rest in order, two and two; there it was the scene opened, and great cistern appeared, and the healths began; first in spoons some time after in silver cups; and tho' the healths were many, and great variety of names were given to them, it was observed after one hour's hot service, the posset did not sink above one inch.

References to posset pepper English literature, diaries and letters from the fifteenth to the early twentieth centuries. An essential and much enjoyed part of everyday English life, it entered folklore. The sixteenth-century puck-like character Robin Goodfellow, on seeing the benediction posset at a wedding, turned himself into a bear to frighten away all the guests so that he could drink it all by himself.

Although it continued to feature in late eighteenth-century recipe books, posset vanished from the fashionable table. The warm, sustaining posset was replaced by the frivolous whipped cold syllabub, whose fluffy peaks rose above the delicate lips of the syllabub glasses and pretty epergne baskets of the eighteenth-century table.

From Hedgerow and Orchard

The orange wine will want our care soon. But in the meantime, for elegance and ease and luxury; the Hattons and Milles dine here today – and I shall eat ice and drink French wine, and be above vulgar economy.

Jane Austen, writing to her sister Cassandra in 1808, shows how for gentry families, homemade brews, wines and distilled spirits were affordable alternatives to imported alcohols. Since the climate of northern Europe did not always favour the sun-loving grapevine, the plentiful fruits of the hedgerow, garden and orchard were employed instead. Austen may well have followed a recipe for orange wine in Eliza Smith's *The Compleat Housewife: or Accomplished Gentlewomen's Companion* (1758), a collection of 300 'Family Receipts' intended for medicinal use in the home. Since there were virtually no proprietary medicines before the eighteenth century, a good stock of homemade herbal remedies was essential. Eliza Smith's wine required Malaga raisins, Seville oranges, sugar and water: 'let it stand two days to clear, then bottle it up; it will keep three years, and is better for keeping.' Even the wealthiest, who could afford continental wines and spirits for pleasure, took pains to make wines and cordials to cure, revive and stimulate from their estate produce.

Cultivating many varieties of fruit was fashionable from the early seventeenth century. Texts such as William Lawson's *The Countrie Houswifes Garden* (1617) and *A New Orchard and Garden* (1618) or nurseryman Ralph Austen's *A Treatise on Fruit-Trees* and *Spiritual Use of an Orchard* (both 1653), looked back to the medieval cottagers and smallholders who enclosed a couple of fruit trees in a garden, alongside beehives for honey. Monasteries were renowned both for extensive orchards and for expertise in healing and often ran lucrative cider mills, as at Battle Abbey in East Sussex. Apples were pressed without sugar and the sharp liquid drunk immediately or left for up to two years if preferred strong and sweet. Like wine-based possets and hypocras, highly spiced cider was enjoyed. Pears were grown alongside apples and although mostly eaten as fruit, they could be pressed and fermented into perry, a cheaper version of cider. Physicians recognized the healing qualities of cider and cider vinegar. In 1652 Nicholas Culpeper commented in *The English Physitian* (later known as *The Complete Herbal*), 'The Juyce of Crabs [crab apples] either Verjuyce or Cider, is of singular good use in the Heat and faintings of the Stomach'. Even as late as the 1930s

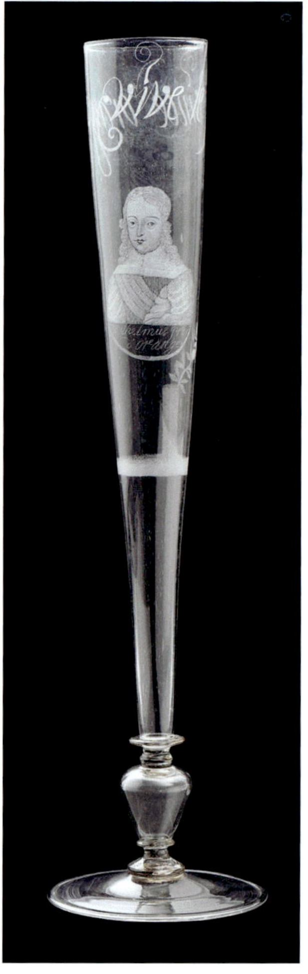

Above: Tall narrow glass for cider, c.1672. The English cider market was boosted by restrictions on French wines, and selective planting of apple varieties.

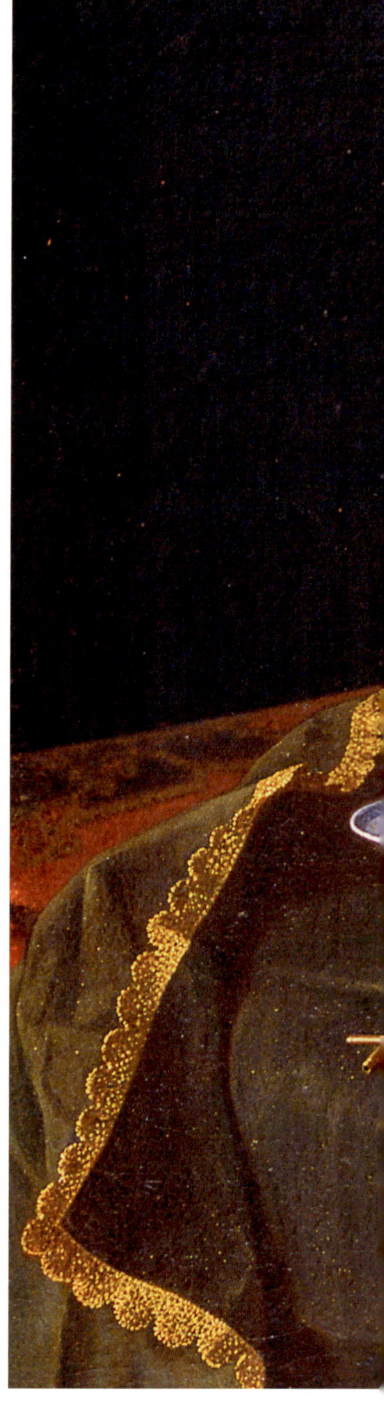

Whiteways Cider Company of Devon advertised 'Woodbine Blend Dry Cider for Rheumatism and Gout'.

Between the 1620s and 1700 cider became widely available in taverns, competing on price as well as quality with imported wine. In North America, where apples are not indigenous, early settlers led by the legendary Johnny Appleseed, quickly planted apple trees for their daily cider and applejack. After Lord Scudamore discovered the special qualities of the Red Streak apple for good cider and mastered the technique of fermenting it in bottles to produce a sparkling drink, commercial production of fine cider in

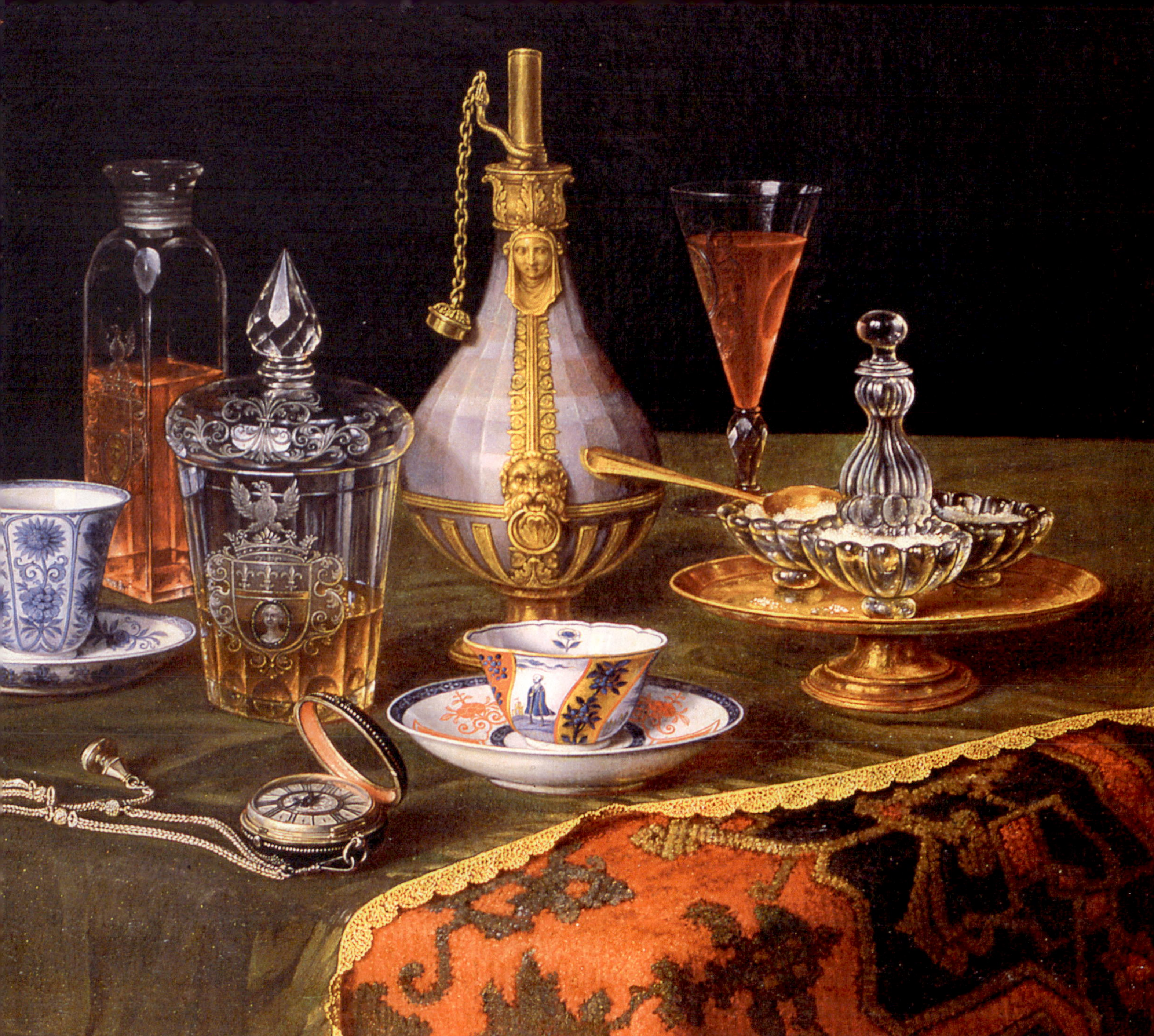

Christian Berentz's sumptuous still life shows flasks and glasses containing a variety of wines and cordials.

Worcestershire grew rapidly. Apples brought to the Woburn Abbey cider mill were supplemented by fruit picked at Thorney, another estate of the dukes of Bedford. J.W. Worlidge's *Vinetum Britannicum, or a Treatise of Cider and such other Wines & Drinks* (1676), extolled Herefordshire, Gloucestershire and Worcestershire cider, which 'is valued above the wines of France, partly from its own excellency and partly from the deterioration of the French Wines which suffer in their exportation and from the sophistication and

adulterations they receive from those that trade in them.' Bottled Herefordshire cider was being sent to London by river in the 1670s and remained a fashionable drink until wine-prices dropped in the early nineteenth century.

Honey, both as a sweetener and as the basis of mead, or 'honey-wine', had an equally ancient history. Some monastic estates produced honey on an industrial scale and sold the surplus. Although cider, ale and then beer usurped mead in popularity as everyday drinks by the fifteenth century, the practice of making mead on a domestic scale survived. In 1758 Eliza Smith gives recipes for standard, 'Strong' and

'Small' mead, each containing different herbs and spices. The Norfolk parson James Woodforde spent the afternoon of 20 October 1794 making Mead Wine:

to fourteen Pounds of Honey, I put four Gallons of Water, boiled it more than an hour with Ginger and two handfuls of dried Elder-Flowers in it, and skimmed it well. Then put it into a small Tub to cool, and when almost cold I put in a large gravey-Spoon full of fresh Yeast, keeping it in a warm place, the Kitchen during night.

Mead infused with spices or herbs was called metheglin, from the Welsh *meddyglyn* meaning 'medicine'. Healing herbs were often preserved in metheglin, which mellowed their sour or bitter taste. Other honey-based alcohols such as clary, a very sweet drink of white wine, honey, sugar and spices, and piment, in which spices such as pepper were infused with red wine and honey, played a similar role.

The English climate meant that medieval wine made from English grapes was often sour. To make it drinkable, spices, fragrant petals and sweet juicy berries were added, such as mulberries, blackberries and elderberries. Herbs such as parsley, hyssop and rue and wild plants like nettles and ivy were added to wine or verjuice (grape or cider vinegar) for medicinal purposes. In 1666, the year plague ravaged

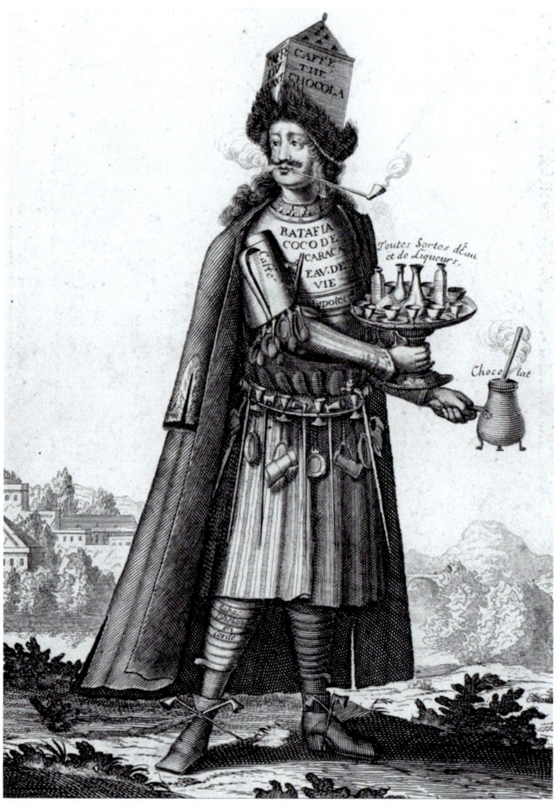

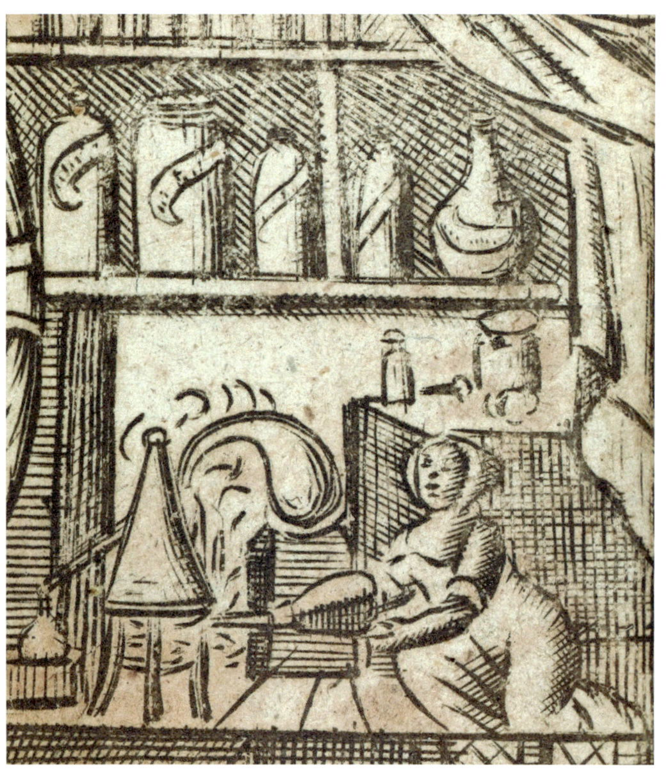

Left: Herb-based distillations were a female speciality; a recipe for lavender wine is printed in *The Queen's Closet Opened* (1655). Hyssop syrup was drunk in warm ale fasting 'to cause an excellent colour and complection'.

Above: A French cafe owner or *caffetier* selling liqueurs alongside tea, coffee, hot chocolate and tobacco.

London, the physician Thomas Willis recommended the benefits of such mixtures: 'In time of Sweating, give the Patient Posset Drink made with Pestilential Vinegar; boyl in the Milk Scordium or Marigold Flowers; if he is very dry, boyl Medesweet, or Wood Sorrel.'

Homemade wines from plants and flowers, such as cowslip and primrose, were flavoured with spices, sugar, home-produced honey and berries. Fragrant gillyflowers were a popular addition to wine and beer. In 1625 James I was believed to have died of poison possibly administered by the Duke of Buckingham in 'a posset smelling of gillyflowers'. Perhaps it was an earlier version of the 'Cordial Caudle' recommended by Thomas Fuller in his *Pharmacopoeia extemporanea* (1710), which combined almonds, egg yolks, 'conserve of red Roses and Gilly Flowers', Aqua Coelestis, Canary wine, damask rose water, 'confection of Alkerms' and

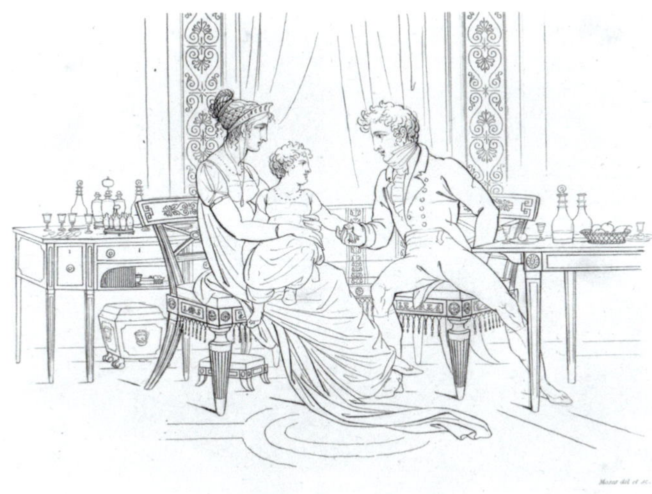

A man and woman enjoying cordials, with decanters and delicate glasses to hand. Around 1820, the artist J.M.W. Turner, entertaining modestly at his Twickenham villa, rejected some currant wine made by his father as being too full of hollands (gin).

oil of cinnamon: 'It greatly Nourisheth, Recruiteth and Reviveth the Spirits, when wasted and low'. Ironically, Culpeper states that gillyflowers 'strengthens nature much, in such as are in consumptions. They are also excellently good in hot pestilent fevers, and expel poison.'

For the Dutch, as in England, weak beer was the recommended everyday drink for children and adults. Usually made from hops and malted barley, homemade beers could also be brewed from fruit and herbs. *De Verstandige Kok of Sorghvuldige Huyschouder* (Amsterdam, 1668) listed at least 18 types of beer including those made with marjoram, rosemary and plums.

Simple fruit and flower waters treated ailments and diluted wine, in preference to plain water, which was 'cold, slow and slack of digestion' according to Tudor physician Andrew Boorde. During the sixteenth century 'cordial waters' made from herbs, flowers, fruits and roots distilled in alcohol grew in popularity, as they could be kept indefinitely. Since they could be made with the dregs of wine, after-wash of brewing and damaged fruit they could be very economical. Strawberry water was thought good for the eyes; gin-based juniper or 'genever' water was taken to cleanse the kidneys and bladder. The *Northumberland Household Book* of 1512 – detailed rules and regulations for the management of the 5th Earl of Northumberland's household – lists 'water of roses, of borage, of fumitory of oakenleaf, of primroses, of parseley of elderflowers' and 'of marigolds'. In the practical manual *The*

London Distiller (1639), the first recipe is for *aqua vitae*, well flavoured with aniseed, often coloured red with rose or poppy petals, or sandalwood and known as 'London redwater'.

By the eighteenth century home-brewing and distillation were considered domestic female skills and a necessary element of household management. Hannah Glasse's manual *The Art of Cookery made Plain and Easy* (1747) gives a recipe for Hysterical Water (to calm the nerves) in which brandy was mixed with a diverse list of ingredients including parsnips, mistletoe, peony root and dried millipedes. Some distilled waters that began as healing drinks evolved into 'fine waters' for entertaining, such as ratafia (an infusion of peach or cherry kernels or almonds in brandy), cherry and damson brandies, and cider royal (cider spirit added to cider).

Hedgerows and orchards also supplied alternatives to alcohol for temperate members of society. On 8 August 1839, the Methodist Mary Ames (1775–1866), wife of stationer Daniel Ames of Lakenham, Norwich, wrote to her teetotal daughter Mary Thompson: 'the bushes are loaded as the first year we took it – and a vast lot of gooseberries – your two bushes (Red and White sweet) by the Bullace tree are loaded. I intend to preserve a few – have made some tea totlers wine – I think it will be superfine.'

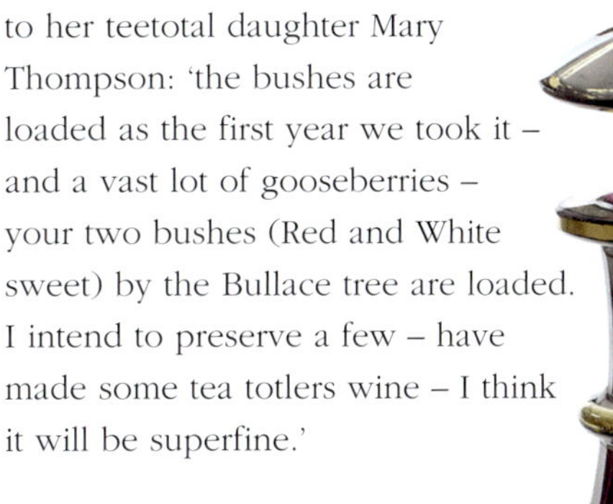

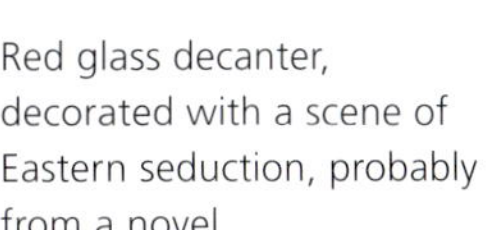

Red glass decanter, decorated with a scene of Eastern seduction, probably from a novel.

Punch-drunk: sour and sweet, strong and weak

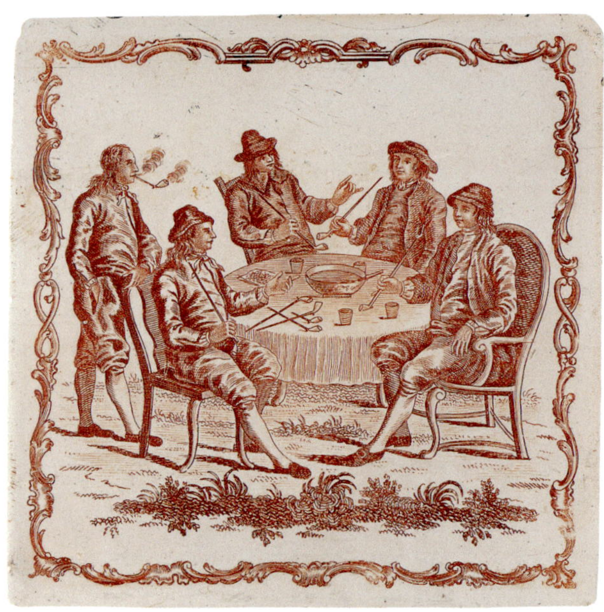

'"You will kill me," said the Doctor. "You will make me drunk," said the Captain. I made no reply, but rang for rum, sugar, and lemons [and] made some punch.' With these words Brillat-Savarin, the famous French writer on gastronomy, described his elderly guests' reaction to a beverage that they had never tasted before. That punch should prove a novelty for Parisian gentlemen in 1801 seems surprising when the drink was familiar to Londoners in the early 1600s and it serves to underline the strength of British interests in the West and East Indies, sugar, rum and spice trade. Punch, the subject of many poems and songs, featured in the name of many London tippling houses. In 1710 punch was even supplied in an extraordinary mechanical vending machine erected at the 'Black Horse Tavern', Hosier Lane, Smithfield, known as the 'new Mathematical Fountain' and advertised by the proprietor Charles Butcher at '6*d* to see't and 2*d* each glass'.

So popular was punch that many of the cases brought before the Old Bailey Sessions were directly attributable to this potent and potentially lethal concoction of spirits, sugar, spice, lemons and water consumed by the bowl-full and drunk either hot or cold. More often than not the defendants had drunk 'too freely by [their] reeling', imbibing three or even five bowls at a sitting. Others, having taken their fill, made off with the vessels themselves and hundreds of cases refer to thefts of ceramic or silver punch-bowls, punch-porringers, punch-cups, punch-ladles, wine strainers and sugar-tongs.

Opinions differ about the origin of the word 'punch', perhaps an Anglo-Indian derivation from *panch*, the Hindi word for five. The actual composition of this mixture, however, was infinitely various, and ingredients differed according to taste, cost and availability. Alexander Radcliffe's *Bacchinalia Coelestia: a Poem in praise of Punch* of 1680 describes a cordial with six ingredients including wine. Before about 1690, the favoured spirit for punch was brandy, but this was supplanted by rum (distilled from molasses) and 'rack' made from arrack (a distillation of palm sap or fermented fruits, grain or sugarcane). Rack punch caused the downfall of Jos Smedley in Thackeray's *Vanity Fair*, and was evidently popular at the Vauxhall Pleasure Gardens. Nutmeg was the main spice, partly for its distinctive flavour as well as for its psychotropic properties as a powerful stimulant.

Above left: A tile depicts fellowship around a punch bowl on a dressed table. There were well-established rituals for smoking and drinking.

Below: 'In London they like everything that is powerful and heavy'. Overwhelmed by punch and tobacco fumes, these men have been drinking till almost 4 am.

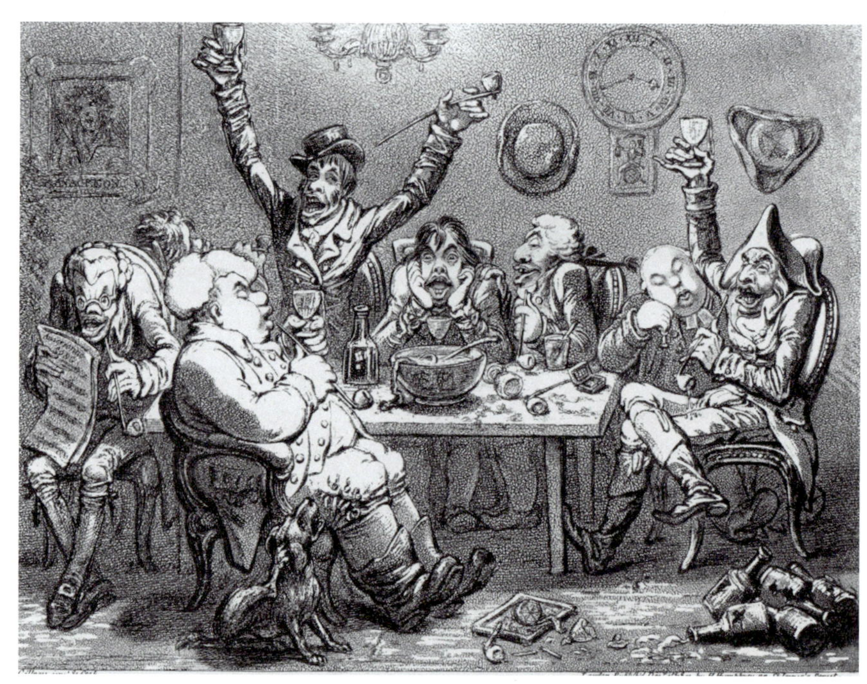

The ingredients were mixed in a simple formula of one part sour (citrus fruit and spice), two parts sweet (sugar), three parts strong (spirit) and four parts weak (water).

Punch and similar alcoholic beverages such as bishop, a mixture of spiced wine oranges and sugar heated with a red-hot poker, remained popular despite government policies of levying taxes both on spirituous liquors and on the manufacture of heavy stout drinking glasses. Drinks were served with a ladle or toddy lifter (a glass tube with holes at each end) which siphoned up the liquid, and the goblets were larger than those for wine. Charles Dickens described them as 'great dropsical, bloated articles … supported on a huge gouty leg'. Substantial glasses for punch were also favoured in Rome. When the composer Hector Berlioz was in residence at the Villa Medici in the 1830s he commented on his preference for punch rather than wine '*voilà le punch! ne bois pas ton vin … il en faut pour le punch; je ne pense pas que vous veuillez le boire dans de petits verres.*' ('I'm not going to drink wine when there's punch … you would not want to drink punch out of little glasses.')

The capacious punch bowl was ideally suited for convivial drinking and the large surface area offered a range of decorative possibilities for the maker and client. Many bowls were made to mark an anniversary or commemorate a particular event and by the nineteenth century this was a common household object.

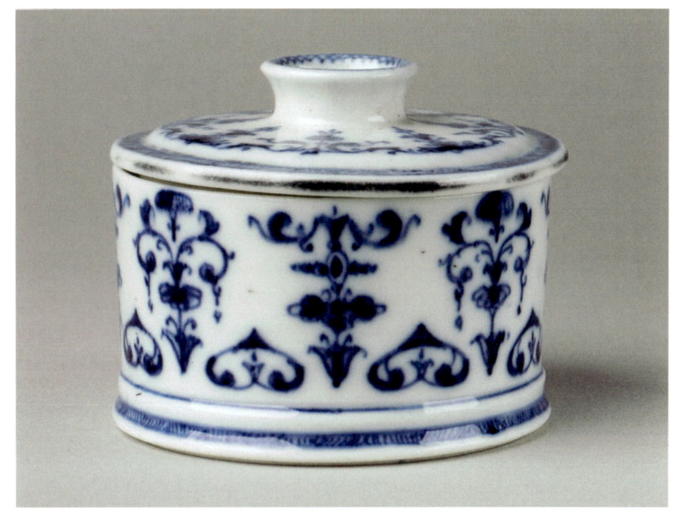

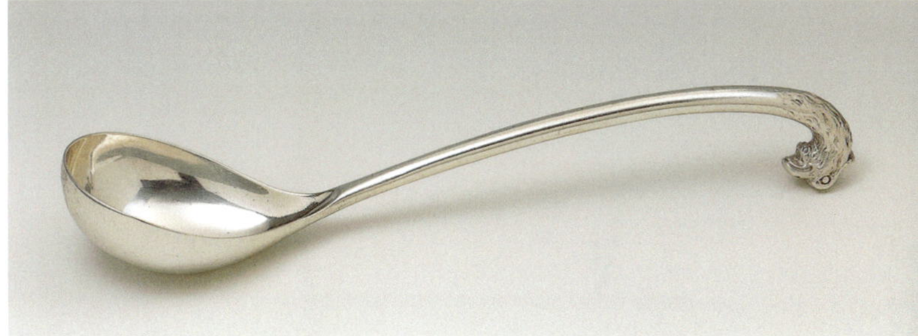

Below: A mocking image of a late night party after Hogarth. A Whig symbol, the punch bowl stood for British support of West Indies rum and sugar rather than the Tories' French claret.

Punch and bishop, a spiced winter warmer based on wine, called for sugar and spice boxes, nutmeg graters, ladles and strainers. Not all were strongly alcoholic. Madeira 'makes the most pleasant whey or negus of all the wines … good for elderly persons when the functions of life have begun to fail'.

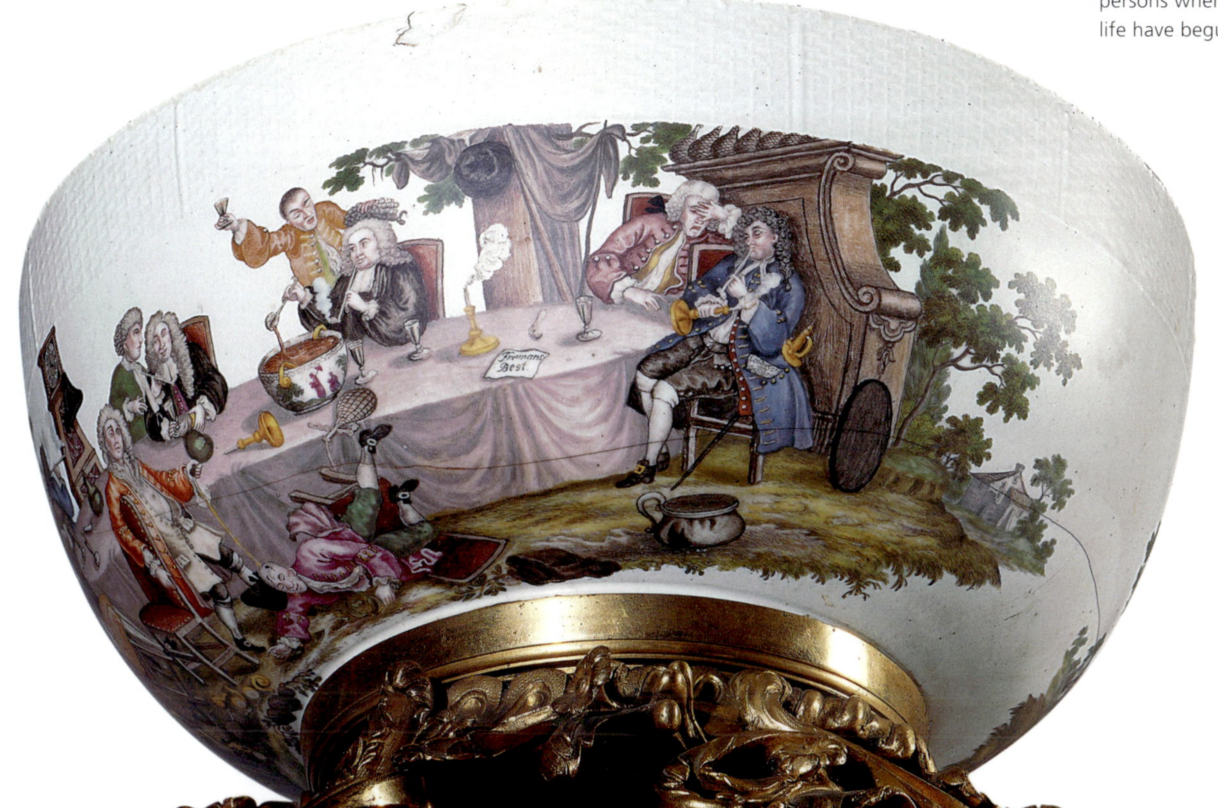

A Toast to Vodka and Russia

While popular lore presents vodka as an eternal and inseparable part of the drinking culture of Russia, its history as an alcoholic beverage is in fact surpisingly recent. Early references to vodka speak of it as medicine. For example, a Russian chronicle for the year 1533 advocates vodka as a sterilizing agent: 'Take *vodki,* apply to the wound and press out'; it is also suggested as a headache remedy: 'For my headache, the master ordered me to take … *vodki* from his medicine chest.'

Medicinal vodkas were infusions of herbs in distilled alcohol that were then diluted with water, just as the spirits or 'hot waters' of northern Europe were medicinal in origin. This usage may explain why a Slavic diminutive for water – vodka – came to signify what we know as vodka today. Aromatic vodkas, including *pertsovka* (pepper vodka), *limonaya* (lemon vodka), *okhotnichaya* (vodka with juniper berries, ginger and cloves), *starka* (apple and pear-leaf vodka) and *zubrovka* (bison grass vodka), are direct descendants of these early medicinal products.

A distilled spirit that can be produced from any starch/sugar-rich plant matter, vodka today is generally made from grains such as rye. Creating spirits first requires the heating of plant matter and water to release starch for conversion into sugar. However, vodka-makers discovered that if this mash is frozen, it permits the separation of the water, which leaves a liquid of far higher alcohol content than would have been possible through fermentation alone. When this fermented liquid is heated in a still, the vapours condense and produce an even purer and more potent form of alcohol.

Vodka may have been introduced through Poland. Edward Keenan suggests that even the word 'vodka' betrays a Polish origin as it places the stress on the penultimate syllable, linguistically indicating a Polish pathway into the Russian language, and the earliest record of vodka being served at the Muscovite court comes from a time of Polish intervention in 1606, when an Augsburg diplomat comments that at the start of the meal 'they served vodka and exceedingly tasty white bread'.

Vodka was like other spirits across northern Europe still a relative novelty in the early seventeenth century and not yet integrated into formal drinking rituals. In Moscow as at other courts the traditional alcoholic drinks of the region, fruit-flavoured meads and beer, with imported wines, remained central to ceremonial drinking, particularly toasts. When the Earl of Carlisle was entertained by Tsar Alexey Mikhailovich in the Facetted Chamber of the Kremlin in 1664, the tsar sent him red mead from his own table in a gold *kovsh*, or cup of friendship. But a small cup of gold set with rubies and emeralds, made for Mikhael Fyodorovich in the Kremlin workshops before 1650, shows that spirits were now part of the tsar's personal diet.

At first vodka drinking was restricted to the upper classes. Isaac Massa, a foreign trader and long-term Moscow resident, mentions that 'Drinking beer and mead is the greatest pleasure [for Muscovites], especially when they are able to drink as much as they wish, and in this they are masters; most of all [they love] vodka which is forbidden to all of them except the gentry and merchant stock.' A late seventeenth-century English diarist, Abraham de la Pryme, cites the thirst for spirits among Russia's élite in an anecdote

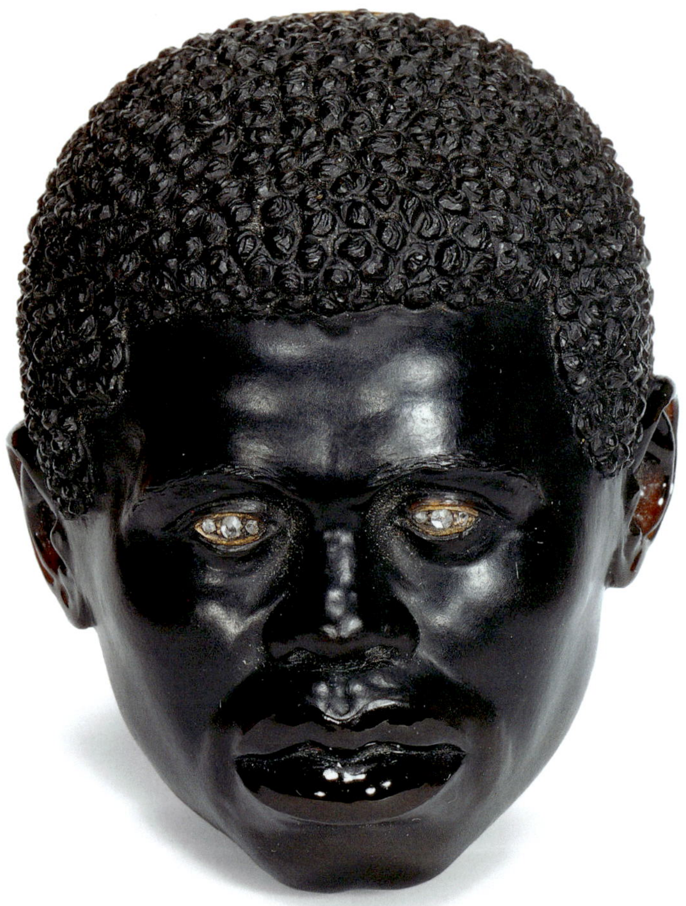

Vodka beaker created for Carl Fabergé, an agate head with diamond eyes, which cannot be securely set down until it is emptied.

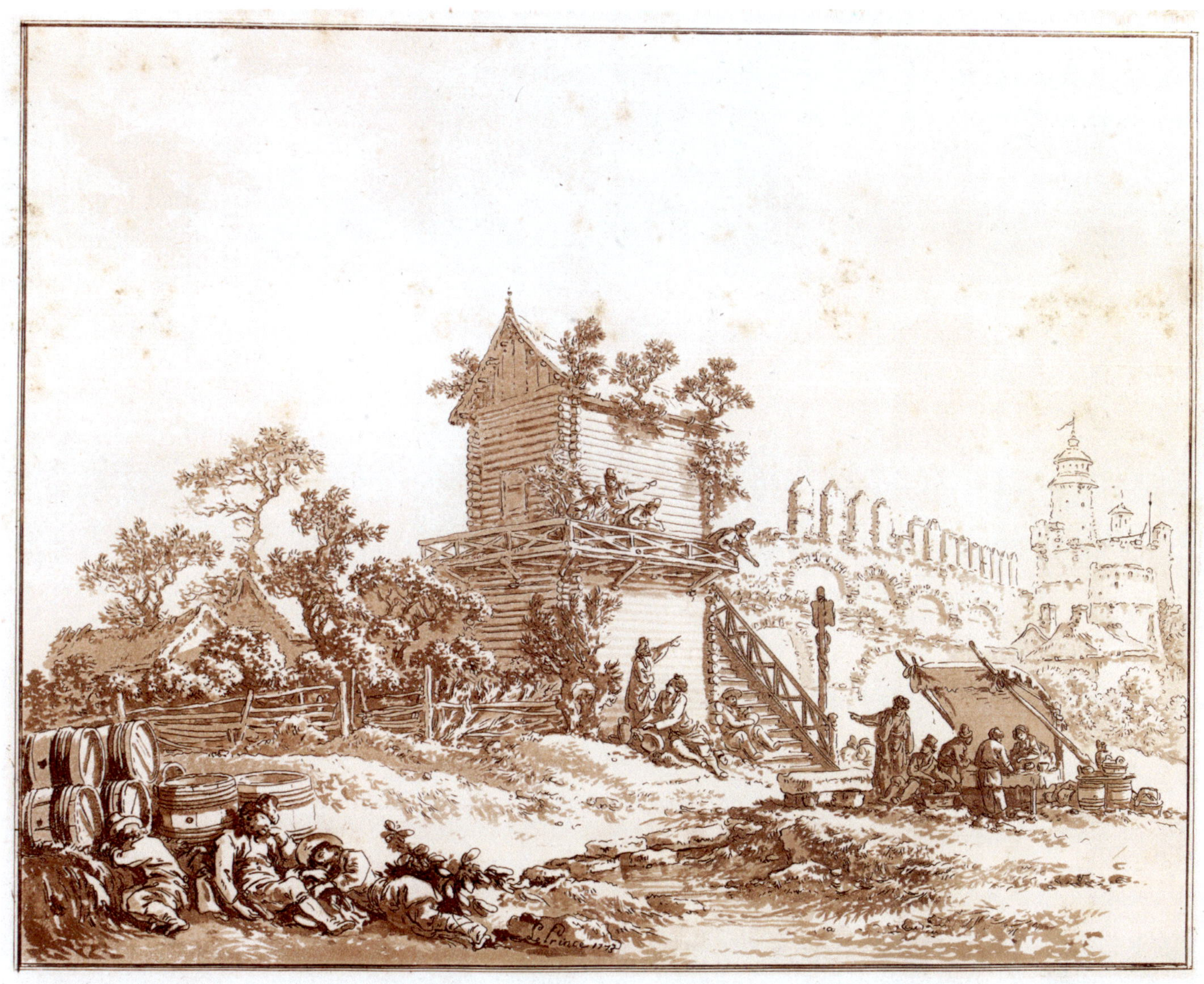

Jean-Baptiste Le Prince, *The Moscow Drink-Shop*, 1765. The 19th-century reality was less picturesque: 'The little drink-shop was in an old, tumbledown little half-rotten hut, shabbier than any poor peasant's hut.… Inside the drink-shop was dirty, dark, full of pipe smoke, cold, crowded and always full.'

about the Russian ambassador to the court of Charles II. For his morning draught he downed a quart glass of spirits mixed with a handful of pepper. Tom Killigrew, courtier and favourite of Charles II, who witnessed and attempted politely to emulate this ritual swore that 'the divel [sic] and hell itself was in it'.

As in western Europe, the Russian government soon discovered that 'trade in spirits provided a large and reliable source of revenue.' Therefore, class restrictions on its sale were lifted and spirits, which profitably used-up surplus grain, became widely available. Vodka was sold through drink-shops, which were state-licensed but often so rustic

they had little more than vodka to attract customers. J.B. Le Prince's 1765 engraving *The Moscow Drink-Shop*, suggests what these drink-shops may have been like. Vodka in Russian culture came to occupy a place of singular importance. Glasses of it were lifted to celebrate church feast days and family events; it functioned as a ceremonial seal when a deal was struck, and it fuelled the village institution known as the 'work party', whereby a meal and some vodka were traded for help with an urgent task, such as gathering in a crop.

Vodka's allure sometimes proved disastrous. An official described the drinking habits of the Russian peasantry in the 1830s:

Everyone drinks—the young, the middle-aged, and the old; the men and the women. In the … [drink-shops] are people of both sexes drinking together with abandon and often forgetting about the children they have left at home with a

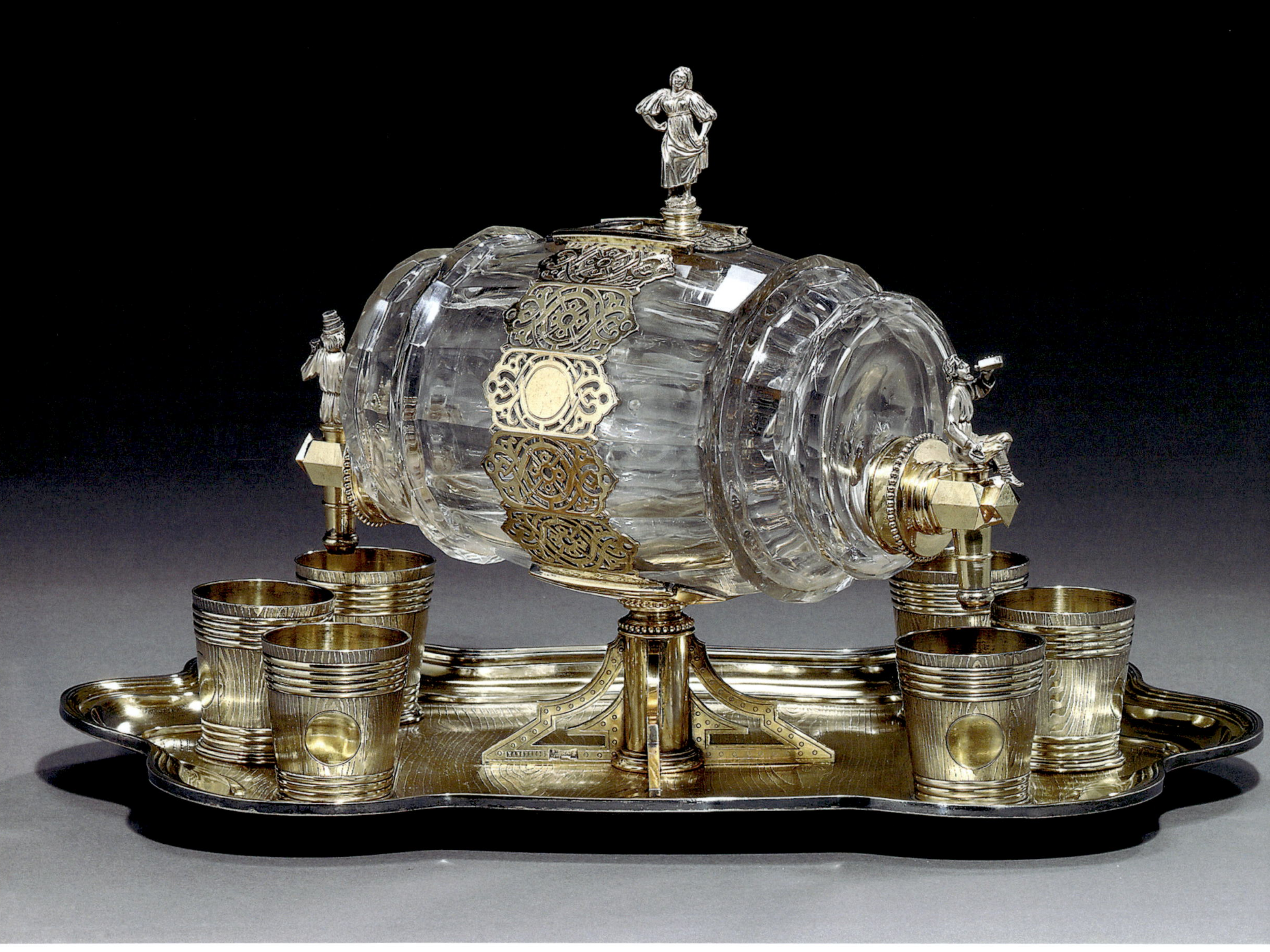

crust of bread. They drink without regard to place or time, whenever and wherever the opportunity arises. They drink because they are rich, and they drink because of their poverty; they drink in happiness and in sorrow; and always drink to excess.

Russian culture embellished vodka drinking with a number of charming customs, as well as distinctive small vessels, both beakers of metal decorated in *niello* and glasses. For example, if one taps the side of one's chin or windpipe, it signifies having a drink. The gesture recalls the story of a peasant who saved the life of Peter the Great. His reward was the right to drink as much vodka as he liked. Fearing

that a written document attesting to this honour would be stolen, he tapped on his throat and begged that the imperial seal be placed there. According to long-established Russian practice, vodka is served ice-cold, neat and in small glasses. A shot is often drunk in one gulp, *do dna* (to the end), and accompanied by a slice of bread or hors-d'oeuvres known in Russian as *zakuski*. Before imbibing it is often customary to blow on the vodka, as this unseats the devil who sits at the top of the glass.

Although vodka may not have originated in Russia, the product we know today owes a great deal to Russian research. The late nineteenth-century Russian scientist D.I.

Opposite: Silver-gilt and glass vodka set by the Moscow silversmith Ivan Klebnikov, 1873. The swivelling glass container harks back to the wooden barrels of peasant drink-shops, also suggested by the beakers. Drunks ornament each spigot, while a peasant girl dances on top.

Right: The Gardner Porcelain Factory, founded in Russia in 1766 by an English entrepreneur, produced this statuette of a wife and her collapsed drunken husband, who clutches a typical Russian four-sided vodka bottle.

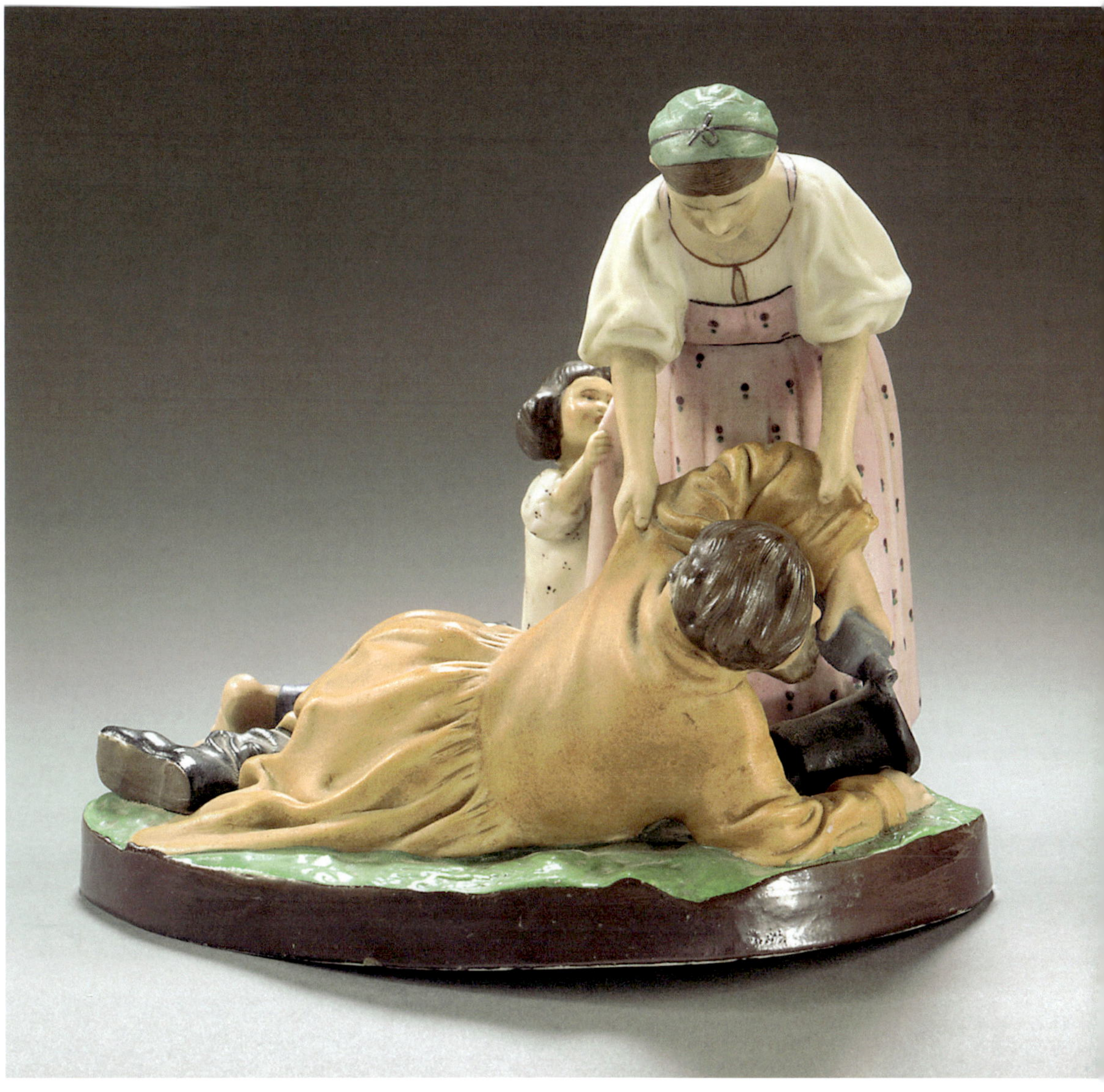

Mendeleyev created the modern template for vodka, which he designated as 'a grain spirit, triple-distilled and then diluted with water to a concentration of 40 per cent by weight.' It was Mendeleyev who insisted that any other terms for the spirit be abandoned in favour of the word 'vodka'. Mendeleyev's formula was adopted in 1894 by the Russian government as the standard for the country's national vodka production.

In modern times, vodka still has its devotees. One notable example is the former Russian president Boris Yeltsin, whose budget 'categorized vodka as an essential commodity like bread or milk.' Affectionately referred to by the Russians as 'our man', Yeltsin was purportedly discovered drunk by one of his ministers during a 1991 political crisis. The quick-thinking minister maintained calm by reassuring waiting crowds that 'I have met the president and he is standing firm in the defence of democracy.'

Thanks to twentieth-century cocktail culture and the advent of the mixed drink, vodka has taken on an entirely new life outside Russia and Eastern Europe. Now one of the most popular spirits in the world, it is associated with urbane sophistication. James Bond's celebrated drink, a vodka martini – shaken not stirred – epitomizes this transformation.

Drink in the Americas

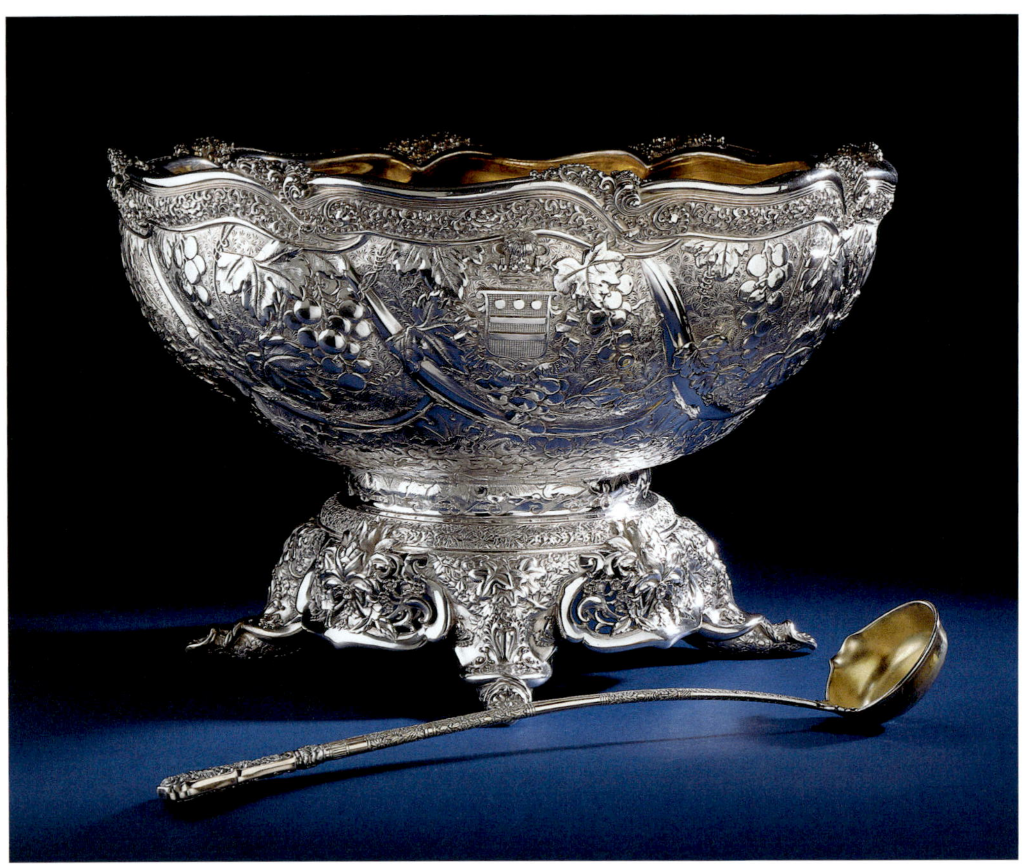

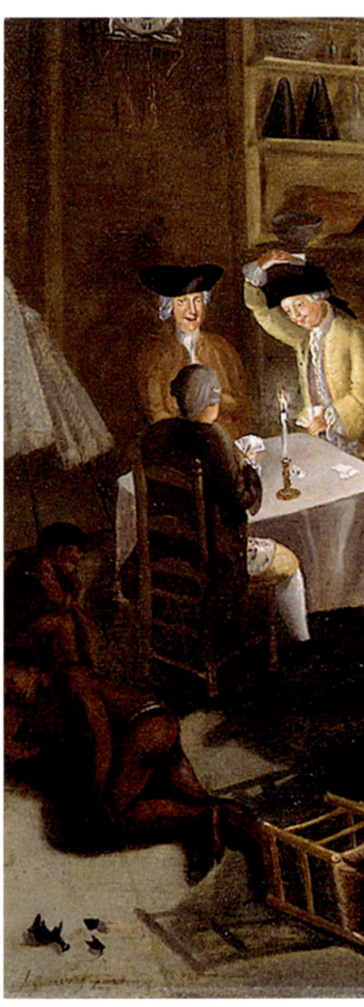

*P*hylloxera, prohibition, rum, mint julep and the cocktail:
these are the stereotypes associated with drink and North
America. Each has had an impact on the visual and material
culture of the past 400 years. America evolved distinctive
drinking customs because of its climate and, as in Britain,
drink and tax became entwined. To assuage hunger and
thirst, to celebrate or commiserate, as medicine and to
escape, alcohol fulfilled the same needs on both sides of the
Atlantic. North America itself is the original cocktail, mixing
ingredients of indigenous, foreign and then a truly American
population whose only constant was change.

Christopher Columbus arrived in Haiti in 1493 and planted
the first sugarcane. Within a century and a half the sugar and
rum trade was to become as valuable as the gold he sought.
Franciscans monks imported European grape vines, as well
as making sacramental wine from native grapes in their quest
for converts. In the seventeenth century the Portuguese,
French, Dutch and English in the West Indies capitalized on
Europe's rampant desire for sugar. Jesuits in Canada
introduced apple trees, just as the Pilgrims did in
Massachusetts. Spruce fir bark and shoots from Canada were
sent back to Britain to flavour beer, and even tested as an
anti-scurvy remedy for the Navy.

Unfamiliar and harsher conditions made brewing beer very
difficult for the settlers. Ploughing for grain and corn was
initially hard, and Indian corn did not produce good malt.
Salted food and heavy starches made for thirsty settlers.
Wine, beer and cider helped them endure or escape from

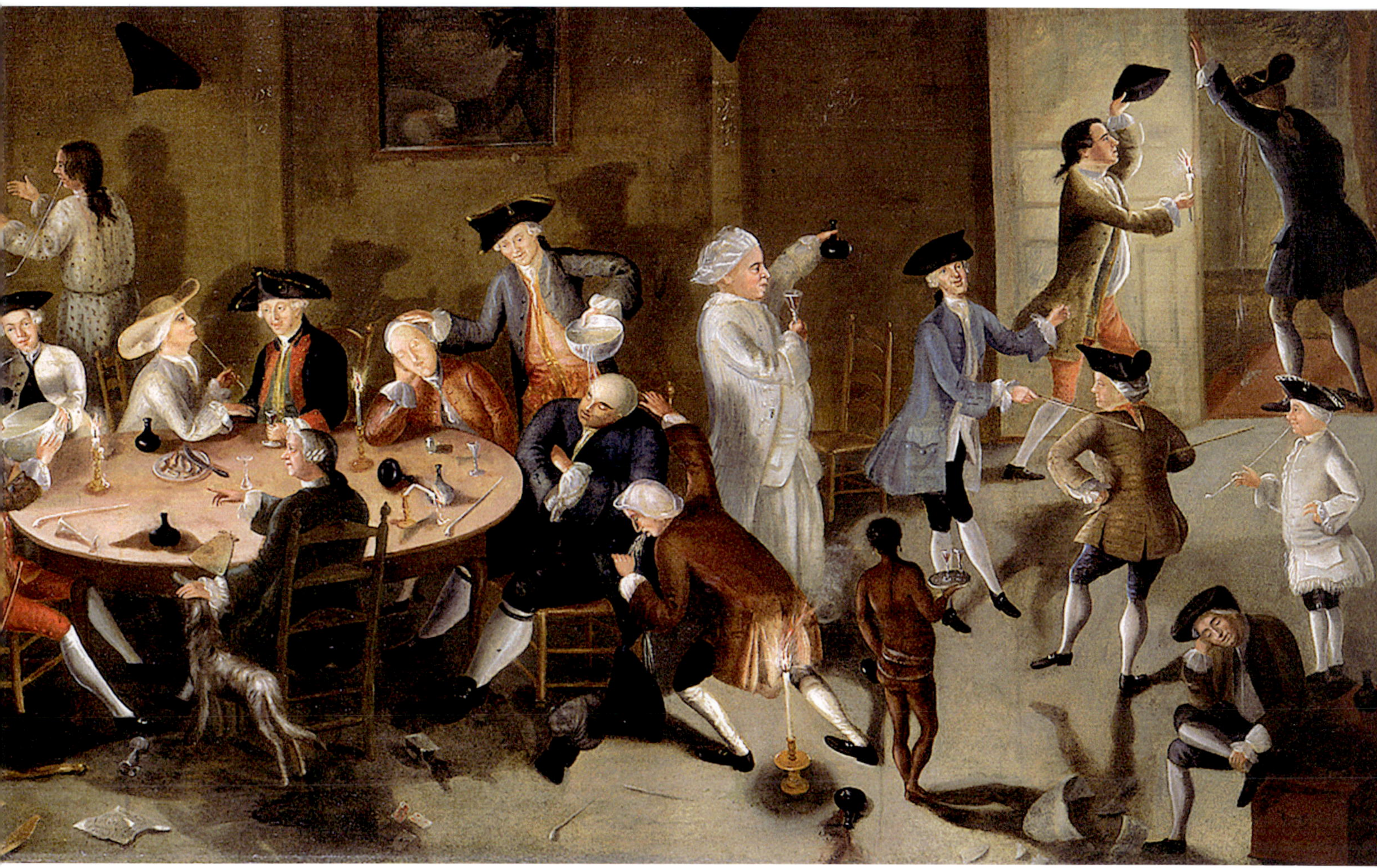

Opposite: A massive punch bowl and ladle made by Tiffany & Co., New York, in 1878, from 20 kg of silver mined in the Comstock Lode, Nevada.

Above: Sea captains carousing on the Dutch sugar island of Surinam, exchanged by the British for Manhattan in 1667. On the northeast coast of South America, Surinam became a frenetic port of call.

their harsh life; Bible verses praising wine and beer satisfied the pious farmer of their goodness.

Apples are not native to America. Before 1625 the eccentric clergyman William Blaxton planted the first apple orchard in Boston. Apple brandy was distilled from apples left in a barrel to ferment. Applejack, lower in alcohol, is concentrated by freezing. Cider, plentiful and cheap, is still much relished.

Even though the use of slaves was legal in New England, religious and political reformers urged consumers to boycott sugar, rum and molasses to stifle this ignominious trade. A moral and cheaper source of stimulus and pleasure, honey was the sweet solution to this stinging problem; mead became a popular and ethical substitute, along with whiskey. However, rum was 'adored by the American English…. 'Tis held as the comforter of their souls, the preserver of their bodies, the remover of their cares, and promoter of their mirth', reported Edward Ward in *A Trip to New England* (1699). This preference for rum was pragmatic. In America's extreme climate beer spoiled easily. Low population density made the constant brewing of beer unprofitable and wasteful; though most settlements had a brewery. It was also difficult to grow hops and barley on the wild land. Wine and brandy were expensive European imports. Distilled spirits were preferred as smaller in volume and easier to transport and store, not to mention the appeal of their kick.

Commerce across the Atlantic was fuelled by the growing demand for alcohol, in a great circle reaching from Britain, Portugal, through the Madeiras to the West Indies, the Carolinas and up the North Atlantic coast. William Cowper, in a poem reflecting on the slave trade in 1768, wrote, 'I pity them greatly but I must be mum, For how could we do without sugar and rum?'

Eager to raise revenues, the British government required colonial-bound cargoes to touch at an English port and pay

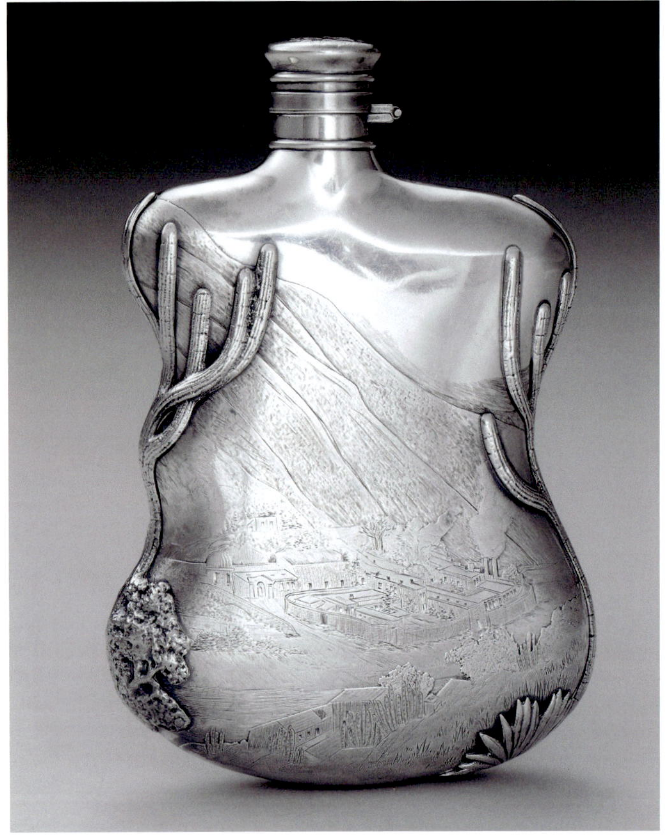

Whisky flask, with scene of the Batopilas silver mine, Sierra Madre, Mexico, one of 10 made as gifts for Alexander Robert Shepherd, head of the silver mine.

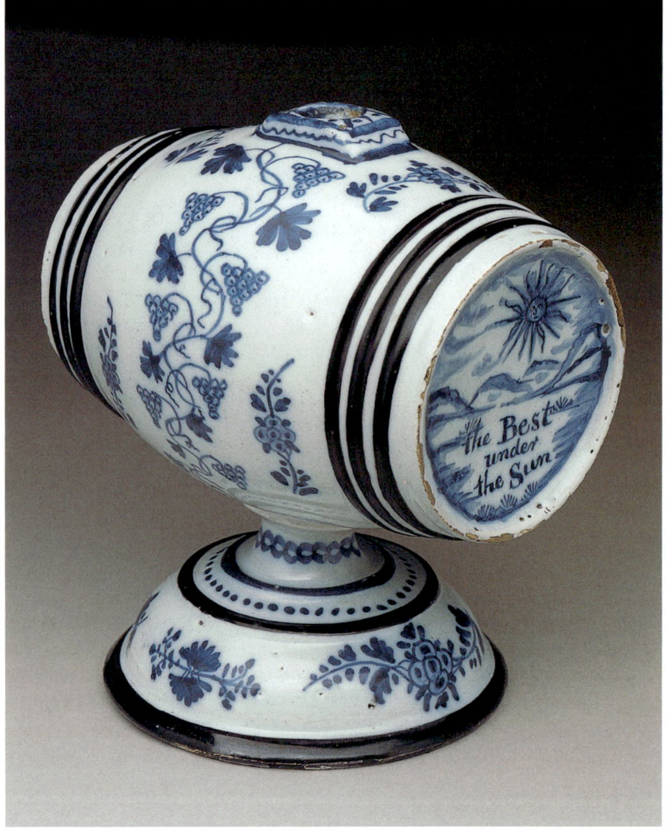

Rum barrel for the sideboard, tin-glazed, late 18th century.

tax before being sold. American ships circumvented the law by buying molasses cheaper from the excise-free French and Spanish controlled islands. England retaliated with the Molasses Act in 1733, imposing heavy import duties, to the anger of northerners who imported the molasses and made and traded the rum. This act was openly flaunted, America's first act of civil disobedience. Eventually, British restrictions on rum and the development of American whiskey made the latter more desirable.

In 1764 Britain, wanting a larger market for English-shipped port, imposed an excise duty on Madeira wine, which had been imported duty free from the Portuguese-owned islands. Boycotting Madeira and port, Americans drank beer and cider instead. Vast amounts of Madeira wine were smuggled; John Hancock's ship *The Liberty* was seized in Boston Harbour. Customs officials had discovered that 100 of the 125 pipes (126 gallons each) of Madeira had been unloaded the night the ship landed. The seizure caused a riot. The British

retaliated by sending 12 warships to Boston, which further incited the populace towards revolution. However, it was the proposed tax on tea and the East India Company monopoly that finally triggered the American colonial revolt. Ironically, in 1791, the very first tax in this new nation was imposed on whiskey by Alexander Hamilton. Rebellion against it raged for years and the rebels in Pennsylvania were threatened militarily. The tax still remains today.

The westward move in the nineteenth century, culminating in the Californian Gold Rush of the 1840s, created many dangerous settlements where conventional society was cast aside for hard living and drinking. This drew condemnation from some religious denominations, such as the Methodists, who spearheaded the early prohibition movement. State prohibition laws gained momentum in the 1850s and by 1855 some 12 states as diverse as New York and Nebraska were dry, at least for a while. The Civil War (1860–65) drew public attention away from prohibition. Liquor was the anaesthetic

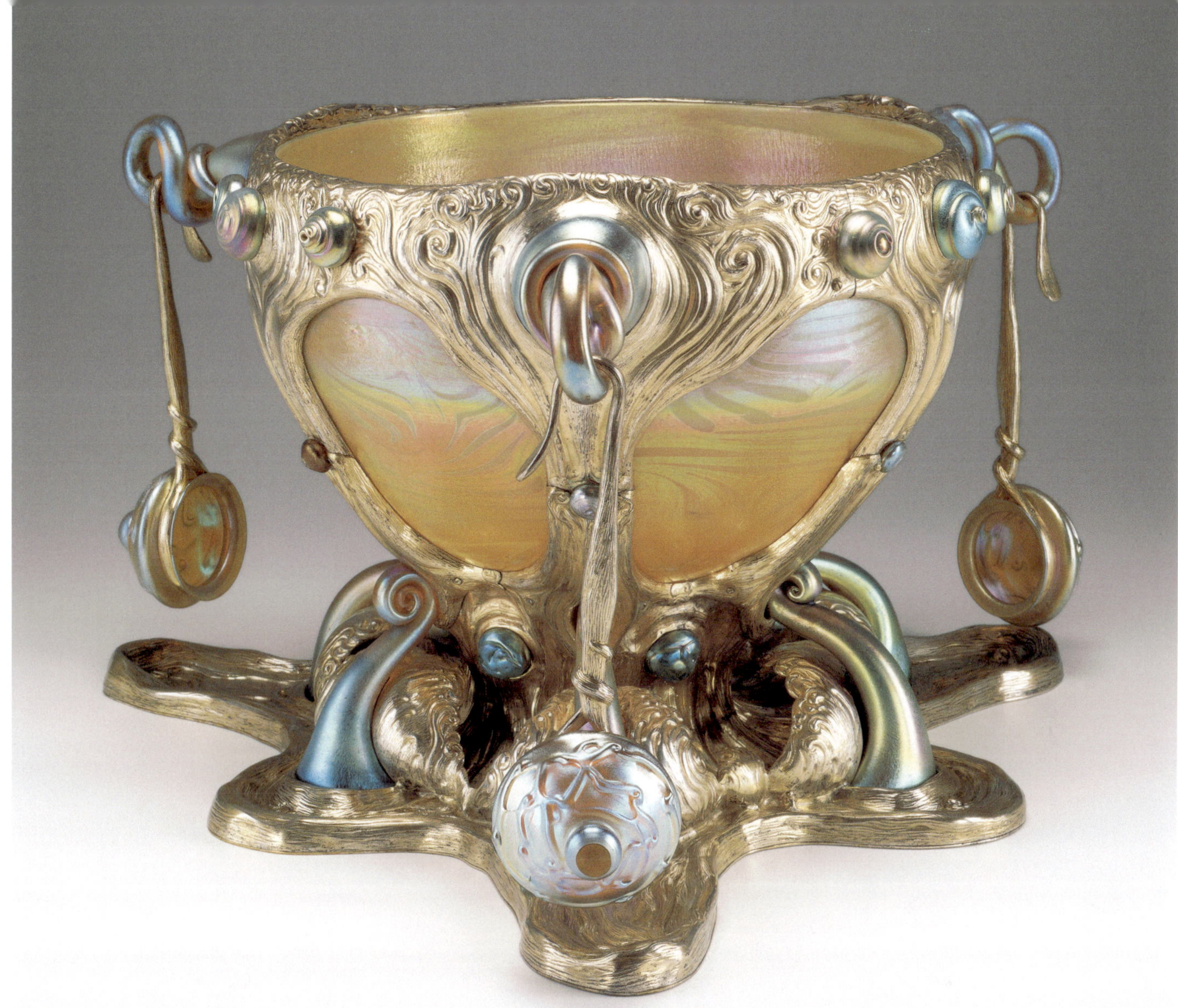

Punch bowl and ladles, iridescent glass, c.1900. Tiffany silver punch ladles were a standard item in 19th-century catalogues. They were made with whalebone or ivory handles to avoid conducting the heat of steaming punch, which was served in winter before and during dinner.

for wounds of the body and the soul, with brother against brother and battlefield amputations common.

In the years that followed, taverns and saloons, once the venue of free thinkers, became seen as dangerous places where new immigrants, especially from southern Europe, gathered to continue this political tradition. The goings-on in hotel spirits bars were also questioned; no respectable woman entered – this was the enclave of the prostitute. The Woman's Christian Temperance Union (founded 1873), the Anti-Saloon League of New York (founded 1893) and male-led Bible-thumping groups trumpeted the virtues of abstinence. It took 69 years from prohibition in 1851 in Maine to the actual implementation of prohibition

countrywide; in 1920 the Eighteenth Amendment to the Constitution prohibited the sale or manufacture of alcohol. Loopholes included a doctor's prescription, tonics and wine for religious services. Some grape juice was labelled 'Warning: will ferment and turn into wine'. Legal commercial production of alcohol was severely curtailed, almost wiping out the burgeoning wine industry, since producers could legally make only sacramental wine, grapes and grape juice.

Ingeniously Americans concealed their alcohol in hollow canes, books and even a fire extinguisher, made in silver by Aspreys. Speakeasies, illegal drinking dens, sprang up across the country, with patrons whispering passwords through unmarked doors. The jazz age was born. Due in fact to economic pressures, President Franklin D. Roosevelt ratified the Twenty-first Amendment, which repealed prohibition, in 1933, declaring, 'I believe this would be a good time for a beer', though he was known for his love of the martini.

The Birth of the Cocktail

The influence of cocktails on modern life cannot be exaggerated…

E.M. DELDERFIELD, *DIARY OF A PROVINCIAL LADY IN AMERICA* (1934)

Many stories circulate about the invention of the cocktail, a drink described as 'excellent for the head' when first mentioned in print in 1803. All are unreliable and the story of choice, rather like the drink itself, is guided by personal preference. The following account combines several flavours. It gives the drink a modicum of antiquity and yet eschews any hint of traditionalism. It underpins the essential racy inventiveness of cocktail culture, combined with a degree of youthful subversiveness that again is often associated with cocktails and their consumption. And it undeniably attributes its invention to the United States, the nation that above all is associated with the cocktail.

Legend has it that the cocktail was invented during the American War of Independence in the early 1770s. A detail of General Washington's troops were relaxing in the 'Four Corners' inn in upstate New York when Betsy Flanagan, the young Irish barmaid serving them, decided to serve a new drink of her own making. Into each glass she poured a mixture of rum, rye whiskey and several fruit juices and in a final flourish, each drink was decorated with a feather, plucked from the plump rooster of a nearby loyalist farmer. Amidst much laughter, as she served the drinks, a young French officer rose and proclaimed, 'Vive le coq tail' and so the cocktail, a mixture of spirits and soft drinks was born.

We can be more certain of the first description of the cocktail. In a May 1806 issue of the local newspaper, *The Balance and Columbian Repository* in Hudson, New York State, after election accounts showed a candidate buying votes with 25 dozen cocktails, a curious reader wrote in:

Sir, I have read your article which appeared on the sixth of this month regarding the accounts submitted by a Democratic candidate … under the title of Loss, 25, do cocktail. Would you be good enough to inform me as to what is understood by this form of refreshment?

To which the editor wrote the following reply.

A cocktail is a stimulating drink made with all manner of spirits, sugar, water and bitters; it is commonly known as bittered sling and is thought to be an excellent potion during the electoral campaign because it emboldens the heart and befuddles the head…. It is also said to be particularly useful

to a Democratic candidate, because anyone who will swallow a glass of it will swallow anything.

Cocktails began to enjoy widespread popularity in the US from the mid-nineteenth century. The basic staples of rye whiskey, brandy and gin were widely available and Americans had been enjoying mixed drinks, often flavoured with herbs, fruit juices or bitters such as wormwood, for two centuries already. More exotic liqueurs for mixes at first could to be found only in the more cosmopolitan centres such as New York, Boston or San Francisco, cities where the cocktail culture began to flourish. But the major influence on its popularity was the invention of artificial refrigeration: in short the ability to manufacture ice, whatever the ambient atmospheric temperature. Dr William Cullen, Professor of

A classic shaped cocktail or Martini glass from the mid-1920s by Lalique. The *Savoy Cocktail Book* recommended icing glasses before using them. The long stem helped keep the cocktail cool.

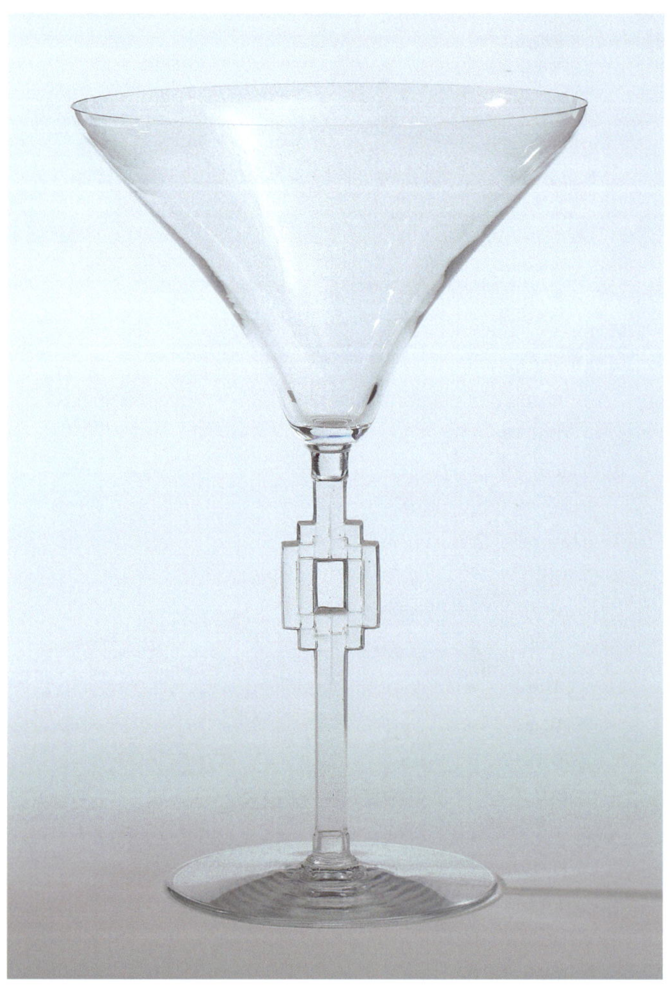

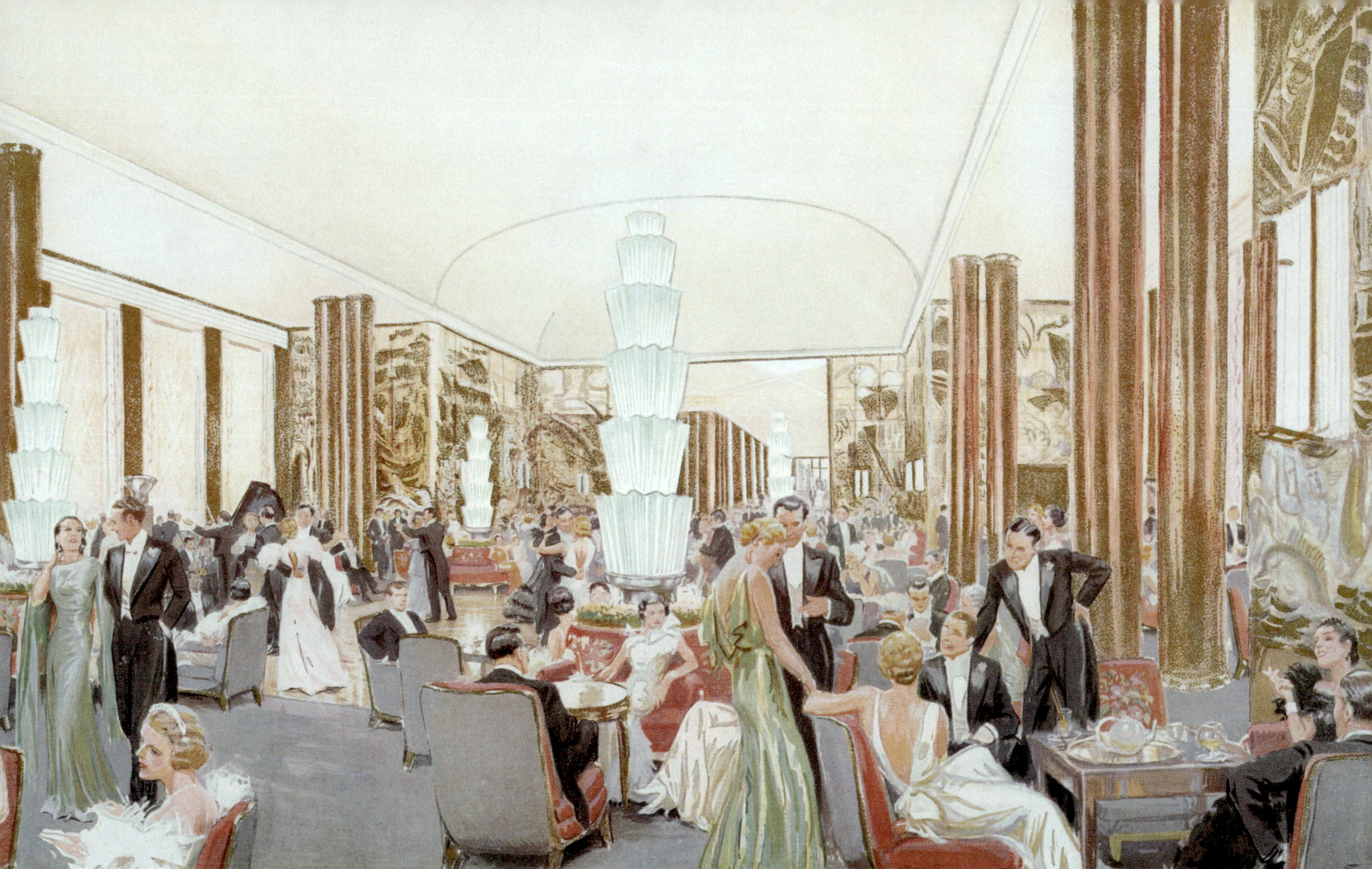

Cocktail hour takes place in the Art Deco Grand Salon of the luxurious French ocean liner SS *Normandie*, launched in 1932.

Medicine and Chemistry at Glasgow University, first made the discovery that ice could be formed by evaporation in 1775, but the practical application was left to others. In 1828, the American Jacob Perkins registered a London patent for a machine that produced low temperatures by the compression and subsequent expansion of a vapour. Further improvements were made by the French brothers Edmund and Ferdinand Carrè who demonstrated their machine at the Great Exhibition of 1851. It was later used extensively throughout North America.

With ice, the vital ingredient, readily and cheaply available, the popularity of cocktails throughout North America surged. In Chicago, in 1866, the first professional association of bartenders, The Bartenders and Waiters Union, was formed, and by 1900 no fewer than 40 books on bar keeping and cocktails had been published. The first and one of the most famous was *The Bon Vivant's Companion or How to Mix Drinks* by the legendary bartender Jerry Thomas in New York in 1862. By the end of the century, the cocktail had become international and the first cocktail bars, whether they were in London, Paris, Berlin or Vienna, inevitably became known as American bars.

In Britain the popularity of the cocktail reflected changing social patterns. Whereas in the Edwardian era dinner was a highly elaborate, formal affair, served over many courses and starting at an early hour, the post-war generation preferred a simplified ritual with fewer courses and starting later in the evening. Suddenly, there was a gap in the day between afternoon tea and dinner that needed to be filled. The cocktail hour was born! Several have claimed the honour of introducing the cocktail party to British society. In 1924, the painter, C.R.W. Nevinson and his wife hosted an 'At Home', reinforced by a large jar of a cocktail; rum seems to have been the chief ingredient, but only two guests arrived. When the writer Alec Waugh hosted a similar occasion the following year, not wanting to alarm his friends, he invited them to tea and served them a rum swizzle, mixed by a friend from New York. According to Waugh, the party was a success and shortly thereafter the cocktail party was an established part of the English social scene. Cocktails became particularly fashionable when the Savoy opened its American Cocktail Bar in 1929 under the management of Harry Craddock, a celebrated New York bartender who since

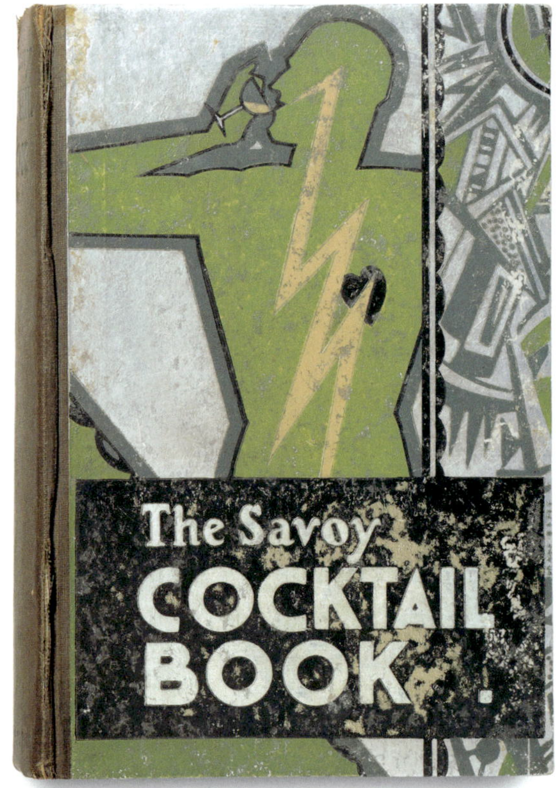

been commonly supposed (although at times it probably tasted as though it was), but was so called because it needed to be diluted before being sold. The bottles did not fit under a sink tap but did fit under those of a bath. Better quality spirits were smuggled in from the West Indies or Canada. The Canadian border was particularly porous. While the Canadian government had to contend with pressures from its own temperance movements, it resolutely refused to introduce similar legislation. It earned far too much money in taxes from the drinks industry. Not until the Great Depression at the end of the decade was the US government forced to accept that this industry was too valuable a source of revenue to forgo. In 1933, the Volstead Act was repealed.

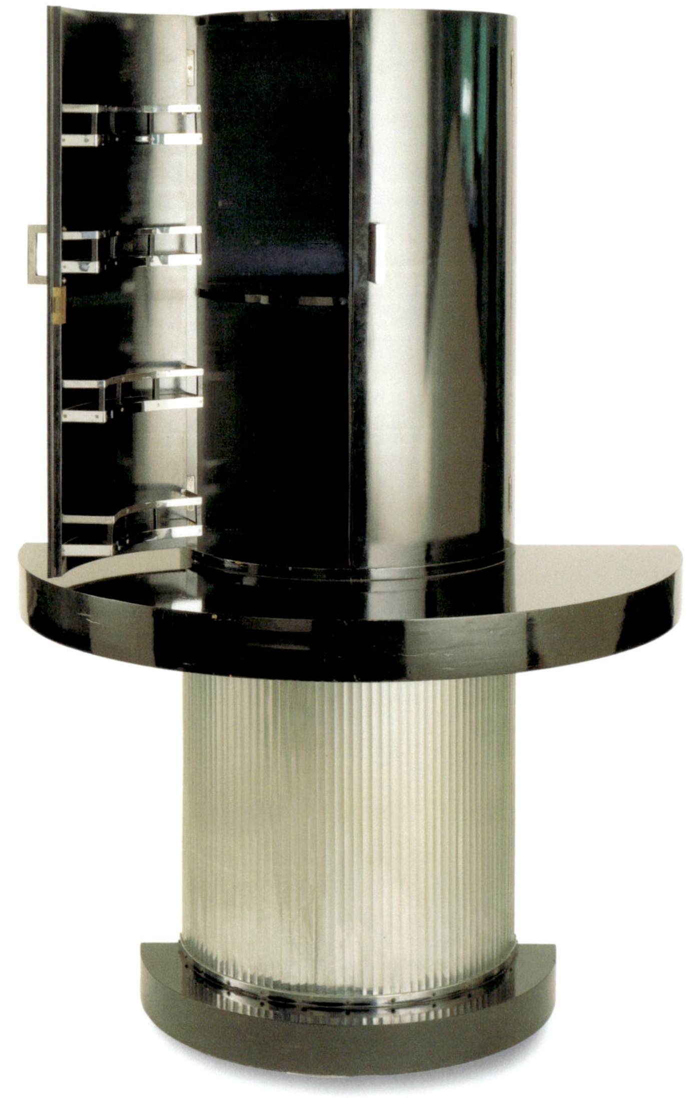

the introduction of prohibition in the US had been having difficulty in finding legitimate work. One year later *The Savoy Cocktail Book*, written by Craddock, was published and has since become the Bible of the British cocktail trade.

Cocktail drinking became the vogue among the bright young things because it was novel, American and outraged their parents, but at least in Britain it was legal. In 1920 the temperance movement, a vocal pressure group in North America since the 1850s, finally gained political victory when Andrew Volstead, the Minnesota Congressman, succeeded in introducing the National Prohibition Act. This measure served as enabling legislation for the total prevention of all manufacture and sale of alcoholic goods within the United States. While it wreaked havoc on the brewing and wine industries, it did little to discourage the spirits trade and therefore cocktail consumption. It simply drove it underground. Cocktails could conveniently disguise the evidence of alcohol until of course they started to be drunk. Americans did not go thirsty; organised crime saw to that.

Spirits were either distilled locally or smuggled in from abroad. Bathtub gin made from distilled corn syrup mixed with glycerine and juniper oil was not mixed in a bath as has

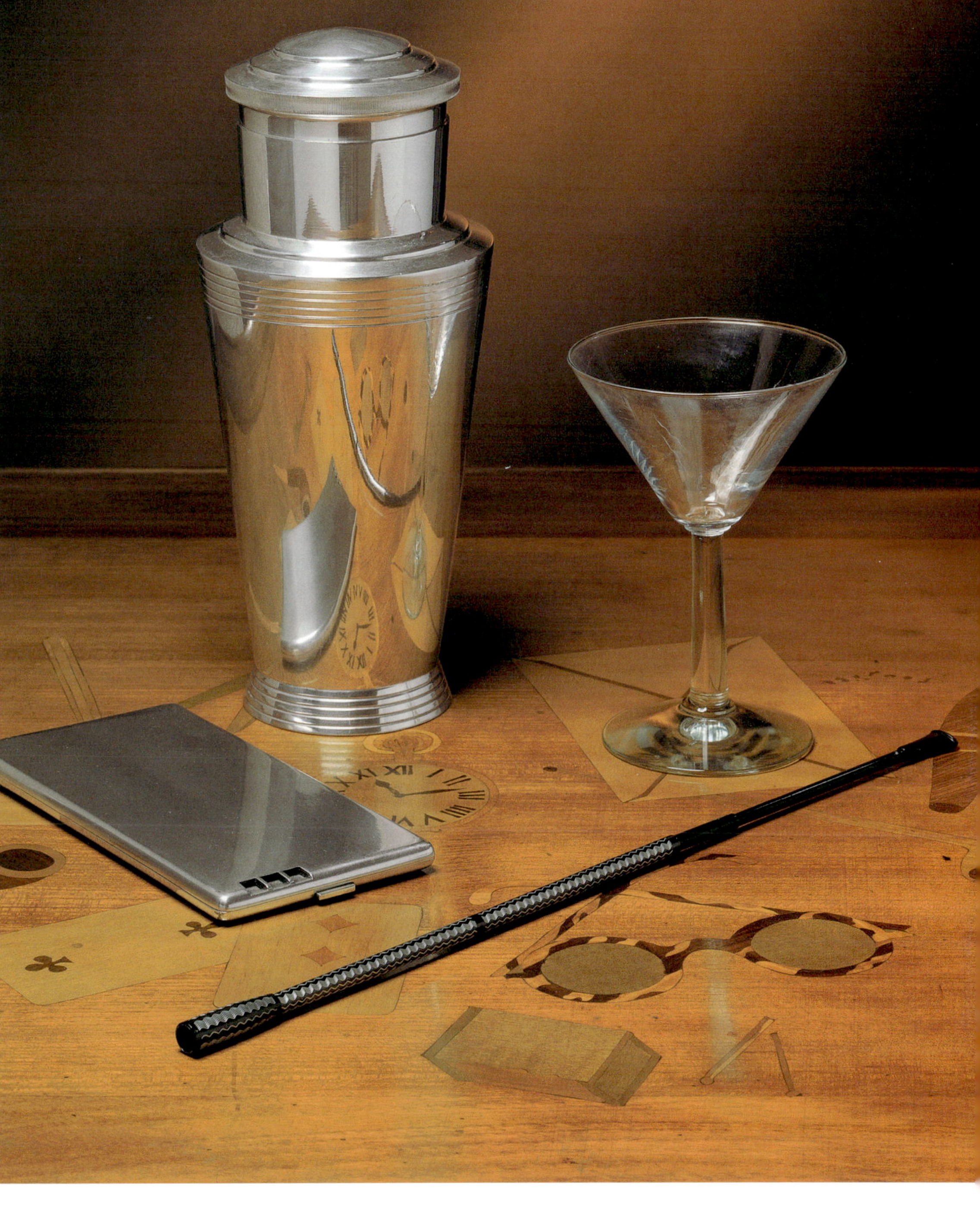

Prohibition acquainted Americans with gin because whisky was so much harder to emulate. The martini, essentially gin with a splash of vermouth and garnished with a Spanish olive, became the classic American drink. Entire books have been devoted to recipes for countless variations and the vodka-based version of that drink has been immortalized by the fictional spy James Bond: 'Martini, shaken not stirred' is, according to one poll, the 90th out of a 100 'most quoted' phrases from film history. But the final word on the subject should go to the American poet, Ogden Nash (1902–1971).

A DRINK WITH SOMETHING IN IT.

There is something about a martini,
A tingle remarkably pleasant;
A yellow, a mellow martini;
I wish I had one at present.
There is something about a martini,
'Ere the dancing and dining begin.
And to tell you the truth,
It is not the vermouth –
I think that perhaps it's the gin.

Further Reading

Arnold, R., 'Emma's Wedding Feast: A Glance at Flaubert's *Madame Bovary*', *Proceedings of Oxford Symposium on Food and Cookery 1990: Feasting & Fasting* (Totnes, 1990)

Barr, A., *Drink: A Social History* (London, 1998)

Banks, F., *Wine drinking in Oxford, 1640–1850: a story revealed by tavern, inn, college and other bottles – with a catalogue of bottles and seals from the collection in the Ashmolean Museum* (Oxford, 1997)

Banks, F., *The Wine bottles of All Souls College* (Oxford, 2002)

Beresford, J. (ed.), *James Woodforde, The Diary of a Country Parson 1758-1802* (Oxford, 1972)

Bickerton, L., *Eighteenth Century Drinking Glasses* (Woodbridge, 1987)

Blair, C., 'Loving Cups and Grace Cups', *Antiquaries Journal* (2004), vol. 84, pp.393-9

Butler, R. and Walkling, G., *The Book of Wine Antiques* (Woodbridge, 1986)

Brears, P.C., *The Collectors' Book of English Pottery* (Newton Abbot, 1974)

Brears, P.C., 'The Christmas wassail in England', *Custom and Ceramics. Essays presented to Kenneth Barton* (Wickham, 1991)

Brown, P. and Schwartz, M., '*Come Drink the Bowl Dry' Alcoholic Liquors and Their Place in Eighteenth Century Society* (York, 1996)

Bruce, P. A,. *Economic History of Virginia in the 17th Century: An Inquiry into the Material Condition of the People, based upon Original and Contemporaneous Records* (New York, 1896)

Clark, P., *British Clubs and Societies 1580-1800: The Origins of an Associational World* (Oxford, 2000)

Charleston, R., *English Glass & the glass used in England, c.400-1940* (London, 1984)

Clifford, H., *A Treasured Inheritance: 600 Years of Oxford College Silver* (Oxford, 2004)

Cornell, M., *Beer: the Story of the Pint* (London, 2003)

Cressy, D., *Birth, Marriage and Death: Ritual, Religion and the Life-Cycle in Tudor and Stuart England* (Oxford, 1997)

Cross, F.L. and Livingston, E.A. (eds), *The Oxford Dictionary of the Christian Church*, 3rd edition (Oxford, 1997)

Dembińska, M., *Food and Drink in Medieval Poland* (Philadelphia, 1999)

Farrington, A., *Trading Places: The East India Company and Asia 1600 -1834* (British Library, exhib. cat., 2002)

Francis, A.D., *The Wine Trade* (London, 1972)

Goldsmiths Company, *The Goldsmith and the Grape Silver in the service of wine* (Goldsmiths Hall, exhib. cat., 1983)

Gourvish, T. R. and Wilson, R. G., *The British Brewing Industry 1830-1980* (Cambridge, 1994)

Harrison, B., *Drink and the Victorians* (London, 1971)

Haydon, P., *The English Pub. A History* (London, 1994)

Haydon, P., *An Inebriated History of Britain* (Stroud, 2005)

Herman, S. and Pascal, J., *Mouton Rothschild: The Museum of Wine in Art* (Paris, 2003)

Hughes-Hallett, P. (ed), *Jane Austen: 'My Dear Cassandra'* (London, 1990) *International Food & Wine Society Journal, passim*

Johnson, H., *Wine, Celebration and Ceremony* (New York, 1985)

Kanof, A., *Jewish Ceremonial Art and Religious Observance* (New York, 1969)

Kettering, Karen L., *Russian Glass at Hillwood* (Washington, D.C., 2001)

Kumin B. and Trusty, B.A. (eds.), *The World of the Tavern: Public Houses in Early Modern Europe* (London, 2002)

Lipski, L.I. and Archer, M., *Dated English Delftware 1600-1800* (London, 1984)

Lynn Martin, A., 'Old People, Alcohol and Identity in Europe, 1300-1700', *Food, Drink and Identity: Cooking, Eating, and Drinking in Europe since the Middle Ages* (New York, 2001)

McConnell, A., *The Decanter: An Illustrated History of Glass from 1650*, (Woodbridge, 2004)

McGovern, P.E., *Wine DNA: Ancient wine Search for the origins of Viticulture* (Princeton, 2003)

McKearin, H., 'Notes on stopping, bottling, binning' and McNulty, R. 'Common Beverage Bottles: production, use and forms in seventeenth & eighteenth century Netherlands', *Corning Journal of Glass Studies* (1971), vol. 13, pp. 120/7; 109-119

Mason, L. (ed.), *Food and the Rites of Passage, Proceedings of the Leeds Symposium on Food History* (Totnes, 1999)

Minchinton, W., 'Cider & folklore', *Folk Life* (1975), vol. 13

Pantin, W.A., *Oxford Life in Oxford Archives* (Oxford, 1972)

Picart, B., *The Ceremonies and Religious Customs of the Reformed, commonly called Calvinists* (London, c.1733)

Pinney, T., *A History of Wine in America: From the Beginnings to Prohibition* (Berkeley, 1989)

Pokhlebkin, V.V., *Chai i Vodka* (Moscow, 1995)

Pokhlebkin, W., *A History of Vodka* (London; New York, 1992)

Pryme, Abraham de la, *The Diary of Abraham de la Pryme: the Yorkshire Antiquary* (Durham, London and Edinburgh, 1870)

Robinson, J.A. (ed.), *The Ames Correspondence. Letters to Mary: A Selection from Letters written by Members of the Ames Family of Lakenham, Norwich, 1837-1847* (Norfolk Record Society, 1962)

Rude, G.F.E., '"Mother Gin" and the London Riots of 1736', *The Guildhall Miscellany* (1959), vol. 10, pp.53 - 63

Salter, J. (ed.), *Wine Labels 1730 to 2003: A Worldwide History* (Woodbridge, 2004)

Scott Thomson, G., *Life in a Noble Household 1641-1700* (London, 1959)

Simon, A., *A History of Champagne* (London, 1962)

Skarzynski, L., *L'alcool et son histoire en Russie: étude economique and sociale* (Paris, 1902)

Smith, E., *The Compleat Housewife: or, Accomplished Gentlewomen's Companion* (London, 1758)

Smith, R.E.F. and Christian, D., *Bread and Salt: A social and economic history of food and drink in Russia* (Cambridge, 1984)

Syndram, D, & Scherner, A. (eds), *Princely Splendour The Dresden Court 1580-1620* (Metropolitan Museum of Art/Staatliche Kunstsammlungen exhib. cat., 2004)

Timbs, J., *The Club Life of London* (London, 1886)

Underdown, D., *Revel, Riot & Rebellion: Popular Politics and Culture in England 1603-1660* (Oxford, 1987)

Willumsen Krog, O., *Kongelige glas. Royal Glass: An Exhibition of four centuries of table glass, glass services and goblets* (Christiansborg Palace Copenhagen exhib. cat., 1995)

Wilson, C.A., '"Burnt Wines & Cordial Waters" The early days of distilling', *Folk Life* (1975), vol. 13

Wilson, C.A., *Liquid Nourishment: Potable Foods and Stimulating Drinks* (Edinburgh, 1993)

Zeldin, T., *Oxford History of Modern Europe, France 1848-1945: Intellect, taste and anxiety*, vol.2 (Oxford, 1977)

Notes on the Illustrations

Jacket

Front: Vanitas, oil on canvas, Willem van Aelst, Netherlands, 1659.
© Johnnie Van Haeften; *front flap:* Decanter for Rum, glass, probably
Bristol, *c.*1790-1800. V&A: C.43-e-1961; *back:* Cocktail Shaker,
electroplated nickel silver, Mappin and Webb, Sheffield, 1933. V&A:
M.226-b-1984; cigarette holder, aluminium, possibly Paris, *c.*1925.
V&A: Circ.40-1972; cigarette case, silver, Margaret Craver Withers,
London, 1946. V&A: M.53-199; *back flap:* Cunha Braga hunting cup,
enamel, gold and rock crystal, German, *c.*1600. © Christie's Images
Ltd. 2006.

Frontispiece : Goblet, blown and engraved glass, Verzelini glasshouse,
London, 1586. V&A: C.226-1983.
page 2 Court wine cup, silver, parcel-gilt, oxidised and patinated,
Kevin Coates for Mr Bruno Schroder, 1990. © The Worshipful
Company of Goldsmiths.
4 Milkmaid cup, silver-gilt, London, *c.*1665. Gans Collection, New
York. Photo by Steve Tucker.
5 Memphis cocktail glass, clear and green plastic, designed by
Richard Holloway for Straight Lines Saltwater Ltd., English, 1983.
V&A: C.338-1983.

Introduction

6 Beer jug, porcelain, Chinese, *c.*1800. V&A: C.43-1951.
7 Menu for Rothschild family dinner, 1885. Reproduced with the
permission of The Rothschild Archive.
8 Beer jug, salt-glazed stoneware and pewter, Cologne, 1525-50.
V&A: 780-1868.
9 'Aristocrat' decanter, glass, designed by Per Lutken for Holmegaard,
Danish, 1955. © Andy McConnell.

ATTITUDES

10-11 Tailors at work, oil on canvas, British School, *c.*1780.
© Museum of London.

Alcohol as Medicine

12 Frontispiece from Johann Sigismund Elsholtz, Georg Wolfgang
Wedel, Werner Rolfink, *Destillatoria curiosa; sive ratio ducendi
liquores coloratos per alembicum,* Berlin, 1674. © The Royal College
of Physicians of London.
13 Illustration of wormwood from Johan Zorn, *Icones Plantarum
medicinaliun,* Nuremberg, 1779-84, vol. II. © The Royal College of
Physicians of London.
14 Cup, cast antimony, English, *c.*1720. V&A: 1370-1900
15 'Our Water Supply' cartoon from J.F. Sullivan, *The British Working
Man,* 1878. V&A: NAL V.G29.G.Box V.

Evils of Alcohol

16 Gin Lane, engraving, William Hogarth, London, 1751.
V&A: F.118.87.
17 *above:* The Temperate and the Intemperate, tempera colours and
ink on parchment, Master of the Dresden Prayerbook, Bruges,
*c.*1475-80. The J. Paul Getty Museum, Los Angeles. 91.MS.81.recto,
bottom: The Ale Bench, earthenware, Staffordshire, late 18th century.
V&A: C.2-2002. Purchased through the Julie and Robert Breckman
Staffordshire Fund.
18 *above:* Lantern slide 'Accidental Death' from *The Temperance
Sketchbook,* no. 29, English, *c.*1875-1900. V&A: Misc.115:22-1977;
bottom: Anti-alcohol poster issued by the National Association Against
the Danger of Schnapps, Zurich, 1925. V&A: E.1110-2004.
19 'Don't ask a man to drink and drive', poster, British, 1966.
V&A: E.216-1982.

Political and Legal Controls of Drink

20 *left:* Parliament Bowl, tin-glazed earthenware, probably Brislington
Pottery, English, 1736. V&A: 3620-1901; *right:* Illustration from
G. Cruikshank, *The Glass and the New Crystal Palace,* London, 1853.
V&A: NAL Box I, G.28.F.
21 Antwerp Market (detail), oil on panel, Flemish School, *c.*1600.
Musées Royaux des Beax-Arts de Belgique, Bruxelles - Koninklijke
Musea voor Schone Kunsten van België, Brussel.
22 *left:* Tavern Mug, pewter, mark of William Hux, English, *c.*1710.
V&A: M.59-1945; *right:* Lid of a baluster measure, pewter, English,
*c.*1550-1600. V&A: M.314-1923.
23 *above:* Behind the Bar, watercolour, J.H. Henshall, English, 1882.
© Museum of London; *bottom:* Brewers Sign, painted and gilded iron,
German, late 18th century. V&A: M.603-1924.

Wine and Religion

24 *above:* The Mass of St Gregory, stained glass, Amsterdam or
Leiden, *c.*1520. V&A: 1015-1905; *bottom:* Kiddush cup, silver and
enamel, Tamar de Vries Winter, English, 2005. V&A: M.19-2005.
25 Passover meal from Bernard Picart, *The Ceremonies and Religious
Customs of the Reformed, commonly called Calvinists,* *c.*1733.
V&A: NAL 24.G.140.
26 The Election II: Canvassing for votes, engraving, William Hogarth,
English, 1757. V&A: F.118.77.
27 *left:* Nonconformist loving cup, pressed glass, English, *c.*1850-90.
The Museum of Methodism and John Wesley's House, London, on
loan to V&A; *right:* Chalice, silver, enamel and ivory, designed by
Henry Wilson, London, 1898. Church of St Bartholomew, Brighton,
on loan to V&A.

Holding Your Drink

28 Dives and Lazarus, engraving, Crispijn van de Passe, Flemish, late
16th century. V&A: 25497 I-14-d.
29 The *Tête-à-tête*, oil on wood, Gabriel Metsu, Amsterdam, late
1650s. Waddesdon, The Rothschild Collection (The National Trust).
Photo by Martin Charles, © The National Trust, Waddesdon Manor.
30 Detail from dining scene, engraving, attributed to Romeyn de
Hooge, Netherlands, *c.*1660-1670. Collection of Ann Eatwell.
31 Tenants' dinner in the Great Hall at Cotehele, Cornwall, lithograph,
Day and Haghe after Nicholas Condy, *c.*1840. From the Collection of
Mt. Edgcumbe House, Plymouth City Museum & Art Gallery:
Plymm.034.
32 *left:* Tray and decanter, papier-mâché, japanned and gilded, with
appliqué of mother-of-pearl, designed by Richard Redgrave,
Birmingham, 1865. V&A: C.108-1992 and C.132-1865; *right:* drinking
after dinner from the Wynnstay Album, watercolour, English, late 18th
century. V&A: E.2387-1948.
33 Service made for the Prince of Wales, cut and engraved glass,
Perrin, Geddes & Co, Warrington, English, *c.*1806-8.
V&A: C.56 & A-1976.

Vintages and Vintners

34 *above:* The Royal College of Art Senior Common Room Wine
Committee, print after a sketch, Edward Ardizzone, English, mid 20th
century. V&A: E.187-1981; *bottom:* Vintners' trade card, ink on paper,
Paris, mid 18th century. Waddesdon, The Rothschild Collection (The
National Trust). Photo by HEDS Digitisation Services, © The National
Trust, Waddesdon Manor.
35 Goblet, blown and engraved glass, Verzelini glasshouse, London,
1586. V&A: C.226-1983.
36 Claret jug, glass and silver, Christopher Dresser, London, 1879-80.
V&A: Circ.416-1967.

37 At the Cafe Royal, oil on canvas, Muriel Minter, English, *c.*1935. Geffrye Museum, London. © Photo Studios Ltd, London.

Alcohol and Pictorial Advertising

38 Poster for Bières de la Meuse, colour lithograph, Alphonse Mucha, Paris, France, 1897. V&A: E.78-1956.
39 Poster for Quinquina Dubonnet, colour lithograph, Jules Cheret, France, 1895. V&A: E.2406-1938.
40 *left:* Poster for Thomas Bräu, colour lithograph, Lucian Bernhard, Germany, 1913. V&A: E.626-1915; *right:* Poster for Guinness, colour lithograph, John Gilroy, British, 1935. V&A: E.129-1973.
41 Poster for Smirnoff vodka, colour lithograph, British, 1979. V&A: E.213-1985.

Nostalgia and the Old English Tavern

42 Poster for Wills' tobacco, colour lithograph, English, 1890-1910. V&A: E.1161-1919.
43 *above:* Interior of Ashopton Inn, watercolour, Kenneth Rowntree, British, 1941. V&A: E.1266-1949; *below left:* Design for a beer jug, pencil and watercolour, London, Henry James Townsend, 1847. V&A: E.5896-1910; *below right:* Beer jug, stoneware, Minton & Co., Stoke-on-Trent, 1848. V&A: 540-1855.

Appreciating Wine

44 Autumnus, engraving, Johann Sadeler I, Dutch, second half of the 16th century. V&A: E.3399-1960.
45 Running Faun, terracotta, after model by Clodion (Claude Michel), French, *c.*1775-1800. V&A: 2627-1856.
46 Wine storage jar, earthenware, Andalucia, Spanish, 18th century. V&A: 295-1970.
47 *left:* Wine glass, Czech, 1950s. V&A: C.101-1984; *right:* Flask for spirits (one of a pair), cut and enamelled glass, Ignaz Preissler, Bohemia, probably 1720s. V&A: C.338-1936.

Bacchus and Ceres

48 Table centrepiece of Bacchus, maiolica, Italy, *c.*1560-75. V&A: C.665-1920.
49 *above:* Beer jug (one of a pair), silver, mark of Charles Frederick Kandler, London, 1739. Chicago Art Institute. © Charles Truman; *below:* Two-handled cup, silver-gilt, mark of Robert Hennell III, London, 1872-73. V&A: M.109-1966.
50 Poster for Champagne, colour lithograph, Walter Crane, English, late 19th century. V&A: E.2444-1938.
51 Costume design for the Spirit of Alcohol, for a production of *The Vintage* at the Empire Theatre, London, 1903. V&A: M. 1557-1925.

RITUAL AND CEREMONY

52-3 Wedding meal at Yport, oil on canvas, Albert-Auguste Fourie, French, 1886. Rouen, Musée des Beaux-Arts. © Musées de la Ville de Rouen.

Rites of Passage

54 *Hansje in de kelder* cup, silver, mark of Focke Raerda, Netherlands, 1669-70. Rijksmuseum, Amsterdam, BK-NM-13029.
55 *left:* Christening tankard, silver, mark of Hans Nieman the Elder, Drammen, Norway, *c.*1680. V&A: M.488-1910; *right:* The Fairfax cup, turquoise glass, enamelled and gilded, Venice, Italy, late 15th century. V&A: C.17-1959.
56 *above:* Many Happy Returns of the Day, oil on canvas, William Powell Frith, English, 1856. Mercer Art Gallery, Harrogate Museums. © Harrogate Museums and Art Gallery, North Yorkshire, UK/The Bridgeman Art Library; London; *below:* Jewish wedding, engraving

from Bernard Picart, *Histoire Generale des Ceremonies Moeurs et Coutumes Religieuses …*, Paris, 1741. V&A: NAL 24.G.148, vol. 1.
57 Guild cup, silver-gilt, mark of Carolus Junger, Hanover, Germany, 1717. V&A: 116&a-1864.

Seasonal Celebrations

58 The Betley Window, painted and stained glass, English, probably 1540s. V&A: C.248-1976.
59 *above, from left to right:* Three tygs or wassail bowls, lead-glazed earthenware, two with slipware decoration, Wiltshire, 1649. V&A: C.35-1987; Staffordshire, 1702. V&A: C.770-1922; Wrotham, 1710. V&A: C.118-1938; *below:* posset pot, salt-glazed stoneware, English, possibly Bristol, about 1700. V&A: C.282-1976.
60 *above:* Fuddling cup, delftware, London, probably Southwark, *c.*1630-40. V&A: 7710-1861; *below:* Pocket flask, white earthenware, John Siggery, Sussex, English, 1794. V&A: C.349-1919.
61 *above:* Harvest Festival at Svartsjö, oil on canvas, Pehr Hillestrom, Swedish, *c.*1780. Nordiska Museet, Stockholm. Photo by Birgit Brånvall © Nordiska museet; *below:* Christmas card, lithograph, Henry Stacy Marks, English, late 19th century. V&A: 15780.38.

Assays and Liveries: drink at court

62 Cunha Braga hunting cup, enamel, gold and rock crystal, German, *c.*1600. © Christie's Images Ltd. 2006.
63 Coronation Banquet of Anne of Brittany, manuscript on vellum, André Delavigne, *c.*1505. Waddesdon, The Rothschild Collection (The National Trust), MS 22, fol. 54v. Photo by Mike Fear, © The National Trust, Waddesdon Manor.
64 *left:* Charles II at a Garter Feast, engraving from Elias Ashmole, *The institution laws and ceremonies of the most noble order of the garter,* London, 1662. V&A: NAL 44.G.58; *right:* Sack bottle, tin-glazed earthenware, London, *c.*1660-65. V&A: C.1042-1922.
65 *left:* Flask containing saffron spirit, porcelain, Japanese, *c.*1860. V&A: 311-1865; *right:* The Lord Keeper's Seal Cup, silver-gilt, London, 1626-27. V&A: M.59:1&2-1993.

Prizes and Presentations

66 *above:* Rummer engraved with Sunderland bridge, cut and engraved glass, Sunderland, 1800-20. V&A: 65-1906; *below:* Beaker, silver and horn beaker, English, *c.*1707. V&A: 917-1864.
68 *left to right:* Guild beer glass (*humpen*), glass painted in enamels, probably Franconia, German, late 17th century. V&A: 1517-1855; humpen, glass painted in enamels, Bohemia, 1616. V&A: 1842-1855.
69 Racing car drivers drinking champagne, gelatine-silver print, Andrew Pitcairn Knowles, 1906. V&A: E.3385-2004.

Drinking in Fellowship: colleges, guilds and societies

70 *above left:* Guild tankard, blue glass with pewter mounts, Saxony, 1705-8. V&A: 5325-1901; *above right:* Punch bowl, tin-glazed earthenware, Cornwall, English, 1731. V&A: 3616-1901; *below:* Flask, pewter, Nuremburg, *c.*1550. V&A: 1333-1872.
71 *above:* Westminster Club dinner, engraving from George Augustus Sala, *Twice Round the Clock*, English, 1859. Collection of Ann Eatwell. © Ann Eatwell; *below:* Parrs Bank dinner at the Savoy Hotel, photograph, English, 1905. Royal Bank of Scotland Archives. Reproduced by kind permission of The Royal Bank of Scotland Group © 2007.
72 Sketches of rummers, ink on paper, from Barnard Ledgers, *c.*1808, pp. 84-85. V&A: Archive of Art and Design.
73 Skull and crossed bones cup, silver, German, *c.*1907. Courtesy of Anthony Marks.

CONTEXTS AND SETTINGS
74-5 Scene in a beer garden, woodcut, Herbert Gurschner, German, 1920s. V&A: E.637-1981.

Alehouses, Taverns and Inns
76 Feast of St Michael, oil on canvas, Pieter Breughel the Elder, Flemish, *c.*1600. © Johnnie Van Haeften.
77 *left:* Gallon measure, pewter, English, 1601. V&A: M.19-1923; *right:* Beer mug, pewter, English, dated 1703. V&A: M.60-1945.
78 Yard of the Oxford Arms, Warwick Lane, photograph, Dixon and Boole, English, mid 19th century. V&A: 63-1958.
79 Public house interior, painting on glass, English, late 18th century. V&A: E.3310-1928
80 *left:* Design for the Rose and Crown, Saffron Walden, pen and ink and watercolour, W. E. Nesfield, English, 1872. V&A: D.1577-1907; *right:* Exterior of the Black Friar pub, London. Photo by Daniel Gulliver, 2006. © Daniel Gulliver.
81 Interior of the Black Friar pub, London. Photo by Daniel Gulliver, 2006. © Daniel Gulliver.

Assemblies
83 *above:* Masquerade at King's Opera House, oil on canvas, attributed to G. Grisoni, British School, 1724. V&A: P.22-1948; *below:* The Great Room at Wynnstay, watercolour, from the Wynnstay Album, English, 1790s. V&A: E.2387-1948.
84 *above:* Designs of modern costume from Henry Moses, *Le Beau Monde*, London, 1823, p. 2, plate 2. V&A: NAL 24.B.19; Illustration from Pierce Egan, *Real Life in London*, London: Jones & Co, 1821, plate opposite p.250. V&A: NAL G.28.C15.
85 Design for door, mirrors, wall decoration, pen and ink, watercolour and wash, James Wyatt, 1760s. V&A: 7231.23.

Clubs, Dens and Political Drinking
86 Meeting of the Hell Fire Club, oil on canvas, James Worsdale, Dublin, *c.*1735. Courtesy of the National Gallery of Ireland. © National Gallery of Ireland.
87 *above:* The Court of Equity, or Convivial City Meeting, mezzotint, after Robert Dighton, English, 1779. V&A: E.540-1976; *below:* Beer Jug, lead-glazed stoneware, John Turner, Staffordshire, *c.*1800. V&A: 2510-1901.
88 The Union Club, engraving, James Gillray, early 19th century. V&A: 1232(2)-1882.
89 Toby jug, earthenware painted with high-temperature oxide colours, probably Staffordshire; *c.*1800. V&A: C.441-1940.

Gardens and Grottoes
90 *above:* Scene of outdoor drinking, engraving, Austrian, *c.*1825. V&A: E.1094-1983; *below:* Still life, oil on canvas, Willem van Aelst, 1660s. © Johnnie Van Haeften.
91 Party by a pool, watercolour drawing, Jean Démosthène Dugourc, French, mid 18th century. Waddesdon, The Rothschild Collection (The National Trust). Photo by Mike Fear, © The National Trust, Waddesdon Manor.

On the Move: hunting and travelling
92 Fox head stirrup cup, silver and silver-gilt, Holland, Aldwinckle and Slater, London, 1904. © Bonhams.
93 *above:* Goblet, engraved glass, 1777-1799, Norwegian. The National Museum of Art, Architecture and Design, Oslo. Photo by Teigens Fotoatelier AS © Nasjonalmuseet for kunst, arkitektur og design, Stiftelsen Kunstindustrimuseet i Oslo; *below, from left to right:* Gin bottles depicting Lord Brougham and Daniel O'Connell, salt-glazed stoneware, London, 1825-50. V&A: 3783-1901; Denby or Codner Park, 1834-47. V&A: 3691-1901.
94 *above:* The Hunt Lunch, oil on canvas, Carle Van Loo, French, 1737. Paris, Musée du Louvre. © Photo RMN/ © Hervé Lewandowski; *below:* Design for fan leaves, gouache on paper, pasted onto wood panel, after David Teniers, possibly France, possibly the Netherlands, probably 18th century. Waddesdon, The Rothschild Collection (The National Trust). Photo by Mike Fear, © The National Trust, Waddesdon Manor.
95 *above:* Shooting party, coloured lithograph, German, mid 19th century. V&A: E.2662-1953; *below:* Spirit flask, silver, London, *c.*1690. V&A: M.10:1-2001.

DRINKS AND VESSELS
96-7 Centrepiece and place setting in the Dining Room, Waddesdon Manor, gilt bronze and porcelain, French, glass Anglo-Irish, *c.*1820-25. Waddesdon Manor, the Rothschild Collection (The National Trust). Photo by Martin Charles, © The National Trust, Waddesdon Manor.

Which Glass?
98 Set of matching glasses, from Silber & Fleming retail catalogue, English, *c.*1882. V&A: Ceramics & Glass Collection Library.
99 *above, left to right:* Wine and water glass, 1760. V&A: C.509-1925; ratafia glass with enamel twist stem, *c.*1765. V&A: C.9-1993; roemer, green glass, *c.*1800. V&A: Circ.872-1924; ale or champagne flute, mid 18th century. V&A: 105-1907; wine glass, mid 18th century, V&A: C.249-1925; *below, left to right:* Spirit glass, *c.*1710-20. V&A: C.537-1936; cordial glass, *c.*1760-80. V&A: C.69-1942; gin glass, *c.*1720-40. V&A: C.158-1925; firing glass, *c.*1740. V&A: C.180-1925.
100 Tavern mug or *gorge*, salt-glazed stoneware, London, *c.*1700. V&A: C.174-1933; hunting jug for beer, salt-glazed stoneware, Mortlake (probably Kishere factory), *c.*1795-1800. V&A: C.137-1982; tavern mug, salt-glazed stoneware and silver, London, *c.*1720-40. V&A: C.199-1939.
101 *above, left to right:* Beer mug or *gorge*, enamelled porcelain, Derby, *c.*1780. V&A: 414:467-1885; pint mug, enamelled porcelain, Derby, *c.*1780. V&A: C.66-1921; beer jug, enamelled porcelain, Liverpool, *c.*1780. V&A: 414:783-1885; *below, left to right:* Dwarf ale glass, English, early 18th century. V&A: C.206-1913; rummer, English, *c.*1810-20. V&A: C.657-1936; tankard, mould-blown glass, English, mid-18th century. V&A: C.599-1936; ale flute, English, *c.*1760. V&A: C.192-1925.

Labelling the Bottle
102 Decanters for Brandy, Rum and Hollands, glass, probably Bristol, *c.*1790-1800. V&A: C.43-e-1961.
103 Wine labels for Chateau Mouton Rothschild vintages featuring artists' works designed by Hugo Jean, colour half-tone and line block, French, *clockwise from left:* Pablo Picasso, 1973. V&A: E. 824-1985; Joan Miro, 1969. V&A: E.820-1985; Andy Warhol, 1975. V&A: E.826-1985; Wassily Kandinsky, 1971. V&A: E.822-1985.
104 *above:* Wine label for Rota, enamel on copper, designed by Ravenet, Battersea, English, 1753-56. V&A: C.5k-1948; *below:* Wine label for claret, tiger claw mounted in silver, Indian, late 19th century. V&A: M.1341-1944.
105 Wine label for Glenmorangie whisky, offset lithograph, British, 20th century. V&A: E.745-1994.

Corkscrews
106 *from left to right:* Wine bottle, glass, English, 1680-90. V&A: C.292-1926; corkscrew, silver, London, *c.*1690. V&A: M.3-1975; wine taster, silver, London, *c.*1690. V&A: M.225-1930.

107 Wine cooler and fountain with buffet pieces, silver, London, early 18th century. © The National Trust, Dunham Massey.
108 *above:* Vauxhall Gardens (detail), watercolour, Thomas Rowlandson, 1784. V&A: P.13-1967; *below, left to right:* Cage corkscrew, iron, English, 18th century. V&A: M.85:1-1993; cage corkscrew, brass and steel, France, early 18th century. V&A: M.72:1-1993.
109 *above, left to right:* corkscrew, base-metal alloy plated with chrome, United States, about 1920. V&A: M.108-1993; corkscrew, engraved steel, Germany, *c.*1730. V&A: M.103-1993; corkscrew, engraved steel, Germany, *c.*1730. V&A: M.70-1993; *below, left to right:* nutmeg grater, silver, English, 1800-50. V&A: M.1065-1927; corkscrew, chiselled and gilt steel, France, about 1770. V&A: M.93-1993.

Drinking Games
110 *above:* Trick glass, Germany or Bohemia, probably 17th century. Waddesdon, The Rothschild Collection (The National Trust). Photo by Mike Fear, © The National Trust, Waddesdon Manor; *below, left to right:* Puzzle jug, tin-glazed earthenware, probably Bristol, *c.*1770-75. V&A: 275-1896; Puzzle cup, maiolica, Italy, *c.*1515. V&A: C.2195-1910.
111 *above:* Page from the Ambras Trinkbuch, manuscript, Austrian, mid to late 16th century. KHM, Wien; *below:* Windmill cup, glass and silver, Netherlands, 1570. V&A: C.416-1936.

Rum and Grog
112 Purser standing by a grog tub, hand-coloured etching, Thomas Rowlandson, London, 1799. © National Maritime Museum, London.
113 Tankard, salt-glazed earthenware, English, 1739. V&A: 414:942-1885.
114 *above:* Nelson recreating with his brave tars, hand-coloured etching, Thomas Rowlandson, London, 1798. © National Maritime Museum, London; *below:* Sailors carousing in the Long Room, engraving, George Cruikshank, *c.*1830. V&A: E.567:1943.
115 Rum issue on board HMS Edinburgh, photograph, 1897. © National Maritime Museum, London.

Liquid Refreshment
116 Title illustration from *The Accomplished ladies rich closet of rarities …*, London, W. & J. Wide, 1691. The Wellcome Trust.
117 *from left to right:* Posset pot, coloured lead-glazed earthenware, Joshua Heath, English, dated 1707. V&A: 4719-1901; posset pot and cover, tin-glazed earthenware painted white, English, *c.*1650-55. V&A: C.589&a-1922; posset pot, tin-glazed earthenware painted blue and white, English, dated 1697. V&A: C.75-1931.
118 *above:* Posset pot, tin-glazed earthenware, English, 1670-1700. V&A: C.98-1944; *below:* Spout cup, silver, Timothy Skottowe, Norwich, 1642-43. V&A: M.84-1920.
119 *from left to right:* Posset pot, lead glass, English, 1740-60. V&A: C.7-1923; posset pot, soda glass, English, probably Duke of Buckingham's glasshouse, London or Greenwich. V&A: C.170-1918; posset pot, lead glass, English, early 18th century. V&A: C.168-1956; posset pot, lead glass, English, mid 18th century. V&A: C.199-1993; posset pot, English, possibly George Ravenscroft at Savoy Glasshouse, London, 1680-90. V&A: C.44-1959.

From Hedgerow and Orchard
120 Flute glass, possibly for cider, Northern Netherlands, *c.* 1665-72. V&A: C.423-1936.
120-121 *Natura Morta: la Mosca*, oil on canvas, Christian Berentz,

German, *c.*1700. Galleria Corsini, Rome. Archivio Fotografico Soprintendenza Speciale per il Polo Museale Romano.
122 *above:* Habit de Caffetier, engraving, Nicolas de Larmessin, French, *c.*1700. V&A: 24686.26.O4c; *below:* Illustration from *The Queen's royal cookery*, London, T. Bates, 1713. The Wellcome Trust.
123 *above:* Illustration from Henry Moses, *Le Beau Monde*, London, 1823. V&A: NAL 24.B.19; *below:* Decanter, red glass, cut, gilt and transfer-printed, Russia, possibly St Petersburg, *c.*1825-50. V&A: C.51&A-1966.

Punch-drunk: sour and sweet, strong and weak
124 *above:* Tile, tin-glazed earthenware, Sadler and Green, Liverpool, *c.* 1765-75. V&A: 3605-1901; *below:* Anacreontick's in full song, aquatint, James Gillray, 1801. V&A: 15490A.JJ11.
125 *above:* Spice box and cover, porcelain, French, *c.* 1695-1710. V&A: C.476&A-1909; *centre:* Punch ladle, silver, English, mid 18th century. V&A: M.16-1972; *below:* Punch bowl, porcelain, Meissen, Germany, *c.* 1755. V&A: C.37&A-1960.

A Toast to Vodka and Russia
126 Vodka cup, agate, Michael Evamplevitch Perchin, Russian, before 1896. Private Collection, on loan to V&A.
127 The Moscow drink shop, from J.B. Le Prince, *The Cries of St Petersburg and Moscow*, French, 1765. V&A: E.1175-1909.
128 Vodka Set, glass and silver gilt, Ivan Khlebnikov, Russian, 1873. © Christie's Images Ltd. 2006.
129 The Vodka Drinker, biscuit porcelain, Gardner Manufactory. Russian, late 19th century. Sotheby's, London. Photo by Chris Halton, © Olympia.

Drink in the Americas
130 Punch bowl and ladle, silver, Tiffany & Co., New York, 1878. Christie's Auctioneers, New York. © Christie's Images Ltd. 2006.
130-1 Captains carousing in Surinam, oil on cotton, John Greenwood, American, 1758. Saint Louis Art Museum, St Louis.
132 *left:* whiskey flask, silver, American, 1888. © Christie's Images Ltd. 2006; *right:* rum barrel, tin glaze, probably Lambeth, London, 1780s. © Jonathan Horne.
133 Punch bowl and ladle, iridescent glass, Tiffany & Co., New York, *c.* 1900. Virginia Museum of Fine Arts, Richmond. The Sydney and Frances Lewis Art Nouveau Fund. Photo by Katherine Wetzel © Virginia Museum of Fine Arts.

The Birth of the Cocktail
134 Martini or cocktail glass, Lalique glassworks, French, *c.* 1925. V&A: Circ.34-1970.
135 The Grand Salon of The Normandie ocean liner, from *l'Illustration*, June 1935. V&A: 08.E.70.
136 *above:* Front Cover of *The Savoy Cocktail Book*, compiled by Henry Craddock, London, Chancellor Press, 1930. V&A: Loan Valerie Mendes Collection; *below:* Cocktail cabinet, ebonised mahogany with chromium mounts and glass, designed by Maurice Adams, British, 1933. V&A: W.96-1978.
137 Cocktail Shaker, electroplated nickel silver, Mappin and Webb, Sheffield, 1933. V&A: M.226-b-1984; cigarette holder, aluminium, possibly Paris, *c.* 1925. V&A: Circ.40-1972; cigarette case, silver, Margaret Craver Withers, London, 1946. V&A: M.53-1996. 'A Drink with Something in it' taken from *Candy is Dandy: The Best of Ogden Nash* (1985), published by André Deutsch.

Index

Numbers in italic identify illustrations. All images are listed and credited in the Notes on the Illustrations, pages 139-42.

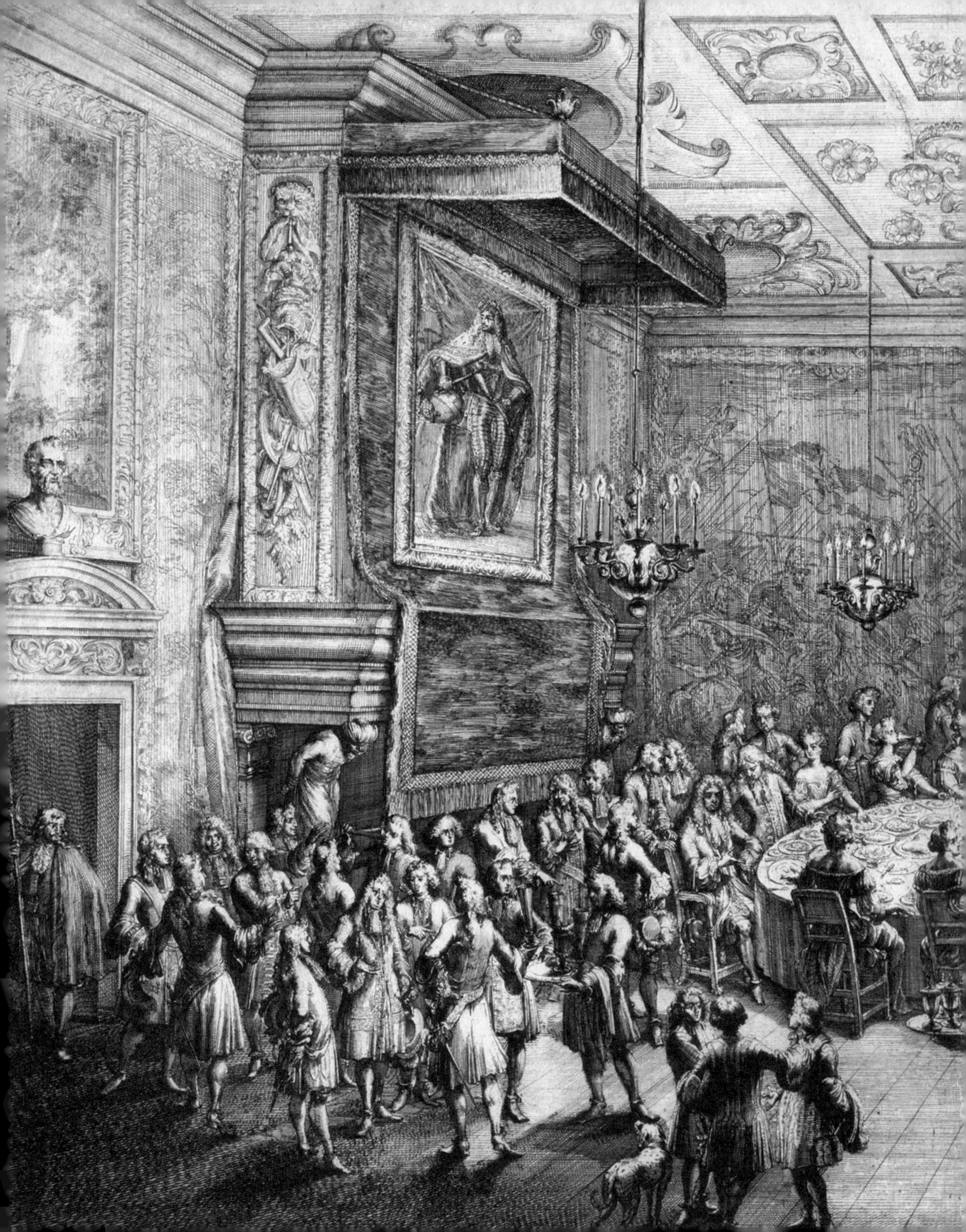